IEE ELECTROMAGNETIC WAVES SERIES 44

Series Editors: Professor P. J. B. Clarricoats
Professor E. V. Jull

Theory of NONUNIFORM WAVEGUIDES

the cross-section method

Other volumes in this series:

Volume 10 **Aperture antennas and diffraction theory** E. V. Jull
Volume 11 **Adaptive array principles** J. E. Hudson
Volume 12 **Microstrip antenna theory and design** J. R. James, P. S. Hall and C. Wood
Volume 13 **Energy in electromagnetism** H. G. Booker
Volume 14 **Leaky feeders and subsurface radio communications** P. Delogne
Volume 15 **The handbook of antenna design, Volume 1** A. W. Rudge, K. Milne, A. D. Olver, P. Knight (Editors)
Volume 16 **The handbook of antenna design, Volume 2** A. W. Rudge, K. Milne, A. D. Olver, P. Knight (Editors)
Volume 17 **Surveillance radar performance prediction** P. Rohan
Volume 18 **Corrugated horns for microwave antennas** P. J. B. Clarricoats and A. D. Olver
Volume 19 **Microwave antenna theory and design** S. Silver (Editor)
Volume 20 **Advances in radar techniques** J. Clarke (Editor)
Volume 21 **Waveguide handbook** N. Marcuvitz
Volume 22 **Target adaptive matched illumination radar** D. T. Gjessing
Volume 23 **Ferrites at microwave frequencies** A. J. Baden Fuller
Volume 24 **Propagation of short radio waves** D. E. Kerr (Editor)
Volume 25 **Principles of microwave circuits** C. G. Montgomery, R. H. Dicke, E. M. Purcell (Editors)
Volume 26 **Spherical near-field antenna measurements** J. E. Hansen (Editor)
Volume 27 **Electromagnetic radiation from cylindrical structures** J. R. Wait
Volume 28 **Handbook of microstrip antennas** J. R. James and P. S. Hall (Editors)
Volume 29 **Satellite-to-ground radiowave propagation** J. E. Allnutt
Volume 30 **Radiowave propagation** M. P. M. Hall and L. W. Barclay (Editors)
Volume 31 **Ionospheric radio** K. Davies
Volume 32 **Electromagnetic waveguides: theory and application** S. F. Mahmoud
Volume 33 **Radio direction finding and superresolution** P. J. D. Gething
Volume 34 **Electrodynamic theory of superconductors** S.-A. Zhou
Volume 35 **VHF and UHF antennas** R. A. Burberry
Volume 36 **Propagation, scattering and dissipation of electromagnetic waves** A. S. Ilyinski, G. Ya. Slepyan and A. Ya. Slepyan
Volume 37 **Geometrical theory of diffraction** V. A. Borovikov and B. Ye. Kinber
Volume 38 **Analysis of metallic antennas and scatterers** B. D. Popović and B. M. Kolundžija
Volume 39 **Microwave horns and feeds** A. D. Olver, P. J. B. Clarricoats, A. Kishk and L. Shafai
Volume 40 **Uniform stationary phase method** V. A. Borovikov
Volume 41 **Approximate boundary conditions in electromagnetics** T. B. A. Senior and J. L. Volakis
Volume 42 **Spectral theory and excitation of open structures** V. P. Shestopalov and Y. V. Shestopalov
Volume 43 **Open electromagnetic waveguides** T. Rozzi and M. Mongiardo

A complete list of publications in this series may be obtained from the IEE

Theory of NONUNIFORM WAVEGUIDES

the cross-section method

B. Z. Katsenelenbaum
L. Mercader del Río
M. Pereyaslavets
M. Sorolla Ayza
M. Thumm

The Institution of Electrical Engineers

Published by: The Institution of Electrical Engineers, London, United Kingdom

The Institution of Electrical Engineers,
Michael Faraday House,
Six Hills Way, Stevenage,
Herts. SG1 2AY, United Kingdom

British Library Cataloguing in Publication Data

A CIP catalogue record for this book
is available from the British Library

IEE ISBN 0 85296 918 X

Printed in England by Redwood Books, Trowbridge

Contents

The authors

B Z Katsenelenbaum graduated in physics from Moscow University in 1941, received his PhD in 1948, the Dr Sc degree in 1960 and his Professorship in 1965. He worked for many years at the Moscow Physical Technical Institute before moving to the Institute of Radio Engineering and Electronics of the Russian Academy of Sciences where he is currently Chief Scientist. He has authored over 115 papers and six books. His main research interest is in theoretical radiophysics, particularly electromagnetics and diffraction theory.

L Mercader del Rio (1923–1998) graduated with an MSc in Electrical Engineering from the Moscow Energy School in 1948 and received his PhD from Moscow University in 1954. In 1965 he became Professor at the Moscow School of Telecommunication Engineers. In 1978 he moved to Barcelona Polytechnic and then on to head the Radiocommunication Department at Madrid Polytechnic where he became Emeritus Professor in 1988. From 1992 he was Emeritus Professor at the Public University of Navarra. One of his main research interests was rain attenuation modelling for microwave radios.

M Pereyaslavets received his PhD in Electrical Engineering from IRE of the Academy of Sciences, Moscow, in 1985 where he worked until 1992. He then joined the Department of Electrical Engineering at Tel-Aviv University, Israel. He has been a visiting scientist at the Fukui University and the National Institute for Fusion Science, Japan. He is a holder of awards from the Max-Planck Society and the Japanese Society for the Promotion of Science. His research interests are in the area of microwave field structure transformers: antennas and mode converters with semitransparent surfaces.

M Sorolla Ayza obtained degrees in telecommunication engineering from Barcelona and Madrid Polytechnics. He worked for Euratom-Ciemat, Madrid, for four years, designing high-power millimetre-wave components and spent 18 months at the Institute for Plasmaresearch in Stuttgart as well as two years at the Ramon Llull University, Barcelona. He also worked in the MIC and MMIC industry. Since 1993 he has been a professor at the Public University of Navarra, Spain, leading the microwave and millimetre wave group. His main research interests are waveguide and antenna components for plasma heating, materials processing and satellite communications.

M Thumm obtained MS and PhD degrees in experimental physics from the University of Tuebingen, Germany. In 1976 he joined the Institute for Plasma Research, Stuttgart, where he worked on RF heating of toroidal pinch plasmas for thermonuclear fusion. From 1982–1990 his research activities were mainly devoted to component development for the transmission of high-power millimetre waves and antenna structures for plasma heating. Since 1990 he has been a professor at the University of Karlsruhe, where the current research projects are the development of high-power gyrotrons, transmission lines and antennas for nuclear fusion plasma heating and materials processing.

Acknowledgments

The authors would like to thank W. Kasparek, H. Kumrić, G.A. Müller and D. Wagner (Institute for Plasmaresearch, University of Stuttgart) and V. Erckmann (Max-Planck Institute for Plasmaphysics, Garching). This work could not have been done without their help, stimulating suggestions and useful discussions. We would also wish to express our deep gratitude to I. Ederra Urzainki (Public University of Navarra, Pamplona) for critical reading of the manuscript and his valuable validation computations using our scattering matrix code. The authors are indebted to Mrs. Feißt (Research Centre Karlsruhe) for her thorough drawing of the figures as well as to G. Michel (Institute of Microwaves and Electronics, University of Karlsruhe) and H. Alpers (Library, University of Stuttgart) for their kind help in finding the English translations of old Russian publications. Finally, we gratefully note the continued encouragement and patience of our respective families, which made completion of this effort possible.

Moscow, Karlsruhe, Pamplona and Fukui

June 1997

B.Z. Katsenelenbaum
L. Mercader Del Río
M. Pereyaslavets
M. Sorolla Ayza
M. Thumm

Foreword

This book is devoted to one of the mathematical methods employed to solve a certain class of electrodynamic problems, considering nonuniform oversized (or overmoded) tubular metallic waveguides. Their properties are monotonically varying along the direction of wave propagation. At present, the most important scientific and technological applications of overmoded transmission lines are in the area of magnetically confined nuclear fusion experiments. The plasma physics community has taken advantage of recent advances in using high-power micro and millimetre waves generated by klystrons and gyrotrons for lower hybrid current drive (1–8 GHz), electron cyclotron resonance heating (ECRH) (28–160 GHz), electron cyclotron current drive (ECCD), plasma production (start up) and plasma diagnostic measurements such as collective Thomson scattering and heat pulse propagation experiments. Electron cyclotron emission (ECE) and microwave reflectometry plasma diagnostics also employ oversized transmission lines. Other possible applications of oversized high-power microwave transmission lines span a wide range of technologies. In materials processing, microwaves are utilised to heat food, textiles, wood products, polymers, minerals, structural and functional ceramics, composites, multilayer structures and chemicals. Other applications are drivers for next-generation high-gradient linear accelerators (TeV-supercolliders), multiply charged ion and soft X-ray sources, high frequency broadband electron spin resonance (ESR) spectroscopy, deep space and specialised satellite communication, high resolution radar ranging and imaging in atmospheric and planetary science. Another technological application, waveguide communications by millimetre waves, was intensely developed 40 years ago. Engineers tried to use the low loss (about 1 dB / km) propagation of the TE_{01} mode in a circular waveguide.

The high ohmic attenuation and low power handling capability of conventional waveguides supporting the fundamental mode are prohibitive for all of these applications. Thus, in order to design low-attenuation and high-power waveguides, oversized structures are preferred. The elements employed in these transmission lines (tapers, waveguide bends etc.) have slowly varying properties compared with the operating wavelength.

The so called cross-section method was formerly developed for the purpose of mathematical analysis and synthesis of overmoded waveguide components for communications. This method allows one to study the main phenomenon of oversized nonuniform waveguides, i.e. mode conversion to parasitic waves. In other words, it provides the necessary tools for maintaining the lowest possible level of all the modes except one, the propagting mode. The cross-section method was employed mainly for the deduction of adiabatic solutions, i.e. when the electromagnetic field of the wave is dominated by one mode.

Waveguide communications were not widely applied mainly due to the emergence of fibre optics. Nevertheless, the elaborated mathematical apparatus has been very useful for the solution of various technical and scientific problems, particularly for fusion plasma heating, materials processing, radar and satellite communication systems, as mentioned above. Electronic devices (e.g. gyrotrons) generating high-power millimetre waves produce electromagnetic fields with a very complex structure. It is necessary to launch this power with minimal losses and with a specific field structure. This makes the use of specifically designed components (e.g. mode converters) in oversized nonuniform waveguides compulsive.

Application of the cross-section method for analysis and synthesis of these devices has been very successful. No additional new ideas were necessary and the following fact is very important: on modern computers the method allows us the exact description of the electromagnetic field in an oversized waveguide with several kinds of propagating coupled wave. This opens the possibility of studying very complex processes of highly efficient power conversion between different modes.

The method is based on the reduction of the three-dimensional (3D) vectorial problem to the form of a simple two-dimensional (2D) problem and to ordinary differential coupled-wave equations. The effectiveness of the method can be explained by this reduction, which is well suited to the physical processes taking place in nonuniform oversized waveguides.

This book is a new improved version of the book of one of the authors, B. Z. Katsenelenbaum (1961c), published in Russian by the former Soviet Union Academy of Sciences under the similar title '*Theory of nonuniform waveguides with slowly varying parameters*'. Several misprints have been corrected. Practically all of the authors of this new version are involved in the application of high-power microwaves to plasma heating and materials processing and because of this the new version is oriented mainly in that direction. Several new complements (written by M. Thumm) have been introduced where the components required for low-loss transmission and highly efficient transformation of electromagnetic waves in oversized waveguides have been designed by means of the cross-section method. This book also supplies many useful formulae for the practical layout of such transmission line components. Because of the large dimensions compared with the wavelength, a fully three-dimensional analysis employing modern numerical methods based on finite element, finite difference, finite integration or transmission line matrix formalisms is practically impossible. However, in the case of simple nonuniform waveguide structures having specific symmetry, the so-called mode matching and generalised scattering matrix method can be employed without too high a requirement for computer memory and computing time. Therefore, if appropiate, we use this method to verify the results derived by the cross-section method.

Today, the use of the cross-section method cannot be limited to oversized tubular metallic waveguides. It is employed intensively in other areas, such as fibre optic communications, antenna synthesis (resonant antennas), natural waveguides (submarine, tropospheric, seismic), microwave radiolinks (Earth or space), the design of absorbing surfaces as well as for the solution of nonlinear problems. Also, it is possible to apply the method to acoustic problems. Several of these problems are presented in this new version of the book in an abridged form. Moreover, all the new theoretical results obtained during the last 35 years are also presented; these are mainly methodological results or problems relating to numerical analysis and the application of this method employing computers. A set of new references has been included.

In order to restrict the volume of the book we have excluded the chapter devoted to acoustical waveguides encountered in the old book.

Propagation of electromagnetic waves (in the air, on the water and Earth's surface) in long transmission lines leads to several very interesting technical and physical problems; the properties along these transmission lines are not invariant. Hence, the mathematical apparatus described in this book and the corresponding physical analysis based on the processes in oversized transmission lines when their length is large compared with the wavelength can be of great interest for specialists in several areas.

The mathematical foundations of the method are simple and they have a mainly geometrical character. However, the development of the method led us to the solution of various nontrivial pure mathematical problems. For this reason the mathematical apparatus described in this book can also be of interest for mathematicians engaged in abstract problems and not only in applied physics and engineering. The same is valid for mathematical analysts and for specialists in numerical analysis.

Introduction

I.1 The contents of the book

I.1.1 General considerations

Uniform waveguide theory for constant cross-section straight waveguides is a very well known topic. There are also many books devoted to analysis methods for the influence of various nonuniformities in a monomode standard rectangular waveguide. These nonuniform elements are usually coupling and tuning elements and have a local distribution, i.e. they are concentrated in a region of the order of the rectangular waveguide dimensions. They provide narrowband matching (no reflections). As a general rule, high-order modes cannot propagate in these waveguides if operated in the nominal frequency band.

In recent years there has emerged the requirement to employ broadband transitions and mode converters in oversized waveguides. In these waveguides it is not feasible to apply local tuning and matching elements such as slots, diaphragms, stubs, etc. A typical nonuniform element is a gradual waveguide taper with a length which is longer than either of the dimensions of the waveguide cross section. The parameters of these waveguides are slowly changing functions of one of the co-ordinates. For example, we can consider in this class of device a bent waveguide with a radius of curvature larger than the cross section dimensions or a long tapering between two waveguides with different cross section or a long waveguide twist etc. These devices operate in a broad frequency band.

As a general rule, a lot of various modes can propagate in the gradual waveguide transitions studied later, usually dozens of modes, since the cross section of these waveguides is essentially larger than the wavelength. It is necessary to design these components in such a way that the main part of the incident wave power is carried away by one specific wave while simultaneously, the amplitudes of the other (i.e. parasitic) modes are maintained very small.

For the most typical cases the electrodynamic problem is the following: a wave with a given transversal field structure (mode) of unitary amplitude is incident on a nonuniform waveguide section placed between two different uniform waveguides; therefore, it is necessary to calculate the amplitudes of all the waves scattered on the nonuniform section in both directions. We will call the set of all those complex amplitudes the scattering matrix of the nonuniform section. The inverse problem is obviously of great interest, i.e. to design waveguide parameters for a desired wave distribution in the operating frequency band. This also includes mode converters for transformation of the incident wave to another specific wave with different transversal structure.

This book is devoted to the systematic description of one of the methods for solving such a class of electrodynamic problems, i.e. the so-called cross-section method.

I.1.2 Description of the book contents

The book consists of an introduction and six chapters. First, in the introduction, a description of the contents of the book and a review of the literature are presented. Additionally, a short summary of the theory of uniform waveguides is given. In Chapter 1, the auxiliary method of small nonuniformities is described. Any complicated waveguide nonuniformity can be considered as the superposition of several basic kinds of nonuniformity: a bend, a variation of the filling medium, a variation of the wall impedance and a variation of the waveguide cross section. In the three sections of Chapter 1 the elementary deformation of each basic kind of nonuniformity are analysed, i.e. the small angle tilt, the small step variation of the filling medium properties and the small step in the cross section. Particularly, determination of the scattered wave amplitudes caused by these nonuniformities allows us to find the required tolerances for the design of these transition waveguides.

A waveguide bent by a finite angle can be considered as the limit of a waveguide consisting of a lot of small tilts. Similarly, the properties of a continuously varying filling medium can be represented by the limit of a waveguide filled with a multilayered medium, and a variable cross-section waveguide is the limit of a lot of small cross-section steps. For this reason, in some problems, the analysis of any of these small nonuniformities allows us to calculate the amplitudes of the waves scattered on a finite (not small) nonuniformity of the same kind. This calculation method is not absolutely rigorous, but it has an obvious physical meaning and in most cases gives the complete solution of the problem. This can be considered as a physical interpretation of the mathematical apparatus elaborated for the fundamental method called by us the cross-section method.

This method is described in Chapter 2. The main idea is based on the fact that the electromagnetic field at any cross section of any nonuniform waveguide section can be represented as an infinite sum of the fields of the modes, including forward and backward propagating waves. These modes can also propagate into the so-called reference waveguide, i.e. into a uniform waveguide of identical cross section and with the same distribution of the electrical permittivity and magnetic permeability as in the considered cross section. The coefficients of this expansion are functions of the longitudinal co-ordinate and they satisfy an infinite number of first order ordinary differential coupled-wave equations. The study of nonuniform waveguides, i.e. the three dimensional (3D) electrodynamic problem, is reduced in such a way to the consideration of both the two-dimensional (2D) problem of the uniform waveguide modes and the one-dimensional problem, that is, the solution of ordinary differential coupled-wave equations.

The great difficulty in the application of the cross-section method arises when we study waveguides with varying cross section. In such cases the nonuniform waveguide fields satisfy boundary conditions different from those for the fields of uniform waveguides of the same cross section, and the series expansions of these wave fields are not convergent towards the desired functions on the cross-section boundary. For this reason, the solution is obtained first for a constant cross-section waveguide filled by a medium with a continuously varying dielectric permittivity distribution $\varepsilon(x, y, z)$. Then, the transition to a discontinuous distribution is performed in such a way that a part of the waveguide remains empty (see Figure 2.5) and, finally, we perform the transition to the complex value of ε for $|\varepsilon| \to \infty$. Proceeding in this way we avoid the need to operate with nonuniformly convergent series expansions that cannot be differentiated term by term. The same results can be obtained without performing the limit calculations mentioned above; in such a case the calculations are more complicated but the method is a more natural one.

The main parameters characterising a waveguide inhomogeneity are the coupling coefficients, i.e. the coefficients of the coupled-wave equations for the amplitudes of the waves. The properties of these coefficients have been extensively studied. These coefficients can also be derived from the scattering matrix elements for a small nonuniformity obtained in Chapter 1. The cross-section method delivers analytical expressions for the coupling coefficients caused by any nonuniformity and, in particular, by combined nonuniformities.

In the case of a gradual nonuniformity, when the parameters characterising a waveguide vary slowly, it is easy to solve the coupled-wave equations and to obtain an explicit expression for the calculation of the waves scattered on a nonuniform waveguide section; this expression is analysed in detail in Chapter 2. There are two special cases considered in Chapter 3 when the solution becomes complicated, including the case of very gradual nonuniformities. In Sections 3.1 and 3.2 of this chapter, the case of a variable cross-section waveguide with the so-called cut-off cross section is studied, i.e. the cross section separating at a given frequency the propagating and the cut-off regions of a given mode. In the neighbourhood of this cross section the values of the coupling coefficients become large independently of the speed of variation of waveguide parameters and, therefore, methods like Wentzel-Kramers-Brillouin (WKB) cannot be employed. In particular, an equivalent boundary condition substituting the cut-off cross section is derived. This boundary condition allows us to obtain the solution of the coupled-wave equations far away from the cut-off cross section.

In Section 3.3 the second special case is considered: the wave incident on a waveguide tilt when the operating frequency is nearly equal to the cut-off value of the coupled parasitic mode. In such a case, there are resonant effects and the parasitic mode amplitude can increase up to relatively large values. These effects depend substantially on the conductivity of the wall material. Hence, to perform the analysis it is necessary to employ the expression for the wavenumber in an imperfect wall waveguide. This expression is applicable in particular at any frequency near the cut-off value; it has been deduced in Section 3.4.

In Chapters 4, 5 and 6 the developed method is used for practical engineering problems. In Chapter 4 it is applied to nonuniform waveguides with a straight axis, whereas in Chapter 5 curved waveguide transitions are described. Chapter 6 summarises the application to straight waveguide mode converters with a periodically nonuniform wall. Special attention is paid to the inverse problems. The cross-section method is employed to design tapers, bends and mode converters with the best parameters. These devices are very helpful for solving many technical problems in various applications.

The material described in these chapters cannot always be employed as a simple cookbook, however, it is complete enough to illustrate the possibilities.

I.2 Review of the literature

In the first four subsections of this section we review the literature published before 1960; these works were driven by research for the waveguide telecommunications projects (Miller, 1954, Kaznachev, 1959), i.e. by the requirement to construct pipes of several kilometres length for information transmission using circular oversized waveguides. The communication was carried by the TE_{01} wave at a wavelength of approximately 4 to 8 mm. Several methods for the mathematical analysis of phenomena in oversized waveguides were proposed. The foundations of the theory of nonuniform waveguides were chiefly established during this time. The present review emphasises mainly the Russian literature, however, the publications of Katsenelenbaum (1953–1961c) are neither mentioned nor referred to later in this book, but are included in the references.

I.2.1 Method of field continuity adjustment

The method of field continuity adjustment is based on field expansion into a sum of waves forming the electromagnetic field of the uniform and nonuniform waveguide and on the continuity requirement for the fields at the boundary (mode matching method); this leads to linear algebraic coupled-wave equations for the coefficients of these expansions. If the nonuniformity is small then a small parameter participates in the problem. Therefore, the coupled-wave equations can be solved by a general procedure. The application of this method is limited to simple nonuniform waveguide structures.

For straight waveguides this method allows us to analyse the transition to a cone with a small aperture angle in a circular waveguide or to a small broadening in a rectangular waveguide. In the works of Lewin (1955a and b), Piefke (1957) and Solymar (1959b) the reflection coefficients of the wave were obtained by this method for rectangular waveguide broadening. An equivalent method was employed by Mar'in (1959a) to determine the coupling and reflection coefficients when a rectangular waveguide is broadened in the *E*-plane. The method was applied by Solymar (1958, 1959d) to circular waveguides for the determination of the coupling coefficient between the TE_{01} and TE_{02} modes. Also, Tanaka (1957) did the same for the calculation of the coupling and reflection coefficients in the general case of arbitrary wave incidence.

In the case of a waveguide bent along a circumference arc, the concept of eigenwaves can be introduced. This represents the total field as a superposition of the eigenwaves of the curved waveguide. By using this development, Jouget (1947a) solved the problem of the interconnection of two rectangular waveguides, one being straight and the other being bent (with constant curvature), by the method of field continuity adjustment. In recent years, a considerable effort has been made to extend the capabilities of the mode-matching method to analyse more complicated structures of practical interest. This has been possible, for example, by means of hybrid mode matching/finite element methods. The state-of-the-art and recent advances are presented in Arndt *et al.*, 1997, and the references given there.

I.2.2 Conformal mapping

Conformal mapping was extensively used by Krasnushkin (1945) for the analysis of planar waveguides. By means of this method the complicated boundary of the nonuniform waveguide is transformed at the limit into two straight parallel lines. Proceeding in such a way, the wave equation describing the field in the waveguide becomes more complicated and adopts a form equivalent to an inhomogeneous medium placed between these two parallel lines. The parameters of this inhomogeneity are related to the conformal mapping function in a well known manner. Rice (1949) as well as Rozhdestvenskiy (1953) and Mar'in (1959b) studied, by means of different methods, a set of problems with wave equations with variable coefficients. Rozhdestvenskiy and Chetayev (1951) employed conformal mapping to solve the problem of matching transitions with dielectric filling. Conformal mapping was also applied by Weinstein (1957) where the problem of a slow varying nonuniformity in a planar waveguide was solved by variational methods.

It may be possible to generalise the conformal mapping method for the case of circular straight waveguides, but we must accept important complications. It is practically impossible to employ it for more complex problems, for example the simultaneous broadening of a rectangular waveguide in two planes.

1.2.3 Matched co-ordinate system

Many authors use a co-ordinate system where the waveguide walls are coincident with one of the co-ordinate surfaces. To elaborate such a system for the more general case (i.e. for nonplanar waveguides) it is impossible to apply conformal mapping and it is therefore necessary to use special methods.

Jouget (1947b) studied a circular waveguide, bent along a circumference with a large radius r, by means of this method. He introduced a co-ordinate system where Maxwell's equations incorporate additional terms proportional to the curvature, in contrast to the Cartesian orthogonal system. These terms represent supplementary currents generated by the propagating wave. The problem reduces to the solution of inhomogeneous Maxwell equations. In Jouguet (1947b) the eigenwaves of the bent waveguide were obtained and, in particular, the fundamental result of the theory of circular waveguides, concerning the conversion of the TE_{01} mode into its degenerate TM_{11} mode caused by a bend, was determined.

Nevertheless, the rest of the scattered wave amplitudes were not determined, i.e. their values are of the order of magnitude of $1/r$. This would require supplementary operations involving mode matching at the input and output of the bent waveguide. The publication by Lewin (1955b) also relates to this method, where the wavenumbers of the eigenwaves for twisted and curved rectangular waveguides are determined. In this paper, neither the operation of field continuity adjustment nor the determination of the scattered wave amplitudes was performed.

In Sveshnikov (1956a, 1956b, 1958) and Viktorova and Sveshnikov (1958) this method was strongly developed and applied to the solution of a general kind of problem: a waveguide bent along a double curvature line and simultaneously having a slow variation of its cross section. The originality of these works lies in the use of uniform methods for the solution of the inhomogeneous Maxwell equations emerging in the problem. The later papers of Sveshnikov (1959a, 1959b) increase the possibilities for the use of this mathematical tool. Thanks to the nonorthogonal curvilinear system of co-ordinates introduced by him, the given waveguide surface is transformed into the surface of a cylinder with unity radius. The solution of Maxwell's equations in that co-ordinate system is obtained, as in Viktorova and Sveshnikov (1958), by a method similar to the cross-section method; the desired electromagnetic field is expanded in a series of products of Bessel and trigonometric functions and to obtain the coefficients of this expansion one derives ordinary differential coupled-wave equations containing, particularly, the metric coefficients of the employed co-ordinate system.

Another very productive idea for the calculation of straight waveguides with variable cross section, based also on the introduction of a special system of co-ordinates, was proposed by Pokrovskii, Ulinich and Savvinykh (1958) in their paper about planar waveguides. The co-ordinate system is orthogonal within an accuracy error smaller than the square of the value of the averaged taper generatrix inclination angle. This value can be used as a smallness parameter in the problem and the wave equation written in this system of co-ordinates can be expanded as a series of this parameter. The equations of zero and first order are solved. A homogeneous equation is obtained in the zero order whereas an inhomogeneous equation results as the first-order equation. The solutions are obtained by the WKB method. The coupling and reflection coefficients depend upon the character of the function that describes the generatrix form; the order of size of these magnitudes is determined by the degree of smoothness of this function.

Later, the method proposed in Pokrovskii, Ulinich and Savvinykh (1958) was applied to the solution of a broad set of questions as in Pokrovskii, Ulinich and Savvinykh (1959a, 1959b), Ulinich (1959), Savvinykh (1959), Lykhne (1959), Dmitriev (1959) and Ryvkin (1959). Discontinous horn transitions between two waveguides, that means waveguides with their generatrices described by analytical functions having discontinuities in the first derivative, and problems relating to cut-off cross-section waveguides were investigated. The main interest was in planar waveguides (Pokrovskii, Ulinich and Savvinykh, 1958, 1959a, 1959b; Ulinich, 1959; Lykhne, 1959; Dmitriev, 1959) and in circular waveguides; Ryvkin (1959) tried to study rectangular waveguides.

I.2.4 Cross-section method

Several dozen articles where different versions of the cross-section method were applied to practical engineering problems has been published, most of which appeared after 1955. The basic idea of this method consists in the fact that the electromagnetic field in a nonuniform waveguide is represented by means of a superposition of the mode fields corresponding to more simple waveguides. The coefficients of this superposition satisfy ordinary differential coupled-wave equations. From the solution of these coupled-wave equations the amplitudes of the waves scattered by the nonuniform waveguide can be determined.

The initial ideas relating to this method belong to the book of Kisun'ko (1949) and the papers of Gutman (1957, 1958, 1959a, 1959b). Emelin (1958) developed similar ideas. Stevenson's article (1951a) was the first one published in western countries. Nevertheless, the intensive application of the method did not start there until after the publication of Schelkunoff (1955).

In the paper of Stevenson (1951a) straight waveguides with variable cross section were discussed. The electromagnetic field was expressed through six functions, each of which was expanded in terms of eigenfunctions of *TE* and *TM* modes corresponding to the constant cross-section uniform waveguide. To obtain the coefficients of these expansions, second-order differential coupled-wave equations were derived and later studied by the WKB method. The mathematical apparatus obtained from this procedure is very large and complicated. The only attempt to apply it to practical problems for the determination of the field scattered on a nonuniform waveguide section was done in

Leonard and Yen (1957). In this work the authors calculated the reflection coefficients for several waves at the connection of a straight circular waveguide with a cone and, moreover, the reflections from the broadening of a rectangular waveguide. The obtained formulae for the rectangular waveguide are still valid, but for the case of circular waveguides they are wrong.

In Schelkunoff (1955) the cross-section method is proposed again and it is illustrated by several examples of broadening planar waveguides and also by waveguides bent along a circumference arc. The work published by Heyn (1955) also presents the expansion of the wave fields of a constant cross section bent waveguide through straight waveguide fields. In both papers, first-order differential equations are established for the determination of the expansion coefficients; but these equations remain unsolved and the expressions for the scattered wave amplitudes are not presented.

Unger (1958a) applies the mathematical apparatus proposed by Schelkunoff to the practical problem of the incidence of a TE_{01} wave into a symmetrical transition between two circular waveguides. The amplitudes of the waves propagating in both directions are used as variables instead of the coefficients of the field Fourier expansion as it was done in Schelkunoff (1955) and Heyn (1955), and therefore the expressions for the amplitudes of the TE_{0n} waves scattered forwards have been obtained. In Iiguchi (1961) the author tries to calculate the complex transition from a rectangular to a circular waveguide.

The publications of Morgan (1957), Shimizu (1957) and Oguchi and Kato (1958) used the same method to obtain the differential equations for the wave amplitudes in a bend and the coupling coefficients between the TE_{01} mode and the modes TE_{1m} and TM_{11} of a curved circular waveguide. Andreasen (1958a) further developed Morgan's work (1957) and calculated the coupling coefficients of the fundamental TE_{11} wave of circular waveguides to the waves TM_{0n} and TE_{2n} (E-plane bend, $n = 1,2,3,...$) and TE_{0n} and TE_{2n} (H-plane bend, $n = 1,2,3,...$) and of the fundamental rectangular waveguide TE_{10} wave to the waves $TE_{0,2n}$ (E-plane bend, $n = 1,2,3,...$) and $TE_{1,2n+1}/TM_{1,2n+1}$ (H-plane bend) ($n = 0,1,2,....$). The problem of variable curvature waveguides was first considered by Unger (1957) and Andreasen (1958b). Nevertheless, the method applied by these authors was limited to the case of a continuous curvature distribution along the whole waveguide, i.e. they excluded the case of curvature discontinuities; later, this case was demonstrated as being very helpful for shortening the bend length.

It seems that Reiter (1959), independently from Schelkunoff but employing approximately the same method, studied variable cross-section straight waveguides. His results were applied later to the calculation of practical cases in Solymar (1959a) in a more comfortable form and by Schnetzler (1960a, 1960b). The latter calculated in this way a rectangular waveguide with torsion and the simple transition from a rectangular to a circular waveguide.

Variable cross-section straight waveguides are also considered in the papers of Gutman (1957, 1958, 1959a, 1959b). Their originality, compared with

Schelkunoff (1955) and Reiter (1959) and to the further works based on these publications, consists in the introduction of a special system of co-ordinates as well as in another method to deduce the differential equations. The paper published by Emelin (1958) generalises this method to the case of simultaneously varying shape of the cross section and the direction of the axis.

The paper of Lyubarski and Povzner (1957) can also be viewed as another version of the cross-section method. In this work the electromagnetic field is described as a series of wave fields not in a uniform waveguide but in a cone formed by the tangents to the considered section boundary. Differential coupled-wave equations are derived for the calculations of the coefficients of the expansion; the smallness parameter of these coupled-wave equations is not the inclination angle of the generatrix but its derivative. If the generatrix inclination angle varies slowly along the z co-ordinate, one can obtain an integral expression for the amplitudes of the waves scattered on this section and, in order to ensure the validity of the first-order approximation, the angle is not required to be small. But this can only be achieved by introducing an essential complication into the coupling coefficients' expressions. The method is only applicable to planar and cylindrical waveguides and no practical results were obtained in the paper in spite of the fact that they solved an acoustic problem less complicated than the electrodynamic one.

The first published paper about waveguides with variable filling medium was Schelkunoff (1952). Strictly speaking, only uniform waveguides with ε and μ depending on the transversal co-ordinates x and y and not on the longitudinal co-ordinate z were studied there. In these waveguides, the field was expanded into a series of the empty waveguide modes and first-order differential coupled-wave equations were derived for the coefficients of this expansion. Operating in this way each term of the expansion represents the electromagnetic field in a very simple system. However, even the field of a uniform waveguide is described by an infinite number of coupled-wave equations. The same idea was applied by Morgan (1957) to a curved waveguide and by Brodskiy (1958) to the constant cross-section straight waveguide with ε and μ depending on the three co-ordinates.

Obviously, there are some papers where the employed methods can be classified in some of the groups described above only with a lot of reservations. They use methods elaborated specifically for application to a given engineering problem. In this class we can consider, for example, the paper of Weinstein (1957) mentioned above or the papers of Barlow (1957a, 1957b) and that of Marié (1959); moreover, in these references the conditions are investigated under which the considered mode (for example, the TE_{01} mode in a circular waveguide) becomes the eigenwave of a bend. By their approach to the problem, the papers of Barlow can be considered nearer to those of Rozhdenstvienskiy and Chetayev (1951).

Several papers published later than 1961 relate to these publications. They develop or apply the cross-section method. Here we will present only some of them and their main contents. Let us remember the works of Shevchenko (1962, 1967) concerning variable impedance waveguides and his book Shevchenko (1971), the paper of Tsimring and Pavelyev (1982), the publication of Kovalev (1986) containing a new method of deduction of the main equations of the cross-section method and the works of Voytovich, Voytovich, Zamorskaya and Katsenelenbaum (1993) and Katsenelenbaum, Kolesnichenko, Frantsesson and Shevchenko (1996).

I.2.5 Integrated optics applications

The second technical problem prompting great developments in the work on nonuniform waveguide theory, mainly for open waveguides, was the emergence of fibre optics telecommunications. The theoretical methods employed for their analysis are related to the cross-section method. There are many of books devoted to this problem and we will indicate only some of them: Marcuse (1974, 1989), Unger (1977), Sohda and Ghatak (1977) and more recently Nishihara *et al.* (1985).

I.2.6 High-power microwaves

Today, interest in the theory of nonuniform waveguides is connected to various applications of high-power microwaves (Gaponov-Grekhov and Granatstein, 1994).

In materials processing, microwaves are used to heat food, textiles, wood products, polymers, minerals, structural ceramics, composites, multilayer structures and chemicals. The plasma physics community has already taken advantage of recent advances in using high-power micro and millimetre (mm) waves in the areas of RF plasma heating for magnetic confinement fusion research, such as the lower hybrid current drive (1–8 GHz), electron cyclotron resonance heating (ECRH) (28–160 GHz), electron cyclotron current drive (ECCD), plasma production (start up) and plasma diagnostic measurements such as collective Thomson scattering and heat pulse propagation experiments (Thumm and Kasparek, 1995). Oversized transmission lines with low attenuation are also utilised in plasma diagnostics systems as electron cyclotron emission (ECE) and reflectometry diagnostics. Other important applications are multiply charged ion and soft X-ray sources (Golubev *et al.*, 1996), high frequency broadband electron spin resonance (ESR) spectroscopy (Tatsukawa *et al.*, 1995), deep space and specialised satellite communication (Uher *et al.*, 1993), high resolution radar ranging and imaging in atmospheric and planetary science and drivers for next-generation high-gradient linear accelerators (Granatstein and Lawson, 1996). The high ohmic attenuation and low power handling capability of conventional metallic waveguides supporting the dominant mode are prohibitive for these applications. Thus, in order to design low attenuation waveguides, oversized and overmoded structures are preferred. Since the gyrotron, the most important high-power millimetre wave source (Edgcombe, 1993, Thumm, 1994), operates in a high-order cavity mode, one needs knowledge of mode conversion and transmission concepts. In these areas of activity, the results of the cross-section method and the coupled-wave formalism were applied by Kovalev *et al.* (1968), Moeller (1982), Doane (1982), Thumm (1984, 1986), Kumrić *et al.* (1988), Buckley *et al.* (1991), Vinogradov and Denisov (1991) and Del Río *et al.* (1994). Transmission line components as waveguide cross-section tapers and bends should conserve high purity of the propagating mode and mode transducers should selectively convert a specific mode into only one desired other mode without exciting large amounts of parasitic waves.

I.3 Uniform waveguides

I.3.1 General remarks

In this section the main properties of uniform waveguides are summarised and the terminology and symbols are defined. Our approach is quite similar to the monograph of Weinstein (1988). Corresponding textbooks in Western literature are the monographs of Ramo *et al.* (1994) and Collin (1991).

We say that a waveguide is uniform if all its properties do not vary along its longitudinal axis. This previous characterisation of the properties of a very general type of uniform waveguide is important for the following study of non-uniform waveguides. Let us assume that the cross section is limited by some arbitrary closed boundary and that the properties of the filling medium inside the waveguide vary across the cross section. Generally speaking, all the reference waveguides used in this book are of this kind.

We use the stationary solutions of the Maxwell equations:

$$\mathrm{rot}\vec{E} = -ik\mu\vec{H} \qquad \text{(I.1}a\text{)}$$

$$\mathrm{rot}\vec{H} = ik\varepsilon\vec{E} \qquad \text{(I.1}b\text{)}$$

where k is the free-space wavenumber.

In eqns. I.1 we have not introduced external current sources. It is assumed that these current sources are located outside of the waveguide section considered, whereas the fields are excited by waves incident on this section.

The character of the problems considered in this book allows the authors to avoid the typical discussion on the system of units best suited for applied electrodynamics problems. Certainly, this is an important question (it is necessary to emphasise, however, that frequently it is exaggerated) when one is interested to find the relationship between the field and its exciting current. In the problems of this book all the waves are excited by other waves and therefore relative values only (ratios of amplitudes) are calculated, i.e. reflection and coupling coefficients. Some use of the Gaussian system of units is slightly preferred in this book only when ε and μ are dimensionless variables equal to unity in vacuum and the electric and magnetic fields have identical dimensional units. The introduction of ε_0 and μ_0 values into the expression of the vacuum wave impedance would cause unnecessary complexity in the formulae. This approach has been compensated for by the introduction of the time dependence factor in the form $\exp(i\omega t)$, i.e. exactly as currently accepted in electrical engineering and not as it is done in theoretical physics.

With the exception of Sections 1.2, 2.2, 3.3 and 3.4, we will consider the waveguide walls as perfect conductors. This yields

$$\vec{E} \bullet \vec{t} = 0 \tag{I.2}$$

Where $\vec{t}$ is a tangential vector to the surface of the waveguide wall. Since we have assumed that ε can vary in an arbitrary way across the cross-section boundaries, eqn. I.2 does not restrict the class of considered waveguides (see subsection 1.2.4).

Let us define a Cartesian co-ordinate system (x, y, z) where the z-axis is parallel to the waveguide axis. We also use a polar co-ordinate system (ρ, β, z) where the angle β is counted from the x-axis. Another 2D (two-dimensional) co-ordinate system, can also be used in the cross-section plane.

The z dependence of the electromagnetic field in a uniform waveguide is $\exp(-ihz)$ where h is the longitudinal wavenumber (i.e. the propagation constant, since the attenuation constant is assumed to be zero) inside the waveguide. If the operating frequency is higher than the cut-off frequency of the given mode then $h^2 > 0$. At the cut-off frequency of the given mode we have $h^2 = 0$ and, finally, if the frequency is below the cut-off value then $h^2 < 0$.

We can define the normalised wavenumber $\hbar = h/k$, which is dimensionless. The relation $0 \leq \hbar \leq 1$ is satisfied for propagating modes. At an operating frequency which is much higher than the cut-off frequency $\hbar \approx 1$, whereas at the cut-off frequency $\hbar = 0$.

Let us introduce the complete set of modes which can exist in the given uniform waveguide, having in mind both propagating modes ($\hbar^2 > 0$) as well as evanescent, nonpropagating modes ($\hbar^2 \leq 0$). Let us numerate all modes with a superscript j, where j is an integer in the range of $-\infty \leq j \leq +\infty$.

There are some cases where the system of modes is not complete. Such cases will be studied later in subsection 2.2.12 and therefore they will not be considered here.

The electromagnetic fields of the usual modes will be written as $\hat{\vec{E}}^j(x, y, z)$ and $\hat{\vec{H}}^j(x, y, z)$ and the factors describing the field dependence on the co-ordinates x and y only, i.e. in the cross-section plane are denoted as $\vec{E}^j(x,y)$ and $\vec{H}^j(x,y)$, respectively, this means

$$\begin{aligned}\hat{\vec{E}}^j(x, y, z) &= \vec{E}^j(x, y)\exp(-ih_j z)\\ \hat{\vec{H}}^j(x, y, z) &= \vec{H}^j(x, y)\exp(-ih_j z)\end{aligned} \tag{I.3}$$

The vectors $\vec{E}^j$ and $\vec{H}^j$ satisfy the six relationships which can be obtained by substituting eqn. I.3 into the Maxwell equations I.1.

$$\begin{aligned} &\frac{\partial E_z^j}{\partial y} + ih_j E_y^j = -ik\mu H_x^j \\ &\frac{\partial E_z^j}{\partial x} + ih_j E_x^j = ik\mu H_y^j \\ &\frac{\partial E_y^j}{\partial x} - \frac{\partial E_x^j}{\partial y} = -ik\mu H_z^j \end{aligned} \tag{I.4a}$$

$$\begin{aligned} &\frac{\partial H_z^j}{\partial y} + ih_j H_y^j = ik\varepsilon E_x^j \\ &\frac{\partial H_z^j}{\partial x} + ih_j H_x^j = -ik\varepsilon E_y^j \\ &\frac{\partial H_y^j}{\partial x} - \frac{\partial H_x^j}{\partial y} = ik\varepsilon E_z^j \end{aligned} \tag{I.4b}$$

Each mode propagates if $h_j^2 > 0$ or is attenuated along the two opposite propagation directions if $h_j^2 < 0$. Waves with opposite propagation directions are denoted with superscripts of different signs. If at the same time we assume that

$$h_{-j} = -h_j \tag{I.5}$$

then eqns. I.3 are satisfied for the waves of mode j in both propagation directions. Let us assume for convenience, if $h_j^2 > 0$, then the wavenumber is positive for $j > 0$, i.e. $h_j > 0$, and if $h_j^2 < 0$ and $j > 0$ then $h_j = -i\left|h_j\right|$. By this notation, the waves with positive superscripts will propagate forwards in the waveguide, that is along the positive direction of the z-axis, whereas the waves with negative superscripts propagate backwards, i.e. in the negative direction. We will call the first waves forward waves and the second ones backward waves. Let us connect the components of the forward and backward waves of mode j by the following relations

$$\begin{aligned} &E_x^{-j} = -E_x^j, E_y^{-j} = -E_y^j, E_z^{-j} = E_z^j \\ &H_x^{-j} = H_x^j, H_y^{-j} = H_y^j, H_z^{-j} = -H_z^j \end{aligned} \tag{I.6}$$

In this case, eqns. I.4 are invariant to the superscript sign, that is, if the fields of the forward wave satisfy eqns. I.4 then the backward wave also satisfies the same equations in agreement with eqns. I.5 and I.6. The wave relations I.6 could be also written with $\vec{E}$ and $\vec{H}$ exchanged.

The fields of different modes are mutually orthogonal. The orthogonality relationship will be formulated for any combination of modes, superscript signs and particularly for forward and backward propagating waves of the same number. The following orthogonality relationships and the superscript sign invariance relations of eqn. I.4 allow us to avoid any restrictions on the propagation directions of the waves.

It is easy to show, applying the Lorentz lemma to the two fields $\hat{\vec{E}}^m$ and $\hat{\vec{H}}^m$ and $\vec{E}^j$ and $\vec{H}^j$ for any (m, j) in a waveguide section defined between two planes $z = z_1$ and $z = z_2$ (see, for instance, Weinstein (1988)), that the integral

$$
\begin{aligned}
&\int\left\{\left|\vec{H}^m \times \vec{E}^j\right|_z + \left|\vec{H}^j \times \vec{E}^m\right|_z\right\}dS \\
&\equiv \int\left(E^j_m H^m_x - E^j_x H^m_y - E^m_x H^j_y + E^m_y H^j_x\right)dS
\end{aligned}
\tag{I.7}
$$

is identically equal to zero if $h_m \neq h_j$. Here, as well as anywhere later, the integral $\int \ldots dS$ is calculated within the limits of the waveguide cross section. The fact that this integral eqn. I.7 is identically equal to zero can be alternatively proven using eqns. I.4 and I.2 directly. As is well known, this identity is also satisfied if we substitute eqn. I.2 by the Leontovich's condition (i.e. the impedance wall condition).

Let us normalise the electromagnetic fields $\vec{E}^j$ and $\vec{H}^j$ in the following way

$$
\int\left(E^j_y H^j_x - E^j_x H^j_y\right)dS = kh_j \tag{I.8}
$$

The right-hand side of eqn. I.8 has been chosen in such a way that the normalisation condition for the transversal eigenfunctions (see eqn. I.16) is as simple and usual as possible.

The orthogonality condition and eqn. I.8 can be written as one equation only in a following way

$$
\int\left(E^j_y H^m_x - E^j_x H^m_y - E^m_x H^j_y + E^m_y H^j_x\right)dS = 2kh_j\delta_{mj} \tag{I.9}
$$

where the Kronecker symbol is $\delta_{mj} = 0$ for $m \neq j$, whereas $\delta_{jj} = 1$. Using eqns. I.4 and I.9, the fields of the normal modes in a uniform waveguide are determined in a unique way although the sign of j is not determined.

The power flux carried through the cross section by a wave with the superscript j is simply h_j, if h_j is real. Here, a constant factor is omitted for all modes; this factor is not essential for further discussion. We will say that the mode amplitude is P_j if its fields are $\vec{E} = P_j\vec{E}^j$ and $\vec{H} = P_j\vec{H}^j$; the power flux being equal to

$$
\left|P_j\right|^2 h_j \tag{I.10}
$$

Here, again, a constant factor is omitted for all modes, since later, when we study the power conversion losses for conversion to another wave of any number, we will always calculate ratios of the amplitudes for eqn. I.10 for this wave and for the incident wave.

I.3.2 Transverse electric (TE) and transverse magnetic (TM) modes

If ε and μ are both equal to 1, i.e. the waveguides are empty, the fields $\vec{E}^j$ and $\vec{H}^j$ can be expressed by the transversal eigenfunctions $\psi^j(x, y)$ and $\varphi^j(x, y)$. These functions are proportional to the longitudinal components of the electric and magnetic Hertz vectors. The functions ψ^j and φ^j satisfy the following equations

$$\nabla^2\psi^j + \alpha_j^2\psi^j = 0 \tag{I.11a}$$

$$\nabla^2\varphi^j + \alpha_j^2\varphi^j = 0 \tag{I.11b}$$

where ∇^2 is the 2D Laplace operator (in the x, y variables). The coefficients α_j are the eigenvalues of eqns. I.11 with the following boundary conditions at the cross-section boundary

$$\frac{\partial\psi^j}{\partial n} = 0 \tag{I.12a}$$

$$\varphi^j = 0 \tag{I.12b}$$

Here, as well as everywhere later, $\vec{n}$ is the external normal to the boundary of the cross section of a uniform waveguide. There is a relationship between the free-space wavenumber k, the longitudinal wavenumber h_j and α_j^2, which depends only on the geometry of the cross section

$$h_j^2 = k^2 - \alpha_j^2 \tag{I.13}$$

showing that α_j is the transversal wavenumber.

In general, the eigenvalues of the systems of eqns. I.11*a*, I.12*a* and I.11*b*, I.12*b* are not equal to each other.

The modes are classified as electric (transverse magnetic: *TM*) and magnetic (transverse electric: *TE*). The fields of *TE* modes are expressed through the Ψ^j functions

$$\begin{aligned} &E_x^j = -ik\frac{\partial\psi^j}{\partial y}, \quad E_y^j = ik\frac{\partial\psi^j}{\partial x}, \qquad E_z^j = 0 \\ &H_x^j = -ih_j\frac{\partial\psi^j}{\partial x}, H_y^j = -ih_j\frac{\partial\psi^j}{\partial y}, H_z^j = \alpha_j^2\psi^j \end{aligned} \tag{I.14a}$$

whereas the fields of *TM* modes are expressed through the φ^j functions

$$\begin{aligned} E_x^j &= -ih_j \frac{\partial \varphi^j}{\partial x}, \quad E_y^j = -ih_j \frac{\partial \varphi^j}{\partial y}, \quad E_z^j = \alpha_j^2 \varphi^j \\ H_x^j &= ik \frac{\partial \varphi^j}{\partial y}, \quad H_y^j = -ik \frac{\partial \varphi^j}{\partial x}, \quad H_z^j = 0 \end{aligned} \tag{I.14b}$$

Let us assume that

$$\varphi^{-j} = \varphi^j, \quad \psi^{-j} = -\psi^j \tag{I.15}$$

Then eqns. I.14 will also be invariant to the sign of the superscript j.

It is easy to prove that eqns. I.14 satisfy the orthogonality condition; i.e. the integral of eqn. I.7 is equal to zero when $m \neq j$. This follows from the usual orthogonality relationships for the different solutions of the systems of eqns. I.11 and I.12. The normalisation introduced for the fields $\vec{E}^j$ and $\vec{H}^j$ leads to the following conditions for the functions ψ^j and φ^j

$$\int \left(\nabla \psi^j\right)^2 dS = 1, \ \int \left(\nabla \varphi^j\right)^2 dS = 1 \tag{I.16}$$

where ∇ is the 2D gradient operator. Normally, instead of the above equations, it is simpler to use the following equivalent equations derived from eqns. I.12

$$\alpha_j^2 \int \left(\psi^j\right)^2 dS = 1, \ \alpha_j^2 \int \left(\varphi^j\right)^2 dS = 1 \tag{I.17}$$

Therefore, ψ^j and φ^j are dimensionless.

A special case for which eqn. I.17 cannot be derived from eqn. I.16 and also cannot be used, is presented in subsection 3.5.2.

Sometimes, it is also possible to express the fields in a waveguide filled with a medium through transversal eigenfunctions when the filling is of a special kind of function. In this case, the mode components are generally expressed by two transversal eigenfunctions.

I.3.3 Transversal eigenfunctions

Now we can write down the functions ψ^j and φ^j for circular and rectangular waveguides. For circular waveguides

$$\psi^j = N^j J_n(\alpha_{nq}\rho)\begin{Bmatrix}\cos(n\beta)\\ or\\ \sin(n\beta)\end{Bmatrix}$$
$$\varphi^j = M^j J_n(\alpha_{nq}\rho)\begin{Bmatrix}\cos(n\beta)\\ or\\ \sin(n\beta)\end{Bmatrix} \tag{I.18}$$

where the mode indices (n, q) define the mode (TE_{nq} or TM_{nq}, respectively), and J_n are the Bessel functions. The eigenvalues α_{nq} are connected with the waveguide radius a by the relations

$$\alpha_{nq} = \frac{\mu_{nq}}{a} \tag{I.19a}$$

$$\alpha_{nq} = \frac{\nu_{nq}}{a} \tag{I.19b}$$

for TE or TM modes, respectively. In the above equations μ_{nq} is the qth positive root of the equation $J_n'(\mu) = 0$ and ν_{nq} is the qth positive root of the equation $J_n(\nu) = 0$. Later, we use the obvious notations μ_j and ν_j. The normalisation factors N^j and M^j are determined from eqn. I.16

$$N^j = \sqrt{\frac{2}{\pi\epsilon_n}}\frac{1}{\sqrt{\mu_j^2 - n^2}\left|J_n(\mu_j)\right|}$$
$$M^j = \sqrt{\frac{2}{\pi\epsilon_n}}\frac{1}{\nu_j\left|J_n'(\nu_j)\right|} \tag{I.20}$$

where $\epsilon_n = 1 + \delta_{0n}$, i.e. $\epsilon_0 = 2$ and $\epsilon_n = 1$ for $n \neq 0$.

To describe the electromagnetic fields in a rectangular waveguide we use the (x, y) co-ordinate system. The co-ordinate origin corresponds to a vertex of the rectangle, the x-axis corresponds to the broad side and the y-axis to the narrow side of this rectangle, as it is shown in Figure I.1.

Then,

$$\psi^j = N^j \cos\left(\frac{\pi n x}{a}\right)\cos\left(\frac{\pi q y}{b}\right)$$
$$\varphi^j = M^j \sin\left(\frac{\pi n x}{a}\right)\sin\left(\frac{\pi q y}{b}\right) \tag{I.21}$$

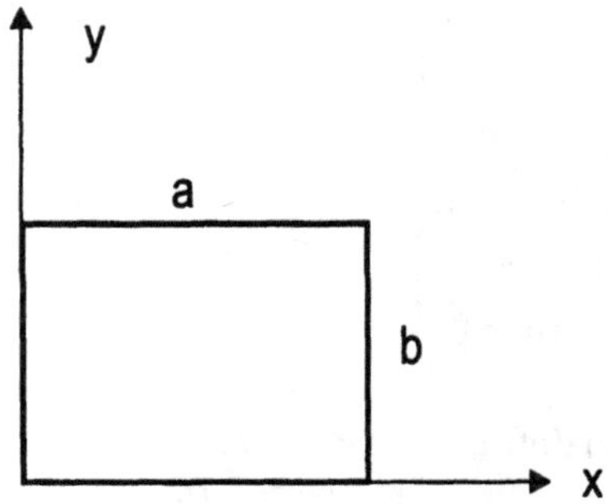

Figure I.1 *Cross section of a rectangular waveguide and definition of the co-ordinate system*

where *a*, *b* are the broad and narrow sides, respectively. The eigenvalues of the TE_{nq} and TM_{nq} modes are described by the same formulae

$$\alpha_{nq}^2 = \pi^2\left(\frac{n^2}{a^2}+\frac{q^2}{b^2}\right) \tag{I.22}$$

The normalisation factors are given by

$$N^j = M^j = \sqrt{\frac{1}{ab\,\epsilon_n\,\epsilon_q}\frac{2}{\alpha_j}} \tag{I.23}$$

Chapter 1

The method of small nonuniformities

The simplest nonuniformity is the small nonuniformity, limited to a small area and having little effect on the waveguiding properties. There are three kinds of principal nonuniformity: waveguide curvature, variation of waveguide filling or impedance and variation of waveguide cross section. The corresponding small nonuniformities are the small angle tilt, a small step of the waveguide filling parameters ε and μ or of impedance and a small step of waveguide cross section, respectively. The electrodynamic problems corresponding to wave scattering from the small nonuniformities mentioned above are considered in the three sections of this Chapter, namely the scattering from a junction of two half-infinite uniform waveguides different by one of the following features: by the directions of their axes, by their filling media, their impedance or by their cross sections. The analysis of the solutions obtained for small nonuniformities leads to the calculation of the field scattered by a nonuniform waveguide section of finite length, where the parameter variations are not small at all.

1.1 Waveguide curvature

1.1.1 Small waveguide tilt

One of the principal kinds of nonuniform waveguide is a bent waveguide with constant cross section. The simplest example of a waveguide bend is a tilt with a small angle. We start from a calculation of the fields of the waves scattered from this elementary tilt. Say that AA and BB are the two planes orthogonal to the axes of the corresponding waveguides connected by the bend. These planes are the boundaries of the bend (Figure 1.1). Let us assume that the angle $\Delta\vartheta$ formed by these two orthogonal planes is so small that the distances between the respective points of both planes AA and BB are less than the wavelength. Later, such a bend with small electrical length is called a waveguide tilt. It is sufficient to study a plane (2D) bend only, because a small 3D bend can be analysed as the superposition of a small 2D bend and a twist by a small angle. A

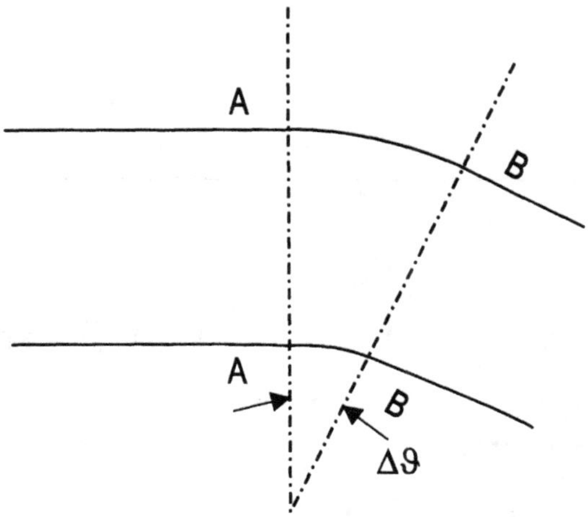

Figure 1.1 *Waveguide tilt*

small twist is a particular case of the deformation studied in Section 1.3. The amplitudes of the modes scattered from a small 3D bend can be derived by summation of the amplitudes of the modes scattered from the two deformations mentioned above.

An *m*th ($m > 0$) waveguide mode with a normalised unitary amplitude of 1 is incident on the tilt from its left-hand side. The fields of the incident wave are $\hat{\vec{E}}^m$ and $\hat{\vec{H}}^m$. The total electromagnetic field at the cross section AA ($z = 0$) is formed by the fields of the incident wave and the superposition of scattered backward waves. This field can be expressed as follows

$$\vec{E} = \vec{E}^m + \sum_{j=1}^{\infty} P_{-j}\vec{E}^{-j}, \quad \vec{H} = \vec{H}^m + \sum_{j=1}^{\infty} P_{-j}\vec{H}^{-j} \tag{1.1}$$

whereas the electromagnetic field at the cross section BB is composed of forward waves only

$$\vec{E} = \sum_{j=1}^{\infty} P_j\vec{E}^j, \quad \vec{H} = \sum_{j=1}^{\infty} P_j\vec{H}^j \tag{1.2}$$

The unknown variables are the amplitudes P_{-j} and P_j.

First, we determine the forward wave amplitudes in the right-hand side (i.e. second) waveguide. Multiplying $\vec{E}$ and $\vec{H}$ in eqn. 1.2 by the vectors $\vec{E}^j$ and $\vec{H}^j$, summing, integrating over the cross section and using the orthogonality condition of eqn. I.9, yields

$$P_j = \frac{1}{2kh_j} \int \left\{ \left[\vec{H}^j \times \vec{E} \right]_z + \left[\vec{H} \times \vec{E}^j \right]_z \right\} dS \tag{1.3}$$

The vectors $\vec{E}$ and $\vec{H}$ in eqn. 1.3 correspond to the cross section BB. Let us express these through the vectors $\vec{E}$ and $\vec{H}$ on the cross section AA. To do this, we orient the x-axis of the Cartesian co-ordinate system on the waveguide cross section in such a way that it points to the curvature centre. The y-axis is normal to the plane of Figure 1.1 and directed from us in order to form a right-hand oriented co-ordinate system (x,y,z). Each pair of points on the cross sections AA and BB corresponding to each other have the same y value in the co-ordinate system for one of the cross sections, and their x co-ordinates differ by terms of the order of $(\Delta\vartheta)^2$, whereas their z co-ordinates differ by Δz proportional to $\Delta\vartheta$. We can neglect the differences in co-ordinate x because the amplitudes of the scattered waves must be determined by the terms of the first order of $\Delta\vartheta$.

The relationships between the transversal components of the fields $\vec{E}$ and $\vec{H}$ in the cross sections AA and BB are

$$H_x \Big|_{BB} = H_x + \Delta z \frac{\partial H_x}{\partial z} - \Delta\vartheta \cdot H_z \Big|_{AA} \tag{1.4a}$$

$$H_y \Big|_{BB} = H_y + \Delta z \frac{\partial H_y}{\partial z} \Big|_{AA} \tag{1.4b}$$

We do not write down here a second similar relation for the transversal $\vec{E}$ components. The first two terms in eqns. 1.4 are the first terms of a Taylor series. The third term corresponds to the rotation of the BB co-ordinate system from the co-ordinate system of the cross section AA. The Δz value in eqns. 1.4 is equal to $r\,\Delta\vartheta$, where r is the distance from the curvature centre to the given cross-section point. Obviously, this distance varies for different points of the cross section.

Let us substitute eqns. 1.4 into eqn. 1.3. Doing this substitution we can use in eqn. 1.4 at the cross section AA only the $\hat{\vec{E}}^m$ and $\hat{\vec{H}}^m$ fields, i.e. the fields of the incident wave instead of the total fields $\vec{E}$ and $\vec{H}$ of eqn. 1.1 containing the incident and reflected waves. Indeed, by substituting the fields of the reflected modes $P_{-j}\vec{E}^{-j}$ and $P_j\vec{H}^{-j}$ ($j > 0$) in the first terms of the sum 1.4, the corresponding terms in eqn. 1.3 disappear because of the orthogonality condition. If these fields are substituted in the second terms of eqns. 1.4 then the corresponding terms of eqn. 1.3 are different from zero; however, these terms are of the order $(\Delta\vartheta)^2$ and therefore can be neglected as well. Hence, it is possible to substitute $\vec{H}$ in eqn. 1.3 by the following field components

$$\begin{aligned} H_x &= H_x^m - \Delta\vartheta\left(ih_m r H_x^m + H_z^m\right) \\ H_y &= H_y^m - \Delta\vartheta i h_m r H_y^m \end{aligned} \tag{1.5}$$

and by the similar components of the $\vec{E}$ field. This is the peculiarity of eqn. 1.3.

Substituting eqn. 1.5 into eqn. 1.3, we obtain the following expressions for the desired values of P_j (according to the above discussion, they are valid including the terms of the order of $\Delta\vartheta$)

$$P_j = \delta_{jm} + \Delta\vartheta \cdot F_{jm} \tag{1.6}$$

The coefficients F_{jm} are given by

$$F_{jm} = \frac{1}{2kh_m}\left\{-ih_m\int r\left(E_y^j H_x^m - E_x^j H_y^m - E_x^m H_y^j + E_y^m H_x^j\right)dS - \int\left(E_y^j H_z^m - E_z^m H_y^j\right)dS\right\} \tag{1.7}$$

As was expected, the amplitudes of the scattered waves with the index $j \neq m$ are proportional to the angle $\Delta\vartheta$ of the waveguide tilt.

The amplitude of the transmitted wave m differs from 1, i.e. it differs from the amplitude of the incident mode by a value of the order of $\Delta\vartheta$. It is equivalent to say that the incident mode transmitted through the tilt acquires the factor

$$\exp(-iq_m\Delta\vartheta) \tag{1.8}$$

where $q_m = iF_{m,m}$, i.e.

$$q_m = \frac{1}{k}\int r\left(E_y^m H_x^m - E_x^m H_y^m\right)dS - \frac{i}{2kh_m}\int\left(E_y^m H_z^m - E_z^m H_y^m\right)dS \tag{1.9}$$

If h_m is a real number, q_m is a real number as well. Therefore, eqn. 1.8 gives us the phase shift of the wave at the waveguide tilt output.

One can always plot a line parallel to the generating line in the curved section of a waveguide in such a way that the phase $q_m\Delta\vartheta$ is equal to the wavenumber h_m multiplied by the arc differential dl_m of this line. In other words, the mode propagates in the curved waveguide section with the same phase velocity as in the uniform waveguide, if the distance between the planes AA and BB is measured along the arc of a definite line. More rigorously, it is correct for an infinitesimal angle only. In the case of a finite curvature, as we will see later, the phase velocity depends on conversion to other modes as well. Generally speaking, the lines l_m are different for various modes. However, e.g. in a circular waveguide, all of these lines are coincident with the waveguide axis (see the calculation in Section 5.3).

To determine the amplitudes of the backward waves propagating in the left-hand side waveguide, we will employ the same technique that was used for the derivation of eqn. 1.6. Multiplying eqn. 1.1 by $\vec{H}^{-j}$ and $\vec{E}^{-j}$ and performing the same transformations as for the derivation of eqn. 1.3, yields

$$P_{-j} = \frac{1}{2kh_{-j}} \int \left\{ \left[\vec{H}^{-j} \times \vec{E} \right]_z + \left[\vec{H} \times \vec{E}^{-j} \right]_z \right\} dS \tag{1.10}$$

The vectors $\vec{E}$ and $\vec{H}$ in eqn. 1.10 correspond to the cross section AA. We express these in terms of the vectors $\vec{E}$ and $\vec{H}$ of the cross section BB; the corresponding formulae similar to eqns. 1.4 have the following form

$$\begin{aligned} H_x\Big|_{AA} &= H_x - \Delta z \frac{\partial H_x}{\partial z} + \Delta\vartheta \cdot H_z\Big|_{BB} \\ H_y\Big|_{AA} &= H_y - \Delta z \frac{\partial H_y}{\partial z}\Big|_{BB} \end{aligned} \tag{1.11}$$

where Δz has the same sense as in eqns. 1.4.

The first term in eqns. 1.11 can be omitted when substituting eqn. 1.10 by eqns. 1.11, since, according to eqn. 1.2, it contains the fields of waves with positive superscripts only. These waves disappear when they are substituted into eqn. 1.10. We can neglect in the other terms of eqns. 1.11 the fields of all waves except the *m*th wave, since eqn. 1.10 is valid up to the terms of the order of $\Delta\vartheta$, inclusively. The *m*th wave amplitude can be assumed there to be equal to 1. Performing the same transformations as for the derivation of eqn. 1.6, we get

$$P_{-j} = -\Delta\vartheta \cdot F_{-jm} \tag{1.12}$$

The coefficient F_{-jm} can be derived from F_{jm} in eqn. 1.7, formally exchanging the subscripts j and $-j$. Eqns. 1.6, 1.12 and 1.7 give the solution of the problem of scattering from a small tilt formulated in this section. The coefficients F_{jm} are studied later, developing the general cross-section method.

1.1.2 Waveguide bend as the limit of a waveguide with a large number of small tilts

The results obtained above allow us to find, in some cases, the amplitudes of the modes excited along a bent waveguide with a finite bend angle ϑ_0. We simulate the bent waveguide by a waveguide with a large number of small tilts, i.e. by a set of interconnected straight waveguide sections forming small angles with each other. The *j*th parasitic output wave is the superposition of the elementary waves of the same number excited by each tilt. A final result for a bend can be obtained as the limit of the sum, i.e. as an integral. As is well known, a similar method is used in transmission-line theory to derive the reflection coefficient of a line with a variable characteristic impedance; in this case the nonuniformity is a very small step of the characteristic impedance.

Following eqn. 1.6, the amplitude of an elementary parasitic wave with the superscript $j \neq m$ excited by the section of a bent waveguide located between the angles ϑ and $\vartheta + \Delta\vartheta$, is equal to the product of $F_{jm}\,\Delta\vartheta$ and the amplitude of the mth wave (i.e. the incident wave) at this point. According to eqn. 1.8 this last amplitude has the value of

$$\exp\left(-i\int_0^{\vartheta} q_m\, d\vartheta\right) = \exp(-ih_m l_m)$$

if we measure the arc length l_m from the beginning of the bend. When the parasitic wave reaches the end of the bend, it gets an additional phase increment

$$\exp\left(-i\int_{\vartheta}^{\vartheta_0} q_j\, d\vartheta\right) = \exp\left(-ih_j\left[l_j(J_0) - l_j(J)\right]\right)$$

Therefore, the amplitude of the elementary parasitic wave at the output of the bend is equal to

$$F_{jm}\,\Delta\vartheta \exp\left(-ih_j l_j(\vartheta_0)\right)\cdot\exp\left(-i\left[h_m l_m(\vartheta) - h_j l_j(\vartheta)\right]\right)$$

The total output amplitude of the parasitic mode is equal to

$$F_{jm}\int_0^{\vartheta_0}\exp\left(-i\left[h_m l_m(\vartheta) - h_j l_j(\vartheta)\right]\right)d\vartheta \qquad (1.13)$$

where the insignificant factor $\exp(-ih_j l_j(\vartheta_0))$ is omitted.

The amplitude of the reflected wave of the parasitic mode is given by the same expression after the exchange of F_{jm} for F_{-jm},as well as h_j for $h_{-j} = -h_j$.

In some cases the integral of eqn. 1.13 can be simply calculated. For example, for a waveguide bend with constant radius, where $l_m(\vartheta)$ and $l_j(\vartheta)$ are proportional to ϑ. For a bend with variable radius (variable curvature), one can assume for the sake of simplicity that $l_m(\vartheta)=l_j(\vartheta)$ and consider the radius r as the distance from the centre of curvature to the line along which the length l is measured. The integral of eqn.1.13 can then be rewritten in the following way

$$F_{jm}\int_0^{L}\exp\left(-i\left(h_m - h_j\right)l\right)\frac{dl}{r(l)} \qquad (1.14)$$

where L is the arc length of the waveguide bend.

The analysis of wave propagation in a bent waveguide with a finite angle described above is based on the solution of the problem of a small tilt. This method has an evident physical explanation. The nature of the coupling coefficient F_{jm} as well as the structure of the exponential factor in eqn. 1.13 become clear and understandable. From this point of view, the excitation of parasitic waves in a bend is the result of two phenomena: the excitation of elementary waves at each tilt and the superposition of all those waves with their corresponding phases. Such an approach is very useful for a qualitative analysis of real circuits. The method, by which the analytical expression eqn. 1.7 for F_{jm} was derived with the help of eqn. 1.4, allows us to distinguish two factors causing the mode conversion in a waveguide tilt and in a bend. The first of these is the difference of the distances between the corresponding points of both cross sections rotated in respect to each other by a certain angle. The second factor has a vector character: transversal components of the fields in one cross section are coupled to longitudinal components in a rotated adjacent cross section.

On the other hand, a rigorous derivation of formulae such as eqn. 1.13 from eqn. 1.6 is a relatively difficult problem, more complicated than their direct derivation from the Maxwell equations. The method developed in the last subsection is limited to the situation where the power of the incident mode does not change significantly as a result of its propagation through the total bend, and therefore the conversion of parasitic modes to each other and their reconversion to the fundamental mode can be neglected. This condition requires a large curvature radius, i.e $kr >> 1$. Besides this, the perturbation method employed to solve the problem of a tilt is not sufficiently accurate near the cutoff frequency, where $h_j \to 0$ and $F_{jm} \to \infty$, as well as in several other cases. Hence, the general analysis of a bend with finite angle is done later by virtue of the coupled-wave differential equations for the coefficients P_j considered as functions of ϑ. These coupled-wave differential equations can be derived from eqn. 1.6 as well; however, a more general and simple technique is the cross-section method developed in Chapter 2.

1.2 Variation of waveguide filling medium

1.2.1 Small step of permittivity and permeability of the waveguide filling medium

The second basic kind of nonuniform waveguide is a straight one with constant cross section, but with permittivity ε and permeability μ of its filling medium varying along the waveguide in such a way that $\varepsilon = \varepsilon(x,y,z)$ and $\mu = \mu(x,y,z)$. To this kind of nonuniformity belong, for instance, all types of insertion with tapered cross section. Normally, ε (or μ) is a step function. Nevertheless, general results can be obtained more easily by first considering ε and μ as continuous functions and approaching the corresponding limits later.

The simplest example of the nonuniformity mentioned above is the small step of the ε and μ values with arbitrary values across the waveguide, i.e. the junction of two semiinfinite uniform waveguides having nearly the same values of ε and μ. Let us evaluate the amplitudes of the waves scattered from this step; we will start from the calculation of forward wave amplitudes.

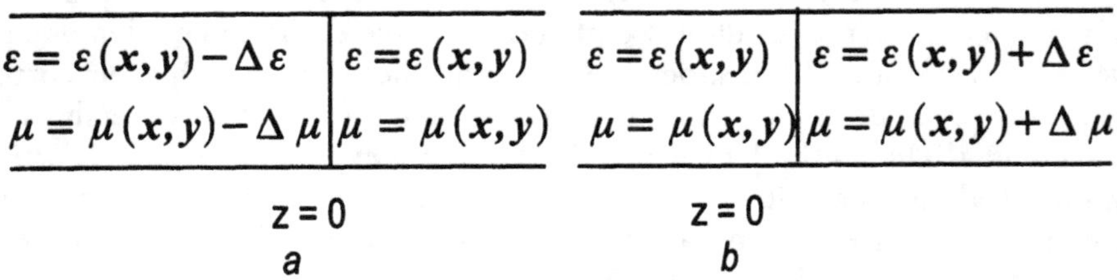

Figure 1.2 Step-wise, small variation of the properties of the waveguide filling medium

Let us assume that $\varepsilon = \varepsilon(x,y)-\Delta\varepsilon$ and $\mu = \mu(x,y)-\Delta\mu$ for $z<0$, whereas for $z>0$ we have $\varepsilon = \varepsilon(x,y)$ and $\mu = \mu(x,y)$. Here, $\Delta\varepsilon$ and $\Delta\mu$ have small magnitudes, and in the general case they are functions of the variables x and y (Figure 1.2*a*). A wave with superscript m and unitary amplitude is incident from the left-hand side and we need to evaluate the amplitudes of the waves propagating away from the junction. A version of the small perturbations method can be applied to solve this problem; here the perturbations are the small steps ($\Delta\varepsilon$ and $\Delta\mu$) of the ε and μ values in the left-hand side waveguide.

If $\Delta\varepsilon = 0$ and $\Delta\mu = 0$, the field is undisturbed along the whole waveguide and equal to the incident wave field $\hat{\vec{E}}^m$ and $\hat{\vec{H}}^m$. If $\varepsilon \neq 0$ and $\Delta\mu \neq 0$, additional fields $\Delta\vec{E}$ and $\Delta\vec{H}$ are excited. These fields satisfy the equations derived by variation of the Maxwell equations (eqns. I.1) written for $\hat{\vec{E}}^m$ and $\hat{\vec{H}}^m$

$$\begin{aligned} \mathrm{rot}\Delta\vec{E} + ik\mu\Delta\vec{H} &= ik\Delta\mu\hat{\vec{H}}^m \\ \mathrm{rot}\Delta\mathrm{H} - ik\varepsilon\Delta\vec{E} &= -ik\Delta\varepsilon\hat{\vec{E}}^m \end{aligned} \tag{1.15}$$

According to eqn. 1.15, the fields $\Delta\vec{E}$ and $\Delta\vec{H}$ are excited by magnetic currents (their density is proportional to $\Delta\mu\hat{\vec{H}}^m$) and by electric currents (their density is proportional to $\Delta\varepsilon\hat{\vec{E}}^m$). These currents are located at $z<0$, and we must determine the fields excited by these currents at $z>0$.

The fields $\Delta\vec{E}$ and $\Delta\vec{H}$ have the following form at $z=+0$

$$\Delta\vec{E} = \sum_{j=1}^{\infty}\left(P_j - \delta_{mj}\right)\vec{E}^j, \quad \Delta\vec{H} = \sum_{j=1}^{\infty}\left(P_j - \delta_{mj}\right)\vec{H}^j \tag{1.16}$$

where P_j are the desired amplitudes of the forward waves.

According to eqns. I.7 and I.6, the orthogonality condition of eqn. I.9 can be rewritten as follows

$$\int\left\{\left[\vec{H}^{-j}\times\vec{E}^m\right]_z - \left[\vec{H}^m\times\vec{E}^{-j}\right]_z\right\}\cdot d\vec{S} = 2kh_j\delta_{jm} \tag{1.17}$$

Employing this expression, we obtain for P_j (when $j \neq m$) an integral expression analogous to eqn. 1.3

$$P_j = \frac{1}{2kh_j}\int\left\{\left[\vec{H}^{-j}\times\Delta\vec{E}\right]_z - \left[\Delta\vec{H}\times\vec{E}^{-j}\right]_z\right\}\cdot d\vec{S} \tag{1.18}$$

On the other hand, from eqn. 1.15 and the Maxwell eqns. I.1 written for the fields $\hat{\vec{E}}^{-j}$ and $\hat{\vec{H}}^{-j}$ in the undisturbed waveguide, it follows that

$$\begin{aligned}&\operatorname{div}\left\{\left[\hat{\vec{H}}^{-j}\times\Delta\vec{E}\right]-\left[\Delta\vec{H}\times\hat{\vec{E}}^{-j}\right]\right\}\\&= ik\left(\Delta\varepsilon\hat{\vec{E}}^{m}\cdot\hat{\vec{E}}^{-j} - \Delta\mu\hat{\vec{H}}^{m}\cdot\hat{\vec{H}}^{-j}\right)\end{aligned} \tag{1.19}$$

Let us integrate this equation over the whole space where $\Delta\varepsilon \neq 0$ and $\Delta\mu \neq 0$. According to the boundary condition eqn. I.2 satisfied by the fields $\vec{E}^{-j}$ and $\Delta\vec{E}$, the components of the vectors $\left[\hat{\vec{H}}^{-j}\times\Delta\vec{E}\right]$ and $\left[\Delta\vec{H}\times\hat{\vec{E}}^{-j}\right]$ normal to the metal are identically equal to zero. Let us add a small complex component to the wavenumber k, then we can guarantee the integral convergence by integrating over z from $-\infty$ to 0 and eliminating the result obtained by substituting the lower limit in the integral. In such a way, we obtain for the desired amplitudes of the forward propagating parasitic waves eqn. 1.18

$$P_j = \frac{1}{2h_j\left(h_j - h_m\right)}\int\left\{\Delta\varepsilon\vec{E}^{m}\cdot\vec{E}^{-j} - \Delta\mu\vec{H}^{m}\cdot\vec{H}^{-j}\right\}dS \tag{1.20}$$

where the integral was calculated at $z = -0$. The physical meaning of the right-hand side of eqns. 1.15 is underlined by comparing eqn. 1.20 with the usual formulae of waveguide excitation theory.

To determine the backward propagating wave amplitudes we will suppose that the perturbation takes place in the right-hand side waveguide, i.e. $\varepsilon = \varepsilon(x,y)$ and $\mu = \mu(x,y)$ for $z < 0$, whereas for $z > 0$ we have $\varepsilon = \varepsilon(x,y) + \Delta\varepsilon$ and $\mu = \mu(x,y) + \Delta\mu$ (Figure. 1.2*b*). In such a way, $\Delta\varepsilon$ and $\Delta\mu$ have the same values as above, however we will derive the electromagnetic field only in the domain without sources, as it was done above by the calculation of the forward wave amplitudes (hence, we already know that this way will be easier). Repeating the same manipulations as for the calculation of the forward wave amplitudes above and employing again eqn. 1.15, yields

$$P_{-j} = \frac{-1}{2h_j\left(h_j + h_m\right)}\int\left\{\Delta\varepsilon\vec{E}^{m}\cdot\vec{E}^{j} - \Delta\mu\vec{H}^{m}\cdot\vec{H}^{j}\right\}dS \tag{1.21}$$

The last two formulae solve the problem of electromagnetic fields excited by a small step in the waveguide filling properties.

1.2.2 *Special cases of* $\Delta\varepsilon$ *and* $\Delta\mu$

In particular, a condition can be obtained from eqns. 1.20 and 1.21, by which a small step-wise variation of ε and μ does not excite parasitic modes of types other than the incident one.

First, we will demonstrate that the fields of different modes satisfy not only the orthogonality condition of eqn. I.9, but also another orthogonality condition containing all the three components of the electromagnetic field

$$\int\left\{\varepsilon\vec{E}^m\cdot\vec{E}^j+\mu\vec{H}^m\cdot\vec{H}^j\right\}dS=-2h_m^2\delta_{mj} \tag{1.22}$$

This condition follows from the identity

$$\begin{aligned}&\operatorname{div}\left\{\left[\hat{\vec{H}}^m\times\hat{\vec{E}}^j\right]+\left[\hat{\vec{H}}^j\times\hat{\vec{E}}^m\right]\right\}\\&=2ik\left(\varepsilon\hat{\vec{E}}^m\cdot\hat{\vec{E}}^j+\mu\hat{\vec{H}}^m\cdot\hat{\vec{H}}^j\right)\end{aligned} \tag{1.23}$$

which can easily be obtained from the Maxwell equations in eqns. I.1 rewritten for the fields of the modes m and j. To obtain eqn. 1.22 from eqn. 1.23, one should find the derivative of eqn. 1.23 with respect to z, integrate the obtained relationship over the whole cross section and apply the boundary condition of eqn. I.2 for $\vec{E}^m$ and $\vec{E}^j$, together with the usual orthogonality condition, eqn. I.9.

From a comparison of eqns. 1.20 and 1.21 with eqn. 1.22 it follows that, if the following condition is satisfied at every point of the border ($z = 0$)

$$\Delta\varepsilon=A\cdot\varepsilon,\ \Delta\mu=-A\cdot\mu \tag{1.24}$$

where A is a constant value independent of x and y, there is no mode conversion and only the transmitted and reflected waves of the same mode are excited.

The condition of eqn. 1.24 is sufficient but not necessary. Nevertheless, it must be remarked that eqn. 1.24 is stronger than the requirement of equal phase velocities in both waveguides. The last requirement can be expressed in the form

$$\Delta(\varepsilon\cdot\mu)=0 \tag{1.25}$$

This condition follows from eqn. 1.24. Generally, the absence of a step-wise variation in the phase velocity, i.e. the condition of eqn. 1.24, is not sufficient to avoid mode conversion.

1.2.3 *Waveguide with continuously varying filling as limit of waveguide with layered filling*

Before considering waveguides with ε and μ being an arbitrary function of the variable z, we have to generalise the expressions eqn. 1.20 and eqn. 1.21 for the case where the transition from a semiinfinite waveguide to another semiinfinite waveguide with nearly the same values of ε and μ occurs on a small length Δz ($k \cdot \Delta z << 1$).

Analysing the method by which these expressions were obtained, it is easy to prove their validity for this case as well. One should only interpret $\Delta\varepsilon$ and $\Delta\mu$ as the full differences of the ε and μ values in both waveguides. For further discussion it is convenient to substitute in the integral $\Delta\varepsilon$ by $\left(\dfrac{\partial\varepsilon}{\partial z}\right)\Delta z$ and $\Delta\mu$ by $\left(\dfrac{\partial\mu}{\partial z}\right)\Delta z$.

Let us define

$$S_{jm} = \frac{1}{2h_j\left(h_j - h_m\right)}\int\left\{\frac{\partial\varepsilon}{\partial z}\vec{E}^m \cdot \vec{E}^{-j} - \frac{\partial\mu}{\partial z}\vec{H}^m \cdot \vec{H}^{-j}\right\}dS \tag{1.26}$$

Then we obtain for the amplitudes of forward and backward propagating parasitic waves scattered by such a nonuniformity

$$P_j = S_{jm} \cdot \Delta z (j \neq m), \quad P_{-j} = -S_{-jm} \cdot \Delta z \tag{1.27}$$

where S_{-jm} is obtained from S_{jm} by a formal substitution of j by $-j$. A rigorous derivation of these formulae will be presented in the next chapter.

Now let ε and μ be arbitrary functions not only of x and y but also of z along a nonuniform section of the waveguide with its length being equal to L. This section can be considered to be composed of a series of short small uniform waveguide sections with ε and μ tapered by small steps $\Delta\varepsilon(x, y)$ and $\Delta\mu(x, y)$. The amplitudes of parasitic waves excited by each step are given by eqn. 1.27.

The total amplitude of the wave scattered from the whole nonuniform waveguide section is the sum of all the elementary parasitic waves, taking into account the corresponding phase shifts. Obviously, this calculation method is valid only if one can neglect secondary wave conversions of the parasitic and the incident waves. Generally, this is possible only in waveguides with slowly varying parameters, i.e. in those waveguides where the variations of ε and μ along distances of the order of a wavelength and of the order of the cross-section linear dimensions are small in comparison with the values of ε and μ and the operating frequency is far away from the values where $h_j = 0$. Similar considerations are employed in the theory of wave propagation along an inhomogeneous infinite medium, Brekhovskikh (1980).

The propagation of the fundamental wave from the beginning of the nonuniform section at $z = 0$ to the layer between z and $z + dz$ as well as the propagation of the parasitic wave excited by this layer to the nonuniform section output (i.e. to $z = L$ for forward propagating waves or to $z = 0$ for the backward waves) is described in a slightly more complicated way than for the similar problem of a curved waveguide, since in this problem the wavenumbers h_m and h_j depend on z. It follows from the representation of the nonuniform waveguide as the limit of the multilayer waveguide that the phase shift corresponding to the propagation along a waveguide section is equal to the integral $\int h_j dz$ over this waveguide section, i.e. it is equal to the difference of the values of the $\gamma_j(z)$ function at both ends of the section, whereas this function is defined by the equation

$$\gamma_j = \int_0^z h_j dz \tag{1.28}$$

According to eqn. I.10, for a wave propagating along the waveguide the absolute value of its amplitude varies in such a way that its power flux is constant; simultaneously, $P_j \approx 1/\sqrt{h_j}$. If $P_m = 1$, then repeating the discussion which led us to eqn. 1.14, we derive the following expression for the amplitudes P_j of the forward parasitic waves excited by the incident wave with the superscript m $(j \neq m)$

$$\sqrt{\frac{h_m(0)}{h_j(L)}}\int_0^L S_{jm}\sqrt{\frac{h_j}{h_m}}\exp\{-i(\gamma_m - \gamma_j)\}dz \tag{1.29}$$

In this expression we have once again omitted an unimportant phase factor.

The backward wave amplitude can be obtained substituting γ_j, S_{jm} and $h_j(L)$ in eqn. 1.29 by γ_{-j}, $-S_{-jm}$ and $h_j(0)$, respectively. Unlike the analogous coefficient F_{jm} in the curved waveguide problem, the coefficient S_{jm} depends on z and therefore it cannot be moved outside of the integral symbol.

We will analyse eqn. 1.29 containing the S_{jm} coefficients later, after deriving them by a more rigorous and consistent, but less obvious, method. Let us only remark here that the sufficient condition for avoiding mode conversion in a waveguide with variable filling is

$$\frac{\partial \varepsilon}{\partial z} = A\varepsilon, \quad \frac{\partial \mu}{\partial z} = -A\mu \text{ (A does not depend on x and y)} \tag{1.30}$$

which is analogous to eqn. 1.24. Once again, this condition is stronger than the requirement of constant phase velocity $\partial(\varepsilon \cdot \mu)/\partial z = 0$. The integral form of eqn. 1.30 implies the existence of such a function $\Phi(z)$ that

$$\varepsilon(x,y,z) = \varepsilon(x,y,z_0)\frac{\Phi(z)}{\Phi(z_0)} \tag{1.31a}$$

and

$$\mu(x,y,z) = \mu(x,y,z_0)\frac{\Phi(z_0)}{\Phi(z)} \tag{1.31b}$$

1.2.4 Small step of the isotropic waveguide impedance

A small step of the waveguide wall impedance can be considered not as an independent kind of nonuniformity, but as a particular case of small step in the dielectric filling. As is well known, the $\vec{E}$ field tangential components on a metallic surface covered with a thin film of a high value of $|\varepsilon|$ are proportional to the tangential components of the $\vec{H}$ field on the surface of this film, i.e. boundary value impedance conditions are satisfied on such a surface. All impedances, independent of the type of surface (e.g. nonideal conductivity, surface corrugations, etc.) can be modelled by means of such a film, where the ε parameter would be a tensor in a more general case. The impedance value depends on the ε value of the film and impedance variations can be considered as variations of ε. Therefore, we can obtain the required scattering matrix on a small impedance step from eqns. 1.20 and 1.21, for a small step in the filling medium.

But we will not implement such a procedure, i.e. we do not model the impedance with an equivalent dielectric film. It is easier to solve the problem of the impedance step variation directly. Nevertheless, the relationship between both these problems allows us to apply properly the results obtained above to achieve the solutions for this and the problems in the rest of subsections in this Section.

In this Section we will consider the simplest impedance boundary condition

$$E_s = wH_z,\ E_z = -wH_s \tag{1.32}$$

We will not assume a small value of w. Leontovich's boundary condition for a nonideal conductor is a particular case of the condition in eqns. 1.32 for $|w| \ll 1$. In eqn. 1.32 w can be a function of the waveguide circumferential coordinate s, i.e. $w = w(s)$, and because of this we do not extract w from the integral symbol for integration over s. The orthogonality condition in eqn. 1.17 employed later is valid for an arbitrary function $w(s)$. In the next subsection we will demonstrate this identity for a boundary condition more general than eqn. 1.32.

Let w be the waveguide wall impedance at $z > 0$ and $w - \Delta w$ its value at $z < 0$ where $\Delta w \ll 1$. The mode m is incident on the impedance step at $z = 0$ from the left-hand side semiinfinite waveguide. Its fields satisfy the condition in eqns. 1.32 where we can substitute for w by $w - \Delta w$. In the semiinfinite waveguide on the right-hand side, i.e. at $z \geq 0$, the electromagnetic field $\vec{E}$, $\vec{H}$ can be written as the expansions in series as in eqn. 1.2

$$\vec{E} = \sum_{j=1}^{\infty} P_j \hat{\vec{E}}^j, \; \vec{H} = \sum_{j=1}^{\infty} P_j \hat{\vec{H}}^j \tag{1.33}$$

The fields $\hat{\vec{E}}^j$ and $\hat{\vec{H}}^j$ satisfy the condition in eqns. 1.32 with the impedance value w for all z. We have to find the amplitudes P_j for $j > 0$ and $j \neq m$.

The fields $\vec{E}$ and $\vec{H}$ as well as the $\hat{\vec{E}}^{-j}$ and $\hat{\vec{H}}^{-j}$ fields satisfy the homogeneous Maxwell equations, hence the following relationship holds

$$\operatorname{div}\left\{\left[\hat{\vec{H}}^{-j} \times \vec{E}\right] - \left[\vec{H} \times \hat{\vec{E}}^{-j}\right]\right\} = 0 \tag{1.34}$$

If the condition $\operatorname{div}\vec{A} = 0$ is valid for a certain vector $\vec{A}$, then integrating this condition over the volume of the semiinfinite waveguide from $z = -\infty$ to $z = +0$ we obtain the following relationship between the integral of A_z over the waveguide cross section and the integral of A_n over the boundary of the cross section

$$\int_{z=+0} A_z dS + \int_{-\infty}^{0} dz \oint A_n \, ds = 0 \tag{1.35}$$

where $\vec{n}$ is the normal to the boundary and three vectors $\vec{n}, \vec{s}, \vec{z}$ are unit vectors of a right-hand side co-ordinate system. Following the same procedure used to deduce eqn. 1.20, it is necessary to add a small complex value to the wavenumber k to derive eqn. 1.35. If $\vec{A}$ is the vector under the div symbol in eqn. 1.34, then according to eqns. 1.33 and 1.17 the first integral in eqn. 1.35 is equal to $2kh_j P_j$.

Now, let us calculate the second term in eqn. 1.35. If $\Delta w = 0$, then this term of the sum will be equal to zero, since the $\vec{E}, \vec{H}$ and $\hat{\vec{E}}^{-j}, \hat{\vec{H}}^{-j}$ fields satisfy the boundary conditions for identical values of w. In the first order of $|\Delta w|$ we can substitute $\vec{E}, \vec{H}$ by the field of the incident mode $\hat{\vec{E}}^m, \hat{\vec{H}}^m$. Replacing the impedance w by $w - \Delta w$ in eqn. 1.32 for the mode with number m we will obtain

$$\left[\hat{\vec{H}}^{-j} \times \hat{\vec{E}}^m\right]_n - \left[\hat{\vec{H}}^m \times \hat{\vec{E}}^{-j}\right]_n = \exp\left(-i\left(h_m - h_j\right)z\right)\left(H_s^m H_s^{-j} + H_z^m H_z^{-j}\right)\Delta w \tag{1.36}$$

Furthermore, calculating the integral over z and taking into account that $\operatorname{Im}(k) \neq 0$, we obtain from eqn. 1.35 the desired expression for the amplitude of the mode j $(j > 0, j \neq m)$

$$P_j = \frac{i}{2kh_j\left(h_j - h_m\right)} \oint \left(H_s^m H_s^j - H_z^m H_z^j\right) \Delta w \, ds \tag{1.37}$$

where, in addition, the relationship in eqn. I.6 between the $\hat{\vec{H}}^{-j}$ and $\hat{\vec{H}}^j$ components has been employed.

This expression is analogous to eqn. 1.20 and could be obtained by interpreting the impedance model described above as the influence of some dielectric film ($\mu = 1$). A similar expression is obtained for backward waves by substituting for j by $-j$.

The physical meaning of eqn. 1.37 is the following: if w is replaced by $w - \Delta w$ on the left-hand side waveguide walls, an additional magnetic current with the components $\vec{H}_s^m \Delta w$ and $\vec{H}_z^m \Delta w$ is excited there by the propagating mode, m. In turn, such a current excites in the right-hand side waveguide modes of all numbers with the amplitudes given by eqn. 1.37.

All the considerations of subsection 1.2.3 and the main formula, eqn. 1.27 are valid for the modes excited along a section of variable impedance $w(s,z)$. According to eqn. 1.37 the coupling coefficients are

$$S_{jm} = \frac{i}{2kh_j\left(h_j - h_m\right)} \oint \frac{\partial w}{\partial z}\left(H_s^m H_s^j - H_z^m H_z^j\right) ds \tag{1.38}$$

1.2.5 Small step of anisotropic waveguide impedance

In a more general case the wall impedance is anisotropic. In technical applications such an impedance is usually produced artificially (for instance, by means of a corrugation) and there is a privileged direction, where the maximum value is reached. Generally, this direction is coincident neither with $\vec{s}$ nor with $\vec{z}$ and therefore the impedance boundary condition has the following matrix form in the local co-ordinates defined by the vectors $\vec{s}$ and $\vec{z}$

$$\begin{aligned} E_s &= w_{ss} H_s + w_{sz} H_z \\ E_z &= w_{zs} H_s + w_{zz} H_z \end{aligned} \tag{1.39}$$

The four elements of the (w) tensor are not independent. One of the relationships between them yields from the reciprocity theorem. This theorem must be valid in any volume limited by the surface, where the conditions of eqn. 1.39 are satisfied. To obtain this relationship let us analyse two fields $\vec{E}^1, \vec{H}^1$ and $\vec{E}^2, \vec{H}^2$ excited by the currents $\vec{j}^1$ and $\vec{j}^2$. As for the proof of the Lorentz lemma, we will employ the identity

$$\operatorname{div}\left\{\left[\vec{E}^1 \times \vec{H}^2\right] - \left[\vec{E}^2 \times \vec{H}^1\right]\right\} = \vec{j}^1 \cdot \vec{E}^2 - \vec{j}^2 \cdot \vec{E}^1 \tag{1.40}$$

following from the inhomogeneous Maxwell equations for both fields. Let us integrate the last expression over the whole volume. The reciprocity theorem will be satisfied if the identity $\int\left(\vec{j}^1 \cdot \vec{E}^2 - \vec{j}^2 \cdot \vec{E}^1\right) dV = 0$ is valid for any currents $\vec{j}^1$ and $\vec{j}^2$, i.e. for any fields. Transforming the integral of the left-hand side of eqn. 1.40 into a surface integral and substituting eqn. 1.39 we obtain the fact that the surface integral of the function $(w_{ss} + w_{zz})(H_s^1 H_z^2 - H_z^1 H_s^2)$ has to be equal to zero. This will be valid for arbitrary fields only if

$$w_{ss} + w_{zz} = 0 \tag{1.41}$$

The trace of the matrix in eqn. 1.39 has to be equal to zero. As can be expected, this condition is invariant on the selection of the directions $\vec{s}$ and $\vec{z}$.

We will only need the condition in eqn. 1.41, and because of this, the other conditions connecting the coefficients in eqn. 1.39 will be given without proof. They are obtained from the requirement that the surface does not generate energy

$$\mathrm{Re}(w_{sz}) \geq 0, \mathrm{Re}(w_{zs}) \leq 0, \mathrm{Re}(w_{ss}) = \mathrm{Re}(w_{zz}) = 0 \tag{1.42}$$

For a nonabsorbing surface, eqns. 1.42 are valid with the equality symbols.

The condition in eqn. 1.41 is necessary and sufficient for the fields of different waveguide modes to be orthogonal to each other under the boundary condition of eqn. 1.39 in the sense of the condition of eqn. 1.17. In fact, eqn. 1.17 is the reciprocity theorem for the fields of m and $-j, (m \neq j)$ eigenmodes in a uniform waveguide, i.e. for the fields excited by sources located at $z = -\infty$ for wave with the number m and located at $z = +\infty$ for wave with the number $-j$.

The simplest way to obtain eqn. 1.17 is to integrate the equation $\mathrm{div}\left\{\left[\hat{\vec{E}}^m \times \hat{\vec{H}}^{-j}\right] - \left[\hat{\vec{E}}^{-j} \times \hat{\vec{H}}^m\right]\right\} = 0$ following the scheme given for eqn. 1.35. The integral over the cross section in eqn. 1.35 coincides with the left-hand side of eqn. 1.17 and it is equal to zero because the line integral, after substituting eqn. 1.39 into it, becomes zero due to eqn. 1.41.

The considerations which led to us to the formulae in eqns 1.37 and 1.38 for the amplitudes of the waves excited by the impedance step, eqn. 1.32, are also valid for a tensorial impedance. The impedance is given by eqn. 1.39 for $z > 0$, whereas for $z < 0$ the impedance is determined by the same formula but having its matrix elements equal to $w_{ss} - \Delta w_{ss}$, and so on. Obviously, we should take into account the fact that $\Delta w_{ss} + \Delta w_{zz} = 0$. The long but elementary calculus realised according to this scheme yields

$$P_j = \frac{i}{2kh_j(h_j - h_m)} \oint \left(-H_z^m H_z^j \Delta w_{sz} - H_s^m H_s^j \Delta w_{zs} + \left(H_z^m H_s^j - H_s^m H_z^j\right) \Delta w_{ss}\right) ds \tag{1.43}$$

For the coefficients S_{jm} describing mode conversion in the waveguide section with monotonically changing impedance, we obtain the obvious expression from eqn. 1.43 containing the derivatives $\partial w_{ss}/\partial z$ instead of Δw_{ss}, and so on. Eqn. 1.38 is its particular case for $w_{sz} = w, w_{zs} = -w$ and $w_{ss} = w_{zz} = 0$.

1.3 Variation of waveguide cross section

1.3.1 The equivalent boundary condition

The straight waveguide with variable cross section is the third and more complicated basic type of nonuniform waveguide. Let us calculate the simplest elementary nonuniformity of this kind, namely a small waveguide step, i.e. the junction of two waveguides with nearly the same cross section.

Let us suppose that for $z < 0$ the cross section of the waveguide is limited by the boundary C_- and for $z > 0$ the boundary is C_+. The boundaries C_- and C_+ can be different either by their position (for instance, by a shift or by a torsion of waveguides) or by their size or by their shape. The waveguide step is characterised by a function $\delta(s)$ defined as the distance between the corresponding points of the boundaries C_- and C_+. This function depends on the curved coordinates counted along the cross-section boundary contour. Since we analyse small steps only, there is no need for a rigorous mathematical definition of correspondence between the points of both boundaries. For a small step, $|\delta(s)|$ is small in comparison with the wavelength and with the linear dimensions of the cross section. We will assume $\delta(s) > 0$ if at a given point the transformation from C_- to C_+ is performed in the direction of the normal vector to the metallic surface, i.e. towards waveguide broadening, and $\delta(s) < 0$ if the opposite is true, as is depicted in Figure 1.3.

For $\delta \equiv 0$ both waveguides are identical and identically positioned; for $\delta \neq 0$ there is a perturbation at the junction and scattered waves are excited. This perturbation from the mathematical point of view is connected not with the equation but with the position of the surface on which the boundary conditions are defined. The perturbation, i.e. a small deformation of the metallic surface, represents a transition from a surface, on which the boundary condition eqn.I.2 is satisfied, to another surface, on which the same boundary condition is satisfied.

To determine the electromagnetic field perturbation, it is worth using instead of the deformation of the surface, on which eqn. I.2 is valid, a small variation of this condition on the undisturbed surface. Let us derive this new boundary condition, which is equivalent to the deformation. We will call it the equivalent boundary condition.

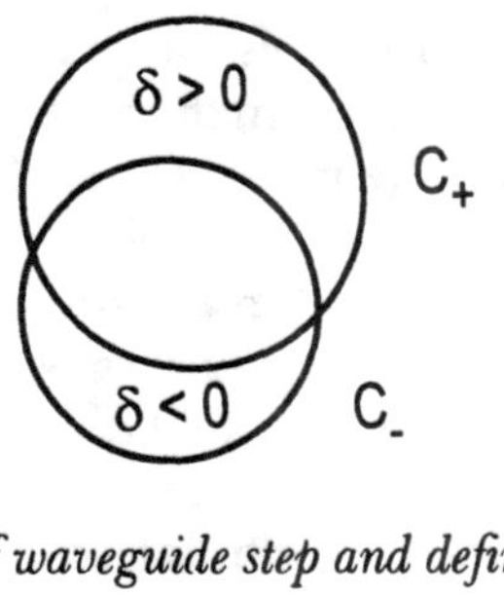

Figure 1.3 Schematic of waveguide step and definition of parameter δ

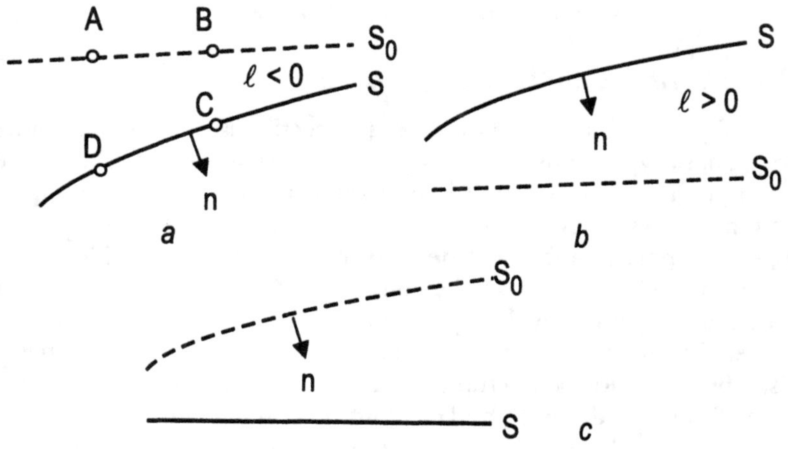

Figure 1.4 Derivation of the equivalent boundary condition, eqn. 1.44

We will assume that before the deformation the surface had occupied the position S_0 (Figure 1.4*a*), here $\vec{n}$ is the normal vector to the metallic surface; whereas after the deformation this surface occupies the position S. It is assumed also that both surfaces are not only situated very close to each other but, moreover, the angle between them is small. The distance between the corresponding points of S_0 and S is denoted as l. The sign of l is negative ($l < 0$) when the surfaces are situated as in Figure 1.4*a*. For the second possible position of S_0 and S shown in Figure 1.4*b* we have $l > 0$.

The tangential component of the electric field $\vec{E}$ on S_0 is small and it is determined by the currents and the electrical charges near the considered point; in other words, for the $\vec{E}$ field there exists a local boundary condition. Applying the integral form of the first Maxwell equation (eqn. I.1) to the boundary $ABCD$ (for $\mu = 1$) in Figure 1.4*a* and taking into account the fact that the usual boundary condition (eqn. I.2) is valid on S, we obtain the desired equivalent boundary condition

$$\vec{E} \cdot \vec{t} = ikl\left[\vec{n} \times \vec{H}\right] \cdot \vec{t} + \vec{t} \cdot \nabla(E_n l) \tag{1.44}$$

where $\vec{t}$ is a unit vector tangential to S_0. This condition has the following meaning: the field excited by arbitrary sources in the area limited by S_0 in the problem with the equivalent condition (eqn. 1.44), satisfied on the undisturbed surface S_0, has the same value as in another problem with the usual boundary condition (eqn. I.2) satisfied on the perturbed surface, S.

Figure 1.4*a* corresponds to the case $l < 0$. As it can be easily proven, the form of the equivalent boundary condition, eqn. 1.44 does not change if the sign of l changes. To derive eqn. 1.44 for the case represented in Figure 1.4*b*, one should start establishing the boundary condition on S in the auxiliary problem depicted in Figure 1.4*c* and then switch to the conditions corresponding to Figure 1.4*b*. In such a way it can be demonstrated that eqn. 1.44 is valid for any sign of l.

The application of the boundary condition eqn. 1.44 instead of the deformation is in fact the introduction of some additional magnetic currents into the problem. From the formal point of view, the field perturbation connected with the deformation is caused by those currents.

1.3.2 Small waveguide step

Let us employ the equivalent boundary condition (eqn. 1.44) to calculate the amplitudes of the waves scattered by the step. Calculating the forward wave amplitudes P_j $(j \neq m)$, we consider the deformation as a deviation of the left-hand side waveguide surface $(z < 0)$ from the continuation of the right-hand side waveguide surface $(z > 0)$; in such a case the magnetic currents exciting a parasitic field are situated outside of the zone where this field is to be found. Obviously, $l(z,s) = \delta(s)$ for $z < 0$ and $l = 0$ for $z > 0$. The undisturbed fields $\hat{\vec{E}}^m$ and $\hat{\vec{H}}^m$ are placed in the right-hand side of eqn. 1.44.

The boundary condition eqn. 1.44 is valid only in the area where the surfaces S_0 and S are not only close to each other, but the angles between them are also small. This condition is not satisfied in the immediate step neighbourhood. Hence, for the distances of the order of δ from the cross section $z = 0$, the component of the electric field $\vec{E}$ tangential to S_0 does not coincide with eqn. 1.44. Therefore, the formulae derived thereafter will have an error of the order of δ^2. Because of this we will determine the scattered mode amplitudes to first-order terms of δ only.

The amplitudes P_j will be derived using eqn. 1.3 as in the problem of a small tilt. To determine the integral over the cross section in eqn. 1.3, we need to integrate the following equation over the whole volume of the left-hand side waveguide

$$\operatorname{div}\left\{\left[\hat{\vec{H}}^{-j} \times \vec{E}\right] - \left[\vec{H} \times \hat{\vec{E}}^{-j}\right]\right\} = 0 \tag{1.45}$$

which is easily derived from the Maxwell equations (eqn. I.1). In eqn. 1.45 $\vec{E}$ and $\vec{H}$ are the complete electromagnetic fields and the $\vec{E}$ field satisfies eqn. 1.44 on the surface S_0. By virtue of its definition, the cross section of the surface S_0 is limited by the boundary C_+ for all z, so the fields $\hat{\vec{E}}^m$, $\hat{\vec{H}}^m$ can be substituted in eqn. 1.44. For the integration we will add a small complex value to k as we have done during the derivation of eqn. 1.20. The fields $\hat{\vec{E}}^{-j}$ and $\hat{\vec{H}}^{-j}$ satisfy the usual condition (eqn. I.2) on the surface S_0 and therefore

$$P_j = \frac{-1}{2kh_j} \oint \left[\hat{\vec{H}}^{-j} \times \vec{E}\right]_n dsdz \tag{1.46}$$

One should calculate this integral using eqn. 1.44. Let us define the direction of $\vec{s}$ in such a way that the two vectors $\vec{n}, \vec{s}$ and the unit vector along the z-axis form a right-hand co-ordinate system. According to eqn. 1.44, the $\vec{E}$ field components on S_0 contained in eqn. 1.46 are equal to

$$E_z = ik\delta \hat{H}_s^m + \frac{\partial}{\partial z}\left(\hat{E}_n^m \delta\right), E_s = -ik\delta \hat{H}_z^m + \frac{\partial}{\partial s}\left(\hat{E}_n^m \delta\right) \quad (1.47)$$

Let us substitute eqn. 1.47 into eqn. 1.46, summing and subtracting the corresponding terms, and pick off under the integral the terms

$$\frac{\partial}{\partial z}\left(\hat{E}_n^m \hat{H}_s^{-j}\delta\right) - \frac{\partial}{\partial s}\left(\hat{E}_n^m \hat{H}_z^{-j}\delta\right) \quad (1.48)$$

Those terms will disappear by the integration: the first one because of the condition $\delta = 0$ at $z = 0$, and the second term disappears as a consequence of the periodicity of all functions versus the argument s. Since the z dependence of the integrand in eqn. 1.46 is known, integration with respect to z is possible. Employing, in addition, the Maxwell equation for the wave fields with the superscript $-j$, we derive the desired expression for the amplitudes of the forward waves scattered by the waveguide step.

$$P_j = \frac{-1}{2h_j\left(h_j - h_m\right)} \oint \delta(s)\left(E_n^m E_n^{-j} + H_z^m H_z^{-j} + H_s^m H_s^{-j}\right) ds \quad (1.49)$$

According to the special theory of relativity, the integrand expression is the scalar product of the four vector currents excited on the wall by the fields of the modes m and $-j$ and is therefore a Lorentz invariant.

Eqn. 1.49 can be rewritten in a more convenient way, taking into account eqn. I.6 connecting the fields of the waves propagating in opposite directions

$$P_j = \frac{1}{2h_j\left(h_j - h_m\right)} \oint \delta(s)\left(E_n^m E_n^{j} + H_z^m H_z^{j} - H_s^m H_s^{j}\right) ds \quad (1.50)$$

The reflected mode amplitudes are obtained by an analogous procedure. We have to choose for all z the surface of the left-hand side waveguide as the surface S_0 and assume $l(z) = 0$ for $z < 0$ and $l(z,l) = -\delta(s)$ for $z > 0$, i.e. we consider a deviation of the right-hand side waveguide surface from the continuation of the left-hand side waveguide surface as a perturbation.

Repeating the same manipulations we yield

$$P_{-j} = \frac{1}{2h_j\left(h_j + h_m\right)} \oint \delta(s)\left(E_n^m E_n^{j} + H_z^m H_z^{j} + H_s^m H_s^{j}\right) ds \quad (1.51)$$

So, the amplitudes of the modes scattered by the waveguide step are expressed through the integrals over the boundary contour of the waveguide cross section; they contain the first power of the step height. All the waves having nonzero values of the integrals eqns. 1.49 – 1.51 for the given $\delta(s)$ are excited by the step, i.e. all those waves are excited by the magnetic currents (eqn. 1.44) corresponding to the propagating wave with the superscript m.

1.3.3 *Reflection at small waveguide step*

There is a simple relationship between P_{-m} and Δh_m, i.e. between the wave amplitude with superscript m reflected from the step and the difference Δh_m of the wavenumbers of this mode between the left-hand side and the right-hand side waveguide. Let us determine this relationship.

The simplest way to derive the difference of the wavenumbers is to apply eqn. 1.45 to the fields of the wave with superscript m of the right-hand side waveguide $\left(\hat{\vec{E}}^m, \hat{\vec{H}}^m\right)$ and to the fields of the wave with superscript $-m$ of the left-hand side waveguide $\left(\hat{\vec{E}}^{-m}, \hat{\vec{H}}^{-m}\right)$. In the left waveguide the fields of the second wave satisfy the simple boundary conditions of eqn. I.2, whereas the fields of the first wave satisfy the equivalent boundary condition in eqn. 1.44, being in this case $l = -\delta(s)$ for all z. Separating in eqn. 1.45 the derivative with respect to z and integrating over the area limited by the boundary C_- we obtain

$$\Delta h = \frac{1}{2ikh_m} \oint \left[\vec{H}^{-m} \times \vec{E}^m\right]_n ds \tag{1.52}$$

Further, substituting eqn. 1.44 into eqn. 1.52 and retaining everywhere only the first-order terms of Δh_m and δ, we derive the expression for the difference of the modal wavenumbers at a small waveguide step

$$\Delta h_m = \frac{1}{2h_m} \oint \delta\left[-(H_s^m)^2 + (H_z^m)^2 - (E_n^m)^2\right] ds \tag{1.53}$$

Comparing this result with the reflected wave amplitude P_{-m} given by eqn. 1.51 we yield the desired relationship

$$P_{-m} = \frac{\Delta h_m}{2h_m} + \frac{1}{2h_m^2} \oint \delta(H_s^m)^2 ds \tag{1.54}$$

In some problems it is easier to calculate the reflection coefficient P_{-m} using eqn. 1.54 instead of the more general formula of eqn. 1.51. For example, the field of a TE_{0q} mode in a circular waveguide does not contain longitudinal currents, i.e. $H_s = 0$ and $P_{-m} = \Delta h_m / 2h_m$ for any shape of the waveguide step, i.e.

for any arbitrary function $\delta(s)$. If the wavenumber is identical in both waveguides, $\Delta h_m = 0$ for example, for the junction of two identical waveguides with a small shift between them, P_{-m} is expressed by the integral of the square of the longitudinal current only. For instance, if an asymmetrical mode is incident on the junction of two identical waveguides, it is easy to find the dependence of the reflection coefficient on the wave polarisation, etc.

The second term in eqn. 1.54 can be expressed for the *TE* and *TM* modes through the eigenfunctions ψ^m and φ^m. According to eqn. I.14, the corresponding formulae have the following form

$$P_{-m} = \frac{\Delta h_m}{2h_m} - \frac{1}{2}\oint \delta \left(\frac{\partial \psi^m}{\partial n} \right)^2 ds$$

$$P_{-m} = \frac{\Delta h_m}{2h_m} - \frac{k^2}{2h_m^2}\oint \delta \left(\frac{\partial \varphi^m}{\partial n} \right)^2 ds \tag{1.55}$$

It follows from eqns. 1.53 and I.14 that the first term of the sum can also be expressed through ψ^m and φ^m. Let us point out that the corresponding formulae can be directly derived from the two-dimensional equations, determining the change of the eigenvalue in α_m eqns. I.11*a* and I.12*a* or in I.11*b* and I.12*b*, caused by a small boundary deformation. The relationship between Δh_m and $\Delta\alpha_m$ is obtained by derivation of eqn. I.13.

1.3.4 Straight, nonuniform waveguide as the limit of a stepped waveguide

Now we will determine the amplitudes of the waves scattered by a finite section of a variable cross-section waveguide. First, let us note that eqns. 1.50 and 1.51 remain valid for the problem, where a junction occurs not in a single plane but occupies a small area Δz and $\Delta z k \ll 1$. In fact, nothing changes in the whole derivation of those formulae for this case. Repeating a discussion similar to the last section we will substitute $\delta(s)$ by the product $\nu(s)\Delta z$. As follows from simple geometrical considerations, $\nu(s)$ is the tangent of the inclination angle of the straight line connecting corresponding points of the boundaries C_- and C_+ towards the z-axis. This line forms an angle approximately equal to $\pi/2$ with the boundaries C_- and C_+; $\nu(s)$ has the same sign as $\delta(s)$. Extracting Δz from the integral sign we obtain the following formulae for $P_j\,(j \neq m)$ and P_{-j}

$$P_j = S_{jm}\Delta z, \quad P_{-j} = -S_{-jm}\Delta z \tag{1.56}$$

Following eqns. 1.50 and 1.51, S_{jm} denotes here the value

$$S_{jm} = \frac{1}{2h_j\left(h_j - h_m\right)}\oint \nu(s)\left(E_n^m E_n^j + H_z^m H_z^j - H_s^m H_s^j\right) ds \tag{1.57}$$

and S_{-jm} is obtained from S_{jm} by the formal substitution of j by $-j$.

For the coefficients in eqn. 1.57 we have employed the same symbols as for the coefficients in eqn. 1.26 in the problem of a dielectric permittivity and magnetic permeability step. As will be shown in Section 2.3, there are very simple relationships between them. Thanks to such representation, eqn. 1.56 is identical to eqn. 1.27.

Now let us consider the problem of a nonuniform waveguide with variable cross section. We may represent such a waveguide as a stepped waveguide as shown in Figure 1.5*a* or Figure 1.5*b*; the result will be exactly the same for both cases. Now the value of $\nu(s)$ depends on z; it is obviously equal to the tangent of the angle formed by the straight line tangential to the metal normal to the cross-section boundary contour towards the z-axis.

The amplitudes of the waves scattered by the analysed nonuniformities are derived by exactly the same method as in the Section 1.2. Eqns. 1.28 and 1.29 for the amplitudes of waves scattered by a finite length waveguide section with a varying filling medium have been derived from eqns. 1.27, related to the amplitudes of the waves scattered by a small waveguide filling step. Eqns. 1.56 are identically equal to eqns. 1.27. Therefore, eqns. 1.28 and 1.29 remain valid for the variable cross-section waveguide as well, with the understanding that S_{jm} is expressed by eqn. 1.57.

The discussion of restrictions of the small nonuniformities method applied to large nonuniformities is also valid here as well as the difficulties of a rigorous derivation of eqns. 1.28 and 1.29 from eqn. 1.27. Hence, the analysis of these formulae and of the S_{jm} coefficients will be performed in Section 2.3, where they will be obtained by a more rigorous method.

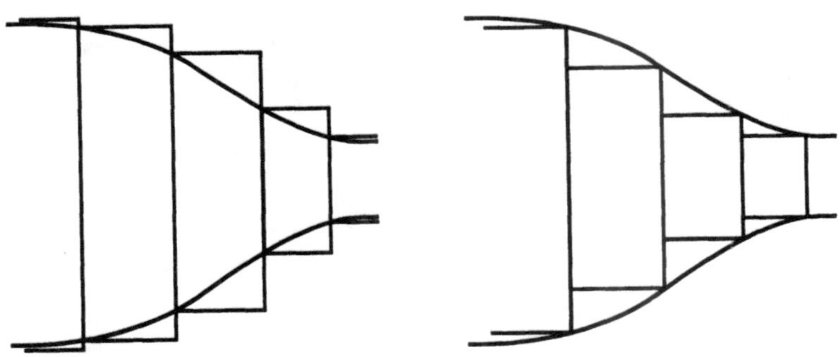

a b

Figure 1.5 Outer (a) and inner (b) stepped waveguide

So, the amplitudes of the waves scattered from the small nonuniformities of three basic types have been derived in Chapter 1. A superposition of those nonuniformities can represent any other kind of waveguide nonuniformity. For finite nonuniformities, which cannot be considered to be small, the parasitic wave excitation can be represented as a superposition of the parasitic waves excited by a series of small nonuniformities approximately reconstructing the studied large nonuniformity. One should take into account the corresponding phase shifts of those parasitic waves and variations of the amplitude corresponding to the power conservation principle. This method is not sufficiently rigorous. It cannot be applied in some problems, but it has a clear physical meaning. Nevertheless, such a method allows us to obtain the mostly interesting features of a nonuniform waveguide section in the simplest way.

Chapter 2
The cross-section method

The basic idea of the cross-section method is that the electromagnetic fields in an arbitrary nonuniform waveguide cross section are represented as a superposition of the waves of different modes propagating in forward and backward directions along an auxiliary straight uniform waveguide of the same cross section and with the identical distribution of ε and μ. The coefficients of this superposition satisfy first-order ordinary differential equations. The general problem of the field derivation in a nonuniform waveguide is reduced in such a way to the problem of the fields in a uniform waveguide and to the solution of coupled-wave ordinary differential equations. Further, this method is employed for the basic types of nonuniform waveguides and for the more general case of combined nonuniformities.

2.1 Calculations on bent waveguides using the cross-section method

2.1.1 Coupled-wave differential equations

We shall start by establishing the basic equations for the fields in a waveguide bent along a circumference arc. We suppose that the considered waveguide section had an arbitrary constant cross section along its full length before the bending. Let us bend this waveguide section along a circular arc. In this way all the generatrices of the straight waveguide become circular arcs with centres which are situated on the same straight line, as shown in Figure 2.1. Let us denote this line as the y-axis in a cylindrical co-ordinate system; the two additional co-ordinates are r and ϑ. The planes $\vartheta = \text{const}$ will intersect the waveguide along the same contour for all angles ϑ. The co-ordinates r and y form a rectangular co-ordinate system in the plane $\vartheta = \text{const}$. We direct the x-axis towards the curvature centre, i.e. $x \equiv -r$, and $\partial/\partial x \equiv -\partial/\partial r$.

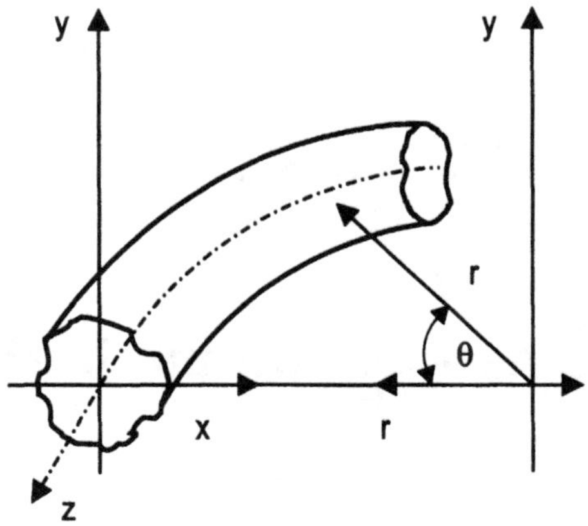

Figure 2.1 Co-ordinate system for description of curved waveguide

Now let us examine a waveguide cross section $\vartheta = \text{const}$ and let us imagine a straight waveguide coming through this cross-section contour and perpendicular to it. The fields satisfy in this reference waveguide and in the bent waveguide the same boundary conditions $\vec{E} \cdot \vec{t} = 0$ along the cross-section contour, because the tangents to both waveguides are the same. If we represent the field in the bent waveguide as a superposition of those fields which can exist in the straight waveguide with the same cross section and with its coefficients depending on ϑ, then each term of this superposition will satisfy the boundary condition of eqn. I.2.

Let us represent

$$E_r = -\sum_{m=1}^{\infty} Q_m E_x^m, \; E_y = \sum_{m=1}^{\infty} Q_m E_y^m$$
$$H_r = -\sum_{m=1}^{\infty} R_m H_x^m, \; H_y = \sum_{m=1}^{\infty} R_m H_y^m \tag{2.1}$$

i.e. we have expanded the transversal electromagnetic field components in the bent waveguide in the transversal field components of all the modes existing in the reference waveguide propagating only in one direction. The coefficients Q_m and R_m depend on ϑ.

Further, we define the new coefficients $P_m(\vartheta)$ and $P_{-m}(\vartheta)$ by the equations

$$P_m - P_{-m} = Q_m, \; P_m + P_{-m} = R_m \tag{2.2}$$

According to our accepted relation eqn. I.6 between the fields of the forward and backward propagating waves, eqn. 2.1 can be rewritten in the following manner

$$E_r = -P_v E_x^v, \quad E_y = P_v E_y^v, \quad H_r = -P_v H_x^v, \quad H_y = P_v H_y^v \tag{2.3}$$

Here and further on, the repeated index v means the series sum from $v = -\infty$ to $v = +\infty$, i.e. the first formula in eqn. 2.3 means, in fact

$$E_r(r, \vartheta, y) = -\sum_{v=-\infty}^{+\infty} P_v(\vartheta) E_x^v(x, y) \tag{2.4}$$

The expansion in eqn. 2.3 unlike eqn. 2.1, contains waves propagating forwards and backwards. Moreover, the coefficients in the expansion of the $\vec{E}$ and $\vec{H}$ fields are identical. The functions employed in the expansion in eqn. 2.3 satisfy the same boundary conditions as the fields $\vec{E}$ and $\vec{H}$, therefore one can differentiate the series term by term. Let us substitute the series into the two Maxwell equations

$$\frac{\partial E_r}{\partial y} - \frac{\partial E_y}{\partial r} = -ik\mu H_\vartheta, \quad \frac{\partial H_r}{\partial y} - \frac{\partial H_y}{\partial r} = ik\varepsilon E_\vartheta \tag{2.5}$$

As follows from a comparison with the last equations in eqn. I.4, there exists an expansion for the longitudinal components of $\vec{E}$ and $\vec{H}$ which is similar to eqn. 2.3, with the same coefficients P_m

$$E_\vartheta = P_v E_z^v, \quad H_\vartheta = P_v H_z^v \tag{2.6}$$

Now let us derive the coupled-wave equations for $P_m(\vartheta)$. The substitution of the series in eqns. 2.4 and 2.6 into the four remaining Maxwell equations

$$\begin{aligned} &\frac{\partial E_y}{\partial \vartheta} - r\frac{\partial E_\vartheta}{\partial y} = -ik\mu r H_r, \quad \frac{\partial (rE_\vartheta)}{\partial r} - \frac{\partial E_r}{\partial \vartheta} = -ik\mu r H_y \\ &\frac{\partial H_y}{\partial \vartheta} - r\frac{\partial H_\vartheta}{\partial y} = ik\varepsilon r E_r, \quad \frac{\partial (rH_\vartheta)}{\partial r} - \frac{\partial H_r}{\partial \vartheta} = ik\varepsilon r E_y \end{aligned} \tag{2.7}$$

gives us

$$\begin{aligned} &P_v' E_x^v = -P_v E_z^v - iP_v h_v E_x^v r, \quad P_v' E_y^v = -iP_v h_v E_y^v r \\ &P_v' H_x^v = -P_v H_z^v - iP_v h_v H_x^v r, \quad P_v' H_y^v = -iP_v h_v H_y^v r \end{aligned} \tag{2.8}$$

The prime symbol represents derivation with respect to ϑ.

Multiplying, now, these equations by $-H_y^j, H_x^j, E_y^j$ and $-E_x^j$, where j can be any positive or negative integer, we add them and further integrate over the cross section. Using the orthogonality condition of eqn. I.9, we obtain the following expression for P_j' as a linear combination of P_m

$$P_j' = F_{jv} P_v \tag{2.9}$$

Since this expression is valid for all j eqn. 2.9 represents the desired first order coupled-wave differential equations for the coefficients $P_j(\vartheta)$.

The expression of the coefficients F_{jm} in eqn. 2.9 is identical to the coefficients F_{jm} in eqn. 1.7 corresponding to the problem of a small waveguide tilt.

According to eqns. 2.3 and 2.6 the value $P_j(\vartheta)$ can be considered as the complex amplitude of the jth mode. It replaces the $\exp(-jh_j z)$ factor in a straight waveguide. However, it varies along the waveguide in a more complicated way. Following eqn. 2.9, the variation of $P_j(\vartheta)$ against the variation of ϑ depends on the amplitudes of all the waves existing at the given value of ϑ. The contribution of each wave amplitude to the formation of the value P_j' increases with the increasing coefficient F_{jm}. Hence, we call the values F_{jm} the coupling coefficients between the modes m and j.

Multiplying eqn. 2.9 by $\Delta\vartheta$, we can underline the following meaning of this equation: the amplitude of any wave propagating along a small angle $\Delta\vartheta$ varies proportionally to $\Delta\vartheta$, whereas this value is the sum of the contributions of all modes. The contribution of a mode to P_j is more important the greater its amplitude and coupling coefficient to this mode are. In this sense eqn. 2.9 generalises eqns. 1.6 and 1.12. As will be demonstrated later, these equations can be derived from eqn. 2.9 supposing that all the modes, except one, are neglected in a given cross section. Following eqn. 2.9, wave propagation in a nonuniform waveguide section is similar to wave propagation in a system of coupled transmission lines with a distributed coupling.

2.1.2 Boundary conditions at the bend ends

Now, let us determine the boundary conditions for $P_j(\vartheta)$ at the ends of a bent waveguide section. We suppose that the bent waveguide section forming a circumference arc is connected to a pair of semiinfinite straight waveguides at $\vartheta = 0$ and $\vartheta = \vartheta_0$. A mode with the superscript m and unitary amplitude is incident on the bend from the left-hand side waveguide.

It follows from the orthogonality of different mode fields and from the continuity of the transversal $\vec{E}$ and $\vec{H}$ components that the values of the coefficients $P_j(\vartheta)$ at the input and at the output of the bend are equal to the amplitudes of the corresponding straight waveguide waves. In other words, $P_j(\vartheta)$ are continuous on the boundaries of the bent waveguide. The boundary conditions are to provide the absence of other waves coming from the straight waveguides, except for the incident wave only. These conditions have the following form

$$P_m(0) = 1, \quad P_j(0) = 0 \text{ for } j > 0, \quad j \neq m, \quad P_j(\vartheta_0) = 0 \text{ for } j < 0 \tag{2.10}$$

The coupled-wave eqns. 2.9 and 2.10 are complete. All the amplitudes $P_j(\vartheta)$ can be found from these equations. The amplitudes of the waves propagating backward in the left-hand side waveguide are equal to $P_j(0)$, with $j < 0$, whereas the amplitudes of the waves propagating forward in the right-hand side waveguide are equal to $P_j(\vartheta_0)$ with $j > 0$; the calculations of those amplitudes do not require any supplementary manipulations to satisfy the field continuity at the ends.

Since the coefficients F_{jm} in the coupled-wave eqns. 2.9 do not depend on ϑ, the solution of these equations can be obtained by the methods for finite number coupled-wave equations with constant coefficients. Retaining in the coupled-wave eqns. 2.9 a finite number N of the variables, we shall obtain a finite number of these coupled-wave equations. Their solution is reduced to characteristic algebraic equations of the Nth order. If some conditions are satisfied, the solution of the infinite number of coupled-wave eqns. 2.9 – 2.10 can be obtained as the limit for $N \to \infty$. Here, we are not interested in the foundations and the development of this method: if the problem does not contain a small parameter, a practical solution of the coupled-wave eqns. 2.9 – 2.10 could be obtained directly by numerical methods, since the structure of eqns. 2.9 is better adapted for computer calculations.

One could face some difficulties in these calculations, since the conditions in eqn. 2.10 should be satisfied at one end of the bent waveguide by a part of the variables only, and should be satisfied by another part of the variables at the other end. However, the well known method can be employed to reduce the considered problem to several similar problems, where conditions like those in eqn. 2.10 could be satisfied by all the variables at one of the waveguide section ends only. An example of this method is given in subsection 4.3.4.

2.1.3 Curvature coupling coefficients

Let us examine in more detail the coupling coefficients; their analytical expressions were given in Section 1.1. If $m \neq j$, the F_{jm} coefficient does not depend on the curvature radius contained in the first sum of eqn. 1.7. This result follows from the comparison of this term with the orthogonality condition of eqn. I.9. Obviously, the coupling coefficients depend on the positions of the curvature centre in relation to the waveguide cross section, because the relative orientation of the x and y-axes in this cross section depends on it.

The coupling coefficients satisfy the condition

$$F_{-j,-m} = -F_{j,m} \tag{2.11}$$

derived directly from eqns. 1.7 and I.6.

These coefficients are also connected by reciprocity relationships. The simplest way to find these relationships is by using the ϑ independence of the integral

$$\int \left\{ E_r^{(1)} H_y^{(2)} - E_y^{(1)} H_r^{(2)} - E_r^{(2)} H_y^{(1)} + E_y^{(2)} H_r^{(1)} \right\} dS \tag{2.12}$$

calculated along the section $\vartheta = \text{const}$. In eqn. 2.12, $\vec{E}^{(1)}, \vec{H}^{(1)}$ and $\vec{E}^{(2)}, \vec{H}^{(2)}$ are two arbitrary fields inside the bend excited by sources situated outside of the bend. The independence of eqn. 2.12 on ϑ follows from the Lorentz lemma applied to the volume between two arbitrary cross sections $\vartheta = \text{const}$. Let us represent the first and the second field components from eqn. 2.12 in the form of eqn. 2.3 with different amplitudes $P_j^{(1)}$ and $P_j^{(2)}$. Further, we substitute them into eqn. 2.12. After simple manipulations employing the orthogonality condition, we will obtain for eqn. 2.12 the expression $h_v P_v^{(1)} P_{-v}^{(2)}$. According to eqn. 2.9, the derivative of this expression with respect to ϑ will be equal to the double sum

$$\left(-F_{v\mu} h_v + F_{-\mu,-v} h_{-\mu}\right) P_{-v}^{(1)} P_{\mu}^{(2)} \tag{2.13}$$

where the summation is performed both in v and in μ. The result must be equal to zero for arbitrary fields $\vec{E}^{(1)}, \vec{H}^{(1)}, \vec{E}^{(2)}$ and $\vec{H}^{(2)}$, i.e. for arbitrary functions $P_j^{(1)}$ and $P_j^{(2)}$. Consequently, the expression included in the parentheses in eqn. 2.13 must be equal to zero for any v and μ. In addition, employing eqn. 2.11, we yield the desired condition connecting F_{jm} with F_{mj}

$$h_m F_{mj} = h_j F_{jm} \tag{2.14}$$

Employing this relationship, eqn. 1.7 for F_{jm} can be simplified. According to eqn. 2.14, there exists the following relationship between the integrals in eqn. 1.7

$$\begin{aligned} &i\left(h_j - h_m\right)\int r\left(E_y^j H_x^m - E_x^j H_y^m + E_y^m H_x^j - E_x^m H_y^j\right)dS \\ &= \int \left(E_y^j H_z^m - E_z^m H_y^j - E_y^m H_z^j + E_z^j H_y^m\right)dS \end{aligned} \tag{2.15}$$

Obviously, this formula can be derived directly from eqns. I.4 and the boundary conditions in eqn. I.2 by elementary (but time consuming) manipulations. Excluding the first of two integrals participating in eqn. 1.7 with the help of eqn. 2.15, we yield a symmetric expression for the coupling coefficients

$$F_{jm} = \frac{1}{2kh_j\left(h_j - h_m\right)}\left(h_m B^{mj} - h_j B^{jm}\right) \tag{2.16}$$

where B^{mj} denotes the integral

$$B^{mj} = \int \left(E_y^m H_z^j - E_z^j H_y^m\right)dS \tag{2.17}$$

All these results are valid not only for waveguides with the walls of infinite conductivity satisfying the boundary condition in eqn. I.2 but also for the walls of finite conductivity, because the orthogonality condition in eqn. I.9 is retained, if one substitutes the Leontovich boundary condition instead of eqn. I.2.

The actual derivation of the coupling coefficients for circular and rectangular waveguides will be realised in Chapter 5. Here we will only express F_{jm} through the eigenfunctions ψ and φ. Those expressions are valid in all problems where eqn. I.14 can be employed and, first of all, for empty waveguides (without dielectric filling).

Let us denote the integrals over the cross section appearing by substitution of eqn. I.14 in the integral eqn. 2.17 by

$$\int \psi^j \frac{\partial \psi^m}{\partial x} dS = K^{jm}, \int \varphi^j \frac{\partial \varphi^m}{\partial x} dS = C^{jm}, \int \psi^j \frac{\partial \varphi^m}{\partial y} dS = M^{jm} \quad (2.18)$$

These magnitudes do not depend on frequency, they depend only on the mode type, on the cross-section shape and on the direction towards the curvature centre in the given cross section. They do not depend on the curvature radius and they are of the order of the linear dimensions of the cross section. It is easy to show that the following pair of relations comes from the boundary condition for the eigenfunction φ along the cross-section boundary contour

$$C^{mj} = -C^{jm}, \int \varphi^m \frac{\partial \psi^j}{\partial y} dS = -M^{jm} \quad (2.19)$$

Substituting the expression for the fields under the form of the eigenfunctions in eqn. 2.16 we obtain, when both m and j areTE modes

$$F_{jm} = \frac{i}{2h_j(h_j - h_m)}\left(h_m \alpha_j^2 K^{jm} - h_j \alpha_m^2 K^{mj}\right) \quad (2.20)$$

When both modes are of *TM* type

$$F_{jm} = \frac{i}{2h_j(h_j - h_m)}\left(h_m \alpha_j^2 + h_j \alpha_m^2\right) C^{jm} \quad (2.21)$$

Finally, if j is a *TE* mode and m is a *TM* mode, then

$$F_{jm} = \frac{ik(h_j + h_m)}{2h_j} M^{jm} \quad (2.22)$$

The coupling coefficient for the possible fourth combination of modes is easily obtained from the last expression and the reciprocity condition eqn. 2.14.

According to eqns. 2.20 – 2.22, the coupling coefficient between two different waves propagating in the same direction is larger than that between the same waves propagating in opposite directions.

2.1.4 Cases of weak curvature coupling

There are two special cases when it is easy to obtain an obvious solution for the coupled-wave eqns. 2.9 – 2.10; these are called weak coupling cases. To study them it is convenient to define the new variables $p_j(\vartheta)$ instead of $P_j(\vartheta)$ as follows

$$P_j(\vartheta) = p_j(\vartheta)\exp\left(-iq_j\vartheta\right) \tag{2.23}$$

where the coefficient $q_j = iF_{jj}$ is given in eqn. 1.9. Unlike for $P_j(\vartheta)$, we will call the variables $p_j(\vartheta)$ reduced amplitudes. The coupled-wave eqns. 2.9 have the following form with these variables

$$p_j' = F_{j\nu}p_\nu\left(1-\delta_{j\nu}\right)\exp\left(-i\left(q_\nu - q_j\right)\vartheta\right) \tag{2.24}$$

As a characteristic feature, the matrix of the coupled-wave equations for these reduced coefficients, p_j, does not contain diagonal elements.

If we can neglect all the wave amplitudes on the right-hand side of eqn. 2.24, except the incident wave m, then the solution of the coupled-wave eqns. 2.24 is obtained in an analytical form.

First, this is possible in the problem of tilt considered in Section 1.1, i.e. a tilt with $\vartheta << 1$. In this case, eqns. 1.6 and 1.12 can be derived immediately from eqn. 2.24 or 2.9 for both forward and backward propagating waves, and one only needs to substitute $\Delta\vartheta$ by ϑ_0. It follows from eqn. 2.24 that the wave index m, $p_m = \text{const} = 1$. Therefore, the phase shift of the mode propagating through the waveguide tilt mentioned above is determined by the exponential factor written separately in eqn. 2.23. Let us study this factor in more detail.

In contrast to F_{jm} the variable q_j contains the curvature radius for $j \neq m$. The ratio q_j/h_j can be called the curvature radius for the wave j. Generally speaking, this value

$$r_j = \frac{q_j}{h_j} \tag{2.25}$$

is different for various waves. However, for example, as the calculations show for a circular waveguide with any symmetrical distribution of ε and μ along the cross section, r_j is the same for all waves and it is equal to the distance between the curvature centre and the waveguide axis. The same is also valid for rectangular waveguides (with or without a homogeneous filling medium).

The general expression for r_j can be obtained from eqns.1.9 and 2.25. This expression has the following form for *TE* and *TM* modes, respectively

$$r_j = \int r\left(\nabla\psi^j\right)^2 dS + \frac{\alpha_j^2}{h_j^2} K^{jj},\ r_j = \int r\left(\nabla\varphi^j\right)^2 dS \tag{2.26}$$

If we measure the distance l_j between two cross sections along an arc of radius r_j as in Section 1.1, then $dl_j = r_j d\vartheta$ and the exponential term in eqn. 2.23 can be rewritten as $\exp\left(-ih_j l_j\right)$. As we have mentioned above, the arc length in a circular waveguide is measured along the waveguide axis. Hence, the exponential term of eqn. 2.23, describing the phase shift variation (and the amplitude variation as well, if h_j is complex or purely imaginary), corresponds to the propagation of the wave with wavenumber h_j along the arc length l_j.

A second simple solution of the coupled-wave eqns. 2.24 can be determined for the problem of a waveguide bend with a very large curvature radius. The simplest way of finding the required solution is to assume at the beginning that we can neglect all the terms in eqn. 2.24 except the term $j = m$ corresponding to the incident mode, because the radius of curvature is so large. Thereafter we need to prove that the solution obtained under this approach satisfies this assumption.

For a large curvature radius the exponential factor in eqn. 2.24 is a fast oscillating function of the angle; this is the core of this assumption. Therefore, the mean value of the derivatives p_j' is small and the values of the variables p_j must be the same as at the end of this interval. The coupled-wave eqns. 2.24 then adopt the following form

$$p_m' = 0,\ p_j' = F_{jm} p_m \exp\left(-i\left(q_m - q_j\right)\vartheta\right),\ j \neq m \tag{2.27}$$

Their solution satisfying the conditions of eqn. 2.10 written for the variables $p_j(\vartheta)$ in the same form as for the variables $P_j(\vartheta)$ has the following form

$$p_m(\vartheta) = 1 \tag{2.28a}$$

$$p_j(\vartheta_0) = F_{jm} \frac{\sin\left(\left(q_m - q_j\right)\frac{\vartheta_0}{2}\right)}{\frac{\left(q_m - q_j\right)}{2}} \exp\left(-i\left(q_m - q_j\right)\frac{\vartheta_0}{2}\right),\ j > 0,\ j \neq m \tag{2.28b}$$

$$p_j(0) = -F_{jm} \frac{\sin\left(\left(q_m - q_j\right)\frac{\vartheta_0}{2}\right)}{\frac{\left(q_m - q_j\right)}{2}} \exp\left(-i\left(q_m - q_j\right)\frac{\vartheta_0}{2}\right),\ j < 0 \tag{2.28c}$$

The first of these equations means that the variation of the incident wave amplitude $P_m(\vartheta)$ is described by the same factor $\exp(-ih_m l_m)$ as in the straight uniform waveguide. The power of this wave does not vary (if h_m is real). This is valid only in the approximation made for the derivation of the coupled-wave eqns. 2.27. As we will see in Section 5.3, the phase shift and the amplitude of the incident wave at the output of a bend with a large curvature radius depend on mode conversion in the second order on the curvature, i.e. in a small degree.

Eqns. 2.28*b* and 2.28*c* provide us with the amplitudes of the parasitic waves scattered forward ($p_j(\vartheta_0)$, for $j > 0$) and backward ($p_j(0)$, for $j < 0$), respectively.

In addition, assuming r_j has the same value for all j and $h_m > 0$, while h_j is real, we can rewrite these formulae in a more convenient form

$$p_j(\vartheta_0) = F_{jm} \frac{\sin\left((h_m - h_j)r\frac{\vartheta_0}{2}\right)}{(h_m - h_j)\frac{r}{2}} \exp\left(-i(h_m - h_j)r\frac{\vartheta_0}{2}\right), \quad j>0 \qquad (2.29a)$$

$$p_j(0) = -F_{jm} \frac{\sin\left((h_m + |h_j|)r\frac{\vartheta_0}{2}\right)}{(h_m + |h_j|)\frac{r}{2}} \exp\left(-i(h_m + |h_j|)r\frac{\vartheta_0}{2}\right), \quad j<0 \qquad (2.29b)$$

The necessary condition for the validity of the whole solution (eqns. 2.28 – 2.29) is a small value of the amplitudes in eqns. 2.29. The curvature radius is contained in the denominator of these formulae. Therefore, the approximation will be accurate for large enough radii of the bend. The validity condition of these formulae may be written as

$$\left| F_{jm} \frac{1}{(h_m \mp |h_j|)r} \right| << 1 \qquad (2.30)$$

For small angles, more exactly when $(h_m \mp |h_j|)r\vartheta_0 << 1$, eqn. 2.29 becomes identical to eqns. 1.6 and 1.12 corresponding to a tilt. A bend with a small angle ϑ_0 is identical to a small tilt for forward propagating waves, if the difference of the bend electric lengths for both waves is small, whereas the sum of the bend electric lengths should be small for the backward propagating waves. Moreover, the requirement of a small ϑ_0 value is more severe for the backward waves.

Consequently, eqns. 2.28 and 2.29 are valid for any angles ϑ_0, if the curvature radius is large enough, but they are valid for all values of r, if the angles ϑ_0

are small. For small r eqns. 2.28 and 2.29 approach eqns. 1.6 and 1.12 for a tilt. Obviously, eqns. 2.28 can also be deduced from eqn. 1.13, the only difference would be an unimportant phase shift factor omitted by manipulations in Section 1.1. Eqn. 1.13 was obtained from elementary considerations.

The amplitude of the parasitic waves at the bend output is a nonmonotonous function of the bend angle, since the parasitic wave is composed by the interference studied in details from this point of view in Section 1.1. The sinus arguments in eqn. 2.29 have a very simple physical meaning: they are equal to the half difference (for the forward waves) and to the half sum (for the backward waves) of the phase shifts for the corresponding waves generated along the whole bend. Let us note that within the accuracy limits of eqns. 2.28, it is possible to reduce the amplitude of any parasitic wave to zero for a fixed frequency choosing the angle ϑ_0 in such a way that parasitic waves excited in different parts of the bend completely compensate for each other.

The forward propagating wave amplitudes of eqn. 2.29*a* are much larger than those of the backward wave amplitudes, eqn. 2.29*b*, since, first, their coupling coefficients are much greater and, secondly, their exponential factor in the integral eqn. 1.13 is a much more slowly oscillating function. Because of this, the denominator of p_j for forward propagating waves contains the difference of their wavenumbers $h_m - |h_j|$, and for backward waves the essentially larger value of their sum $h_m + |h_j|$. Usually, it is possible to neglect the power losses for conversion into the backward propagating waves. This cannot be done only within the frequency band where $|h_j|$ is small; those specific conditions will be studied in Section 3.3.

The coupling coefficient F_{-mm} to the backward wave of the same number, i.e. for the reflected mode, is equal to zero for *TM* modes, since $C^{mm} = 0$ and for *TE* modes it is equal to

$$F_{-mm} = \frac{-i\alpha_m^2}{2h_m} K^{mm} \tag{2.31}$$

If $F_{-mm} = 0$, then the reflection coefficient of the given wave from the bend is equal to zero as well, i.e. $p_{-m}(0) = 0$. Therefore, the reflection coefficient is equal to zero for any *TM* mode. For *TE* modes in circular and rectangular waveguides $K^{mm} = 0$ and the reflection coefficient is also equal to zero, however $K^{mm} \neq 0$ for certain waveguides cross sections. Obviously, from the condition $F_{-mm} = 0$ we conclude that the reflection coefficient $p_{-m}(0)$ is equal to zero only until the term of the first order of $1/r$ (for the bend with a large curvature radius) or of ϑ_0 (for a small bend angle). Generally speaking, other parasitic waves excited in the bend are coupled to the $-m$ wave and the amplitude $p_{-m}(0)$ is different from zero, but it has a small value of the order of either $1/r_0^2$ or ϑ_0^2.

Expr. 2.30, determining the validity condition for the solution of eqn. 2.29, is also not satisfied for an arbitrarily large radius r, if $h_j = h_m$ for a value of j and simultaneously $F_{jm} \neq 0$[1]. This is a so-called degeneracy, when two modes coupled by curvature have the same phase velocities. In this case all the elementary parasitic waves excited at various sections along the bend interfere with identical phases and the result of this superposition has a finite value for finite values of ϑ_0 despite an arbitrarily large value of r. In such problems one cannot neglect the inverse conversion of the wave j into the wave m. Hence, the reduced amplitude of this wave will vary and eqn. 2.28*a* becomes invalid. For example, such a problem arises for the TE_{01} wave propagation in a circular waveguide bend. This practical problem will be considered in Section 5.3.

The mathematical technique based on the coupled-wave eqns. 2.9 or 2.27 allows us to study not only all aspects of the case of complete degeneracy $h_j = h_m$ but also a more general case when the wavenumbers, generally complex, are only approximately equal to each other.

2.1.5 Bends with curvature variations

It is easy to generalise the main results of this section for the problem of a variable curvature bend. It is more comfortable to employ the arc length l_j as a variable in eqns. 2.9 and 2.27 instead of the angle ϑ. To abridge the notation we will suppose that the length l_j is measured along the same circumference arc for all j, so we will omit the subscript of l_j. Then, eqns. 2.9 and 2.27 adopt the following form

$$\frac{dP_j}{dl} = \frac{F_{j\nu}}{r} P_\nu, \quad \frac{dp_j}{dl} = \frac{F_{j\nu}}{r} p_\nu \left(1 - \delta_{\nu j}\right) \exp\left(-i\left(h_\nu - h_j\right) l\right) \tag{2.32}$$

and the relation between P_j and p_j will be

$$P_j = p_j \exp\left(-ih_j l\right) \tag{2.33}$$

The equations in this form can be employed in the case of a variable curvature bend as well, i.e. when r depends on l. In fact, the amplitudes P_j are continuous on the border for the junction of two waveguide sections with different but constant values of r. This follows from the considerations used to deduce the boundary condition, i.e. from the requirement of the field's continuity and from the orthogonality condition. The reduced amplitudes p_j will be continuous as well. The amplitudes satisfy eqns. 2.32 along each of the sections, therefore these equations will be valid along the whole bent waveguide composed of these two sections with different r. This discussion can be extended to an arbitrarily large number of waveguide sections. Approaching the infinite limit,

---

[1]Eqn. 2.16 for the coupling coefficient is undefined for $\left|h_j - h_m\right| \to 0$. In this case, eqn. 1.7 should be used. In particular, from eqn. 1.7 follows that F_{jm} remains finite for $\left|h_j - h_m\right| \to 0$.

we will prove that r in eqn. 2.32 can be an arbitrary (e.g. discontinuous) function of the variable l.

A direct analytical proof of the validity of the coupled-wave eqns. 2.32 could be offered, employing a more general co-ordinate system instead of the cylindrical one, where l is one of the axes; such a co-ordinate system was proposed e.g. in Sveshnikov (1956*b*). The discussion above emphasises the local character of the coupling coefficient, in other words, the independence of the mode conversion on the curvature in neighbouring points. As a clear physical consequence of this fact, eqns. 2.32 can be generalised for the case of a variable curvature radius and the derivation of the curvature does not appear in eqns. 2.32.

Eqns. 2.32 are also valid for straight sections ($r = \infty$). They are then, simply, $p'_j = 0$, i.e. p_j = const or $P_j(l) = P_j(0)\exp(-ih_j l)$. If the waveguide is bent as a 3D curve, the rectangular co-ordinate system orientation varies in each cross section along the waveguide, therefore the coupling coefficients F_{jm} depend on the variable l as well.

A general solution of the coupled-wave eqns. 2.32 can also be obtained for large curvature radii (more exactly, if all the parasitic mode amplitudes are small). Once again, one can neglect all terms in the right-hand side of the second coupled-wave eqns. 2.32, except the term corresponding to the incident mode, in a similar way to the manipulations for the constant curvature bend. Further, one should assume $p_m(\vartheta) = \text{const} = 1$. The parasitic mode amplitudes scattered from the bend are determined then by the equation

$$\frac{dp_j}{dl} = \frac{F_{jm}}{r}\exp\left(-i\left(h_m - h_j\right)l\right) \tag{2.34}$$

Its solution with the boundary conditions, eqn. 2.10, will be

$$\begin{aligned} p_j(\vartheta_0) &= \int_0^L \frac{F_{jm}}{r}\exp\left(-i\left(h_m - h_j\right)l\right)dl, \quad j > 0, j \neq m \\ p_j(0) &= -\int_0^L \frac{F_{jm}}{r}\exp\left(-i\left(h_m - h_j\right)l\right)dl, \quad j < 0 \end{aligned} \tag{2.35}$$

Obviously, this solution is identical to the solution encountered in Section 1.1.

In the case of complete degeneracy ($h_j = h_m$), or in its neighbourhood, one has to retain several terms in the right-hand side of the coupled-wave eqns. 2.32 as well as of eqns. 2.24. These terms must be retained for arbitrarily large curvature radii, hence one has to solve simultaneously several coupled-wave equations (see Section 5.3 for more details).

When r and F_{jm} have constant values, the solution of eqn. 2.35 becomes equal to eqn. 2.28. However, it is also possible to derive the analytical expression of p_j from the integral eqn. 2.35 for a more general case. Let us suppose that $r(l)$ varies slowly with l and its values are of the same order of magnitude along the whole bend, whereas $r(l)$ is discontinuous at both bend ends and becomes infinite in such a way that at the points of the junctions between the bent and the straight waveguides, the curvature of the waveguide bend is of the same order of smallness as along the whole bend and becomes step-wise zero. Integrating eqn. 2.35 by parts under this assumption

$$p_j = \pm \frac{i}{h_m - h_j} \left| \frac{F_{jm}}{r} \exp\left(-i\left(h_m - h_j\right)l\right) \right|_0^L \mp \frac{i}{h_m - h_j} \int_0^L \exp\left(-i\left(h_m - h_j\right)l\right) \frac{d}{dl}\left(\frac{F_{jm}}{r}\right) dl \tag{2.36}$$

we can neglect the second term containing the derivative of the curvature and the product of the curvature by torsion. Generally speaking, the neglected term has the same order of smallness as the terms appearing in eqn. 2.32 when we retain all terms of the sum. The obtained two term formula is a generalisation of eqns. 2.29.

Hence, in this case, analytical expressions for the parasitic mode amplitudes in the higher-order terms depend on the curvature values at the planes where there are curvature discontinuities and on the coupling coefficients at those planes only. Evidently, this does not mean that the parasitic waves are excited only at those planes, really they are excited along the whole waveguide bend; but the curvature discontinuity planes are distinguished, since the mutual compensation of the waves excited along different sections is very weak in the neighbourhood of these points. In Chapter 5 we shall come once again to the calculation of eqn. 2.35. We will also encounter a similar expression for calculating the amplitudes of the waves scattered by waveguide tapers with variable cross section.

If the mean curvature of the waveguide bend with variable radius is not small, it will be necessary to employ numerical methods. The coupled-wave eqns. 2.32 are suitable for relatively simple computer calculations, and it is better to operate with the variables $P_j(\vartheta)$ instead of the $p_j(\vartheta)$.

2.1.6 *Eigenmodes of constant curvature bend*

The electromagnetic field in a constant curvature waveguide bend can also be described in terms of eigenmodes, i.e. the modes propagating without conversion along the bend. The eigenmodes can be considered as a certain linear combination of straight waveguide modes. They can be obtained from the general coupled-wave eqns. 2.9 with the additional requirement that all the amplitudes $P_j(\vartheta)$ depend on ϑ in the same way

$$P_j(\vartheta) = P_j(0)\exp(-iq\vartheta) \tag{2.37}$$

The substitution of eqn. 2.37 into eqn. 2.9 leads us to coupled-wave algebraic equations for $P_j(0)$

$$\left(F_{jv} + iq\delta_{jv}\right)P_v(0) = 0 \tag{2.38}$$

These coupled-wave equations have a solution, if q is equal to one of the roots of the characteristic equation

$$\det\left|F_{jm} + iq\delta_{jm}\right| = 0 \tag{2.39}$$

Each root of this equation corresponds to one of the eigenmodes. The relationships between the coefficients $P_j(0)$ of each eigenmode are determined by substituting the corresponding root into the coupled-wave eqns. 2.38 and solving these equations.

Knowing the fields of the eigenmodes of the bend, it is now possible to determine the scattered wave amplitudes linking the electromagnetic fields on the borders of the bent part. In such a way we obtain coupled-wave algebraic equations for the bend eigenmode amplitudes and for the amplitudes of the waves propagating from the bend into the straight waveguide.

Generally speaking, such a procedure is a little more complicated than the method for the direct determination of the scattered mode amplitudes described above. However, the mathematical apparatus of the eigenmodes is more convenient for some problems (see, for instance, the last point of Section 5.3).

The use of the eigenmode concept was first proposed in the paper of Huang (1960) for the case of a bend with variable curvature. It is possible to define a set of eigenmodes at any cross section. These are the eigenmodes of the bent waveguide with a constant curvature equal to the curvature of the cross section considered. These waves are not independent, there is a coupling between them caused by the variation of the curvature. It is easy to develop a regular method for determining this coupling, if the coefficient matrix of eqns. 2.9 is known.

Let us denote by the symbol O the matrix transforming the matrix F_{jm} to its diagonal form containing only the roots of eqn. 2.39. The eigenmode amplitudes denoted by W_j can be obtained at any cross section from P_j by the matrix O^{-1}. In matrix notation

$$W = O^{-1}P \tag{2.40}$$

where W and P are column matrices with the elements of W_j and P_j, respectively. The amplitudes W_j satisfy coupled-wave differential equations. They can be easily derived from eqns. 2.9 and 2.40

$$W' = NW \tag{2.41a}$$

$$N = O^{-1}FO - O^{-1}O' \tag{2.41b}$$

If the curvature radius is constant, the elements of the matrices F and O are constant as well, more exactly, they do not depend on ϑ and the second term of the N matrix, i.e. $(-O^{-1}O')$, will disappear. Therefore, N will be a diagonal matrix, i.e. the eigenmodes are mutually uncoupled. As it is shown in Huang (1960), the second term of N does not contain diagonal members in a variable curvature bend, i.e. when this term is different from zero; its elements are proportional to the curvature derivative and they describe coupling between the normal modes. If the curvature varies slowly, the coupled-wave differential eqns. 2.41*a* contain a small parameter and therefore their solution can be obtained in an analytical form.

The transformation of the variables P_j into the variables W_j represents, in fact, the application of other reference waveguides to the cross-section method. The reference waveguide bent along the circumference arc is employed instead of a straight reference waveguide tangent to the given bent waveguide. Because of this, the small parameter of such a problem is not the curvature but its derivative. However, the mathematical technique is essentially more complicated, and one has to define the N matrix instead of F and the W_j variables instead of P_j etc.

Lyudbarski and Povzner (1957) proposed an analogous generalisation of the method for the theory of straight waveguide with constant cross section, using a cone as the reference waveguide. However, such reference waveguides can only be applied to particular problems, whereas the mathematical method based on eqns. 2.40 and 2.41 can be employed in principle for any problem with a known original F matrix.

Let us make, in conclusion, a remark about a problem related not to waveguides but to toroidal resonant cavities made from constant cross-section waveguides bent along a circumference. If the radius of this circumference is very large compared to the cross-section dimensions, then the eigenmode fields of these resonators will coincide with the fields of the waves propagating in a straight waveguide, whereas the resonance eigenfrequencies can be derived from an obvious asymptotic condition. According to this condition, the resonator electrical length rh_j has to be equal to $2\pi n$, where $n = \pm 1, \pm 2, \ldots$ and h_j is the wavenumber of the straight waveguide. The resonance eigenfrequencies can be calculated for the arbitrary curvature radius using eqn. 2.39, if we substitute for q by $2\pi n$ and consider it as an equation for determining the frequency. In particular, an analytical expression can be derived in this way for a very large curvature radius to calculate the first-order correction of the resonance eigenfrequency asymptotic value. This correction is proportional to the curvature. The asymptotic value is derived from eqn. 2.39 neglecting the coupling coefficients between different waves, i.e. neglecting all the non- diagonal elements in eqn. 2.39.

2.2 Calculation of waveguides with variable filling medium using the cross-section method

2.2.1 Coupled-wave differential equations and boundary conditions

Let us consider a straight waveguide with constant cross section, where ε and μ are functions of all three co-ordinates. At the beginning we will assume that ε and μ are continuous functions. According to the cross-section method, to describe the electromagnetic field in such a waveguide it is necessary to expand the field at each cross section (z=const) in a Fourier-like series of the modal fields of a reference waveguide. For this problem the reference waveguide will be a uniform waveguide where ε and μ at each cross section are the same functions of the co-ordinates x and y as in the corresponding cross section of the nonuniform waveguide. Obviously, the reference waveguides are different for different values of the co-ordinate z. Let us denote the co-ordinate system in this reference waveguide as x,y and ζ. Obviously, the fields of the reference waveguide eigenmodes depend on ζ appearing in the exponential factor $\exp(-ih_j\zeta)$ but they depend also on z, since simultaneously the functions $\varepsilon(x,y)$ and $\mu(x,y)$ depend on z.

As in the case of bent waveguide we use eqn. 2.1

$$
\begin{aligned}
E_x &= \sum_{m=1}^{\infty} Q_m E_x^m, \quad E_y = \sum_{m=1}^{\infty} Q_m E_y^m \\
H_x &= \sum_{m=1}^{\infty} R_m H_x^m, \quad H_y = \sum_{m=1}^{\infty} R_m H_y^m
\end{aligned}
\tag{2.42}
$$

where the coefficients Q_m and R_m are functions of z. Now we will introduce the variables $P_m(z)$ and $P_{-m}(z)$ using eqns. 2.42. Then the expansions can be written as follows

$$E_x = P_v E_x^v, \quad E_y = P_v E_y^v, \quad H_x = P_v H_x^v, \quad H_y = P_v H_y^v \tag{2.43}$$

The $\vec{E}$ and $\vec{H}$ fields satisfy, on the walls of the nonuniform waveguide, the same boundary conditions eqn. I.2 as the $\vec{E}^m$ and $\vec{H}^m$ fields on the walls of the reference waveguide. The term by term differentiation of eqn. 2.43 can be performed. Substituting eqn. 2.43 into two of the Maxwell equations and comparing the result with the last two formulae in eqns. I.4*a* and I.4*b* we obtain expansions for the longitudinal field components which are similar to the expansions of the transversal components

$$E_z = P_v E_z^v, \quad H_z = P_v H_z^v \tag{2.44}$$

In this way we can consider $P_j(z)$ as mode amplitudes in the given nonuniform waveguide.

Now we will deduce the equations for the functions $P_j(z)$. Let us substitute eqns. 2.43 and 2.44 into the remaining four Maxwell equations. By using eqn. I.4 several times we obtain, after simple manipulations, the following four equations

$$E_x^v(P_v' + ih_v P_v) = -P_v E_x^{v'}, E_y^v(P_v' + ih_v P_v) = -P_v E_y^{v'}$$
$$H_x^v(P_v' + ih_v P_v) = -P_v H_x^{v'}, H_y^v(P_v' + ih_v P_v) = -P_v H_y^{v'} \tag{2.45}$$

where the prime symbol denotes derivation by the variable z.

To deduce the equation required for P_j' let us multiply eqns. 2.45 by $-H_x^j$, H_y^j, E_x^j and $-E_y^j$, respectively. Further, we add the results and integrate over the cross section. Taking into account the orthogonality conditions we obtain the desired expression for P_j' as the following linear combination of P_m

$$P_j' + ih_j P_j = S_{jv} P_v \tag{2.46}$$

This expression is valid for any j, therefore eqn. 2.46 represents the desired first-order ordinary differential coupled-wave equations for $P_j(z)$.

We will call the values S_{jm} coupling coefficients since they describe the interference of the waves scattered from the waveguide nonuniformity. We yield the following expression

$$S_{jm} = \frac{1}{2kh_j}\int\left(E_x^j H_y^{m'} - E_y^j H_x^{m'} + E_x^{m'} H_y^j - E_y^{m'} H_x^j\right)dS \tag{2.47}$$

As in the case of a bent waveguide, the boundary conditions for P_j are deduced considering the continuity of the electromagnetic fields at the junction of the nonuniform and uniform waveguides. If the nonuniform waveguide of length L is interconnected between two uniform waveguides at $z = 0$ and $z = L$ and there is an incident mode with number m and unitary amplitude from the left-hand side uniform waveguide, then these conditions will have the following form

$$P_m(0) = 1,\ P_j(0) = 0 \text{ for } j > 0 \text{ and } j \neq m,\ P_j(L) = 0 \text{ for } j < 0 \tag{2.48}$$

The physical meaning of the coupled-wave eqns. 2.46 is the same as for the coupled-wave eqns. 2.9 and the interpretation given to eqns. 2.9 at the end of Section 2.1.1 can be applied to eqns. 2.46. One needs only to substitute for $\Delta\vartheta$ by Δz.

2.2.2 *Coupling coefficients in variable filling medium*

Several simple relationships can be derived from the general expression in eqn. 2.47 for the coupling coefficients. First, let us compare eqn. 2.47 with the formula resulting from deriving by z the identity eqn. I.8

$$\int \left(E_y^m H_x^m - E_x^m H_y^m\right) dS = kh_m \tag{2.49}$$

This derivation has to be performed for the integrand only, since the area of integration is the same for all z. Proceeding in this way we obtain

$$S_{mm} = -h'_m / 2h_m \tag{2.50}$$

Further, performing the derivation by z of the orthogonality condition eqn. I.9 and expressing the obtained integrals through the coupling coefficients according to eqn. 2.47, we obtain the reciprocity relationship between the coupling coefficients

$$h_m S_{mj} = -h_j S_{jm} \, (m \neq j) \tag{2.51}$$

Obviously, the last two formulae could be deduced by using only the coupled-wave eqns. 2.46 in a manner similar to that in which eqn. 2.14 was obtained. We should only take into account the fact that, according to eqn. 2.47, the value of S_{jm} unlike in eqn. 2.11 does not change its sign when the propagation direction of both waves is reversed, i.e. $S_{-j,-m} = S_{jm}$.

The eqns. 2.50 and 2.51 allow us to verify that the coupled-wave eqns. 2.46 satisfy the requirements of the energy conservation principle. As this principle states that if ε and μ are real, then for any electromagnetic field in a non-uniform waveguide the following sum has to be constant

$$\sum_j \left|P_j\right|^2 h_j \tag{2.52}$$

where all the waves propagating forwards and backwards are included in the sum (h_j should be real). Now, let us transform the derivative by z of eqn. 2.52 substituting P'_j following eqn. 2.46. Then this derivative takes the form

$$\sum_j \left[h'_j + h_j\left(S_{jj} + S_{jj}^*\right)\right]\left|P_j\right|^2 + \sum_{j \neq m}\sum \left[h_j S_{jm}^* + h_m S_{mj}\right] P_j P_m^* \tag{2.53}$$

This expression has to be equal to zero for any P_j. Therefore, the expressions included in the brackets should be equal to zero for both sums. Because S_{jm} is real when ε, μ, h_j and h_m are real, this requirement, according to eqns. 2.50 and 2.51, is also satisfied.

It is possible to demonstrate in a similar manner that the coupled-wave eqns. 2.9 describing the electromagnetic field in a bent waveguide with constant cross section also satisfy the energy conservation principle. The equality to zero of the expressions included in the brackets in eqn. 2.53 follows from the reciprocity relationship eqn. 2.14 and we have also to remember that for the modes propagating without attenuation the coupling coefficients F_{jm} are pure imaginary and $h_j' \equiv 0$ in this case.

2.2.3 *Practical form of the coupling coefficients*

It is not convenient to employ eqn. 2.47 for practical engineering problems. Further, we will derive some alternative formulae for S_{jm} transforming the basic expression in eqn. 2.47. In a uniform waveguide where ε and μ are independent of z, i.e. when $\partial\varepsilon / \partial z = 0$ and $\partial\mu / \partial z = 0$ all the coupling coefficients are equal to zero. Let us deduce the expression for S_{jm} containing the z derivatives of ε and μ instead of the fields derivatives contained in eqn. 2.47.

Let us write the Maxwell equations for the $\hat{\vec{H}}^m$ and $\hat{\vec{E}}^m$ fields in the reference waveguide employing the variables x, y and ζ. In these equations $\hat{\vec{H}}^m$, $\hat{\vec{E}}^m$, h_m as well as $\varepsilon(x, y)$ and $\mu(x, y)$ depend on z, this means that they depend on the kind of uniform waveguide to which they correspond. Let us find z derivatives of these equations

$$\mathrm{rot}\hat{\vec{E}}^{m\prime} = -ik\mu\hat{\vec{H}}^{m\prime} - ik\mu'\hat{\vec{H}}^m,\ \mathrm{rot}\hat{\vec{H}}^{m\prime} = ik\varepsilon\hat{\vec{E}}^{m\prime} + ik\varepsilon'\hat{\vec{E}}^m \tag{2.54}$$

These derivatives can be interpreted as the transition from one reference waveguide to another nearly the same waveguide. For example

$$\left(\hat{\vec{H}}^m\right)' = \left(\vec{H}^{m\prime} - i\zeta h_m' \vec{H}^m\right)\exp(-ih_m\zeta) \tag{2.55}$$

Let us multiply eqns. 2.54 by $-\hat{\vec{H}}^{-j}$ and $-\hat{\vec{E}}^{-j}$; the Maxwell eqns. I.1 written for the wave with the number $(-j)$ will be multiplied by $\left(\hat{\vec{H}}^m\right)'$ and $\left(\hat{\vec{E}}^m\right)'$ and then the results added together. We then obtain the identity

$$div\left\{\left[\hat{\vec{H}}^{-j} \times \left(\hat{\vec{E}}^m\right)'\right] - \left[\left(\hat{\vec{H}}^m\right)' \times \hat{\vec{E}}^{-j}\right]\right\} = ik\left(\mu'\hat{\vec{H}}^m \cdot \hat{\vec{H}}^{-j} - \varepsilon'\hat{\vec{E}}^m \cdot \hat{\vec{E}}^{-j}\right) \tag{2.56}$$

Let us separate the derivative by ζ on the left-hand side and integrate eqn. 2.56 over the cross section assuming $j \neq$ m. $\left(\hat{\vec{E}}^m\right)'$ satisfies the same boundary condition eqn. I.2 on the metal as $\hat{\vec{E}}^m$ and, hence, only the integral of the ζ derivative from the ζ component of the vector for that divergence has been taken remains on the left-hand side. The second terms in eqn. 2.55 will disappear by substituting $\left(\hat{\vec{H}}^m\right)'$ and $\left(\hat{\vec{E}}^j\right)'$ because of the orthogonality condition, whereas the remaining integral in the left-hand side formula will be identical to the integral in eqn. 2.47. This manipulation gives us the following expression for the coupling coefficient eqn. 2.47.

$$S_{jm} = \frac{1}{2h_j\left(h_j - h_m\right)}\int\left(\frac{\partial\varepsilon}{\partial z}\vec{E}^{-j}\cdot\vec{E}^{m} - \frac{\partial\mu}{\partial z}\vec{H}^{-j}\cdot\vec{H}^{m}\right)dS \tag{2.57a}$$

or the equivalent expression

$$S_{jm} = \frac{1}{2h_j\left(h_m - h_j\right)}\int\left(\frac{\partial\varepsilon}{\partial z}\left(E_x^jE_x^m + E_y^jE_y^m - E_z^jE_z^m\right) + \frac{\partial\mu}{\partial z}\left(H_x^jH_x^m + H_y^jH_y^m - H_z^jH_z^m\right)\right)dS \tag{2.57b}$$

The difference $h_j - h_m$ appearing in the denominator of the obtained formula does not mean that the coupling coefficient of these two modes could be very large when their wavenumbers become nearly equal to each other. Following eqn. 2.47, S_{jm} reaches a large value only for small values of h_j, whereas for $\left|h_j - h_m\right| \to 0$ the coefficient S_{jm} has a finite value. Indeed, the value of the integral in the nominator of eqn. 2.57 becomes small when the values of h_j and h_m are nearly the same. This explanation is also valid for several other expressions for S_{jm} containing the same difference $h_j - h_m$ in their denominators (see eqns. 2.53 and 2.3). Those expressions will be derived further.

The expression for the coupling coefficients given in eqn. 2.57 is identical to eqn. 1.26 obtained from the study of small nonuniformities. On the other hand, applying the coupled-wave eqns. 2.46, 2.48 to a small nonuniformity, it is easy to obtain eqns. 1.20, 1.21 for the amplitudes of the parasitic modes scattered by this nonuniformity.

The expression for h_m' can be obtained in a similar manner. To do this we will assume $j = m$ in eqn. 2.56. The ζ dependence of the left-hand side of this equation is determined only by the second term of eqn. 2.55. Separating again the ζ derivative and taking into account the orthogonality condition we obtain

$$h_m' = \frac{1}{2h_m}\int\left(\varepsilon'\vec{E}^{m}\cdot\vec{E}^{-m} - \mu'\vec{H}^{m}\cdot\vec{H}^{-m}\right)dS \tag{2.58}$$

2.2.4 Piecewise constant functions ε and μ

We will employ eqn. 2.57 for the coupling coefficients as an intermediate result only to find another expression for S_{jm} which can be manipulated more comfortably at the usual conditions when ε and μ are piecewise constant functions of the co-ordinates and ε' and μ' are different from zero only on the border of an area filled by a medium.

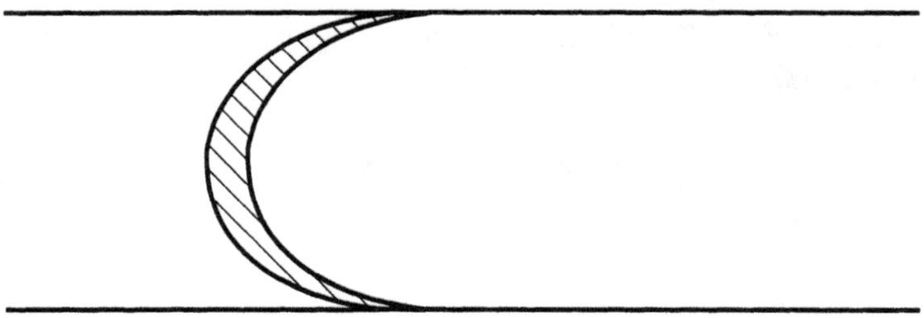

Figure 2.2 Thin intermediate layer at the border of the waveguide filling medium

To calculate S_{jm} for this case, let us first substitute the filling medium border by a thin layer where ε and μ will vary from the proper values of the filling medium to $\varepsilon = 1$ and $\mu = 1$. Then we will calculate the integral in eqn. 2.57 for this area only (see the dashed part in Figure 2.2). This is the substantial difference between eqns. 2.57 and 2.47, since the integral in eqn. 2.47 has to be calculated over the whole cross section for this case. For the sake of simplicity, let us assume that there is only one separating border. Naturally, the result can be generalised for other cases when there are several such borders.

Let us introduce a local co-ordinate system with the axes s and n in this thin layer over which the integral eqn. 2.57 is calculated in such a way that s is tangential to the contour where the separating border and the cross section intersect; n is normal to s and is situated in the cross-section plane and is directed from the area with $\varepsilon = 1$ and $\mu = 1$ towards the filling medium. The vectors $\vec{n}, \vec{s}$ and $\vec{z}$ form a right-hand side system of axes. Note that the vector $\vec{n}$ is not orthogonal to the intermediate layer in the nonuniform waveguide. To find a relationship between $d\varepsilon/dz$ and $d\varepsilon/dn$ and between $d\mu/dz$ and $d\mu/dn$ we should first express these variables through the gradients of ε and μ. Geometrical considerations give us the following expressions

$$\frac{\partial \varepsilon}{\partial z} = -\nu \frac{\partial \varepsilon}{\partial n}, \quad \frac{\partial \mu}{\partial z} = -\nu \frac{\partial \mu}{\partial n} \tag{2.59}$$

where ν is the tangent of the angle α defined between the z-axis and the line tangential to the separating border, orthogonal to the cross-section boundary contour. Eqns. 2.59 are illustrated in Figure 2.3 for the simplest case when the vectors $\vec{n}$, grad ε and the $\vec{z}$-axis are coplanar.

According to Figure 2.3 for the case under consideration, we will have simply

$$\frac{\partial \varepsilon}{\partial z} = -|\text{grad}\,\varepsilon|\sin\alpha, \quad \frac{\partial \varepsilon}{\partial n} = |\text{grad}\,\varepsilon|\cos\alpha \tag{2.60}$$

and eqn. 2.59 follows from these formulae. In a more general case, the factors associated with grad ε in eqn. 2.60 will have a more complex form but their ratio will be equal to $-\nu$ in any case.

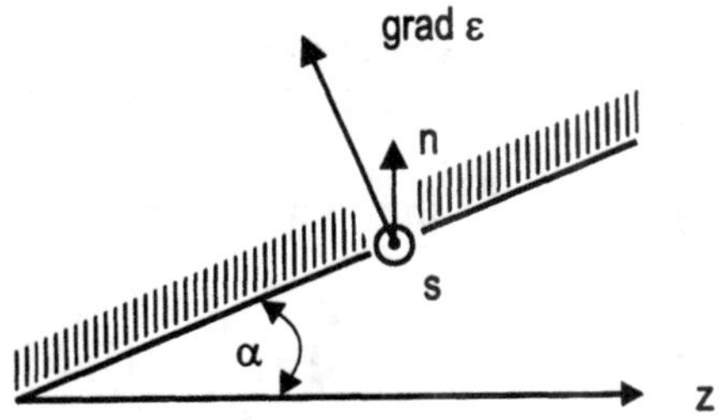

Figure 2.3 Sketch for the derivation of eqn. 2.59

More strictly, eqn. 2.59 is valid only for the limiting case (we will consider this further) when the thin layer with varying ε and μ is reduced to a surface (i.e. the separating border). Then grad ε and grad μ become normal to this surface. ν in eqn. 2.59 has some value before the limit calculation and in the limit it has the geometrical meaning mentioned above.

Substituting eqn. 2.59 into eqn. 2.57 for the calculation of the coupling coefficient, we will obtain

$$S_{jm} = \frac{1}{2h_j\left(h_j - h_m\right)} \oint \nu \left\{ \int \left[\frac{\partial \varepsilon}{\partial n} \left(E_n^j E_n^m + E_s^j E_s^m - E_z^j E_z^m \right) + \frac{\partial \mu}{\partial n} \left(H_n^j H_n^m + H_s^j H_s^m - H_z^j H_z^m \right) \right] dn \right\} ds \quad (2.61)$$

Remember that the electromagnetic fields appearing in eqn. 2.61 correspond to the uniform reference waveguide; now the magnitude ν is the only characteristic of the nonuniform waveguide. The cross section of the reference waveguide is composed of two areas where the parameters ε and μ have constant values and of a thin transition region to separate areas where ε and μ depend on x and y.

In this intermediate layer the tangential components of $\vec{E}^m, \vec{H}^m$ and the normal components of $\varepsilon\vec{E}^m$, $\mu\vec{H}^m$ are continuous in such a way that

$$\begin{aligned} E_n^m(n) &= E_n^m / \varepsilon(n), \quad E_s^m(n) = E_s^m, \quad E_z^m(n) = E_z^m \\ H_n^m(n) &= H_n^m / \mu(n), \quad H_s^m(n) = H_s^m, \quad H_z^m(n) = H_z^m \end{aligned} \quad (2.62)$$

and for the mode number j we have the same relations. Here, $E_n^j(n)$ etc. are the intermediate layer fields; the variables E_n^m etc. on the right-hand side are the field values on the side of the border where $\varepsilon = 1$ and $\mu = 1$. The thinner the transition layer is, the more accurate eqns. 2.62 are. They can be deduced from the integral form of Maxwell's equations in the same way as can the usual conditions on the separating border between two dielectric media.

Substituting eqn. 2.62 into eqn. 2.61 we can perform the integration along n. Now S_{jm} will contain only the integral calculated along the C boundary contour.

In this way the desired expression for the coupling coefficient is obtained

$$S_{jm} = \frac{1}{2h_j(h_j - h_m)} \oint \nu(s) \Big[(\varepsilon - 1)\big(E_s^j E_s^m - E_z^j E_z^m\big) - \Big(\frac{1}{\varepsilon} - 1\Big) E_n^j E_n^m + (\mu - 1)\big(H_s^j H_s^m - H_z^j H_z^m\big) - \Big(\frac{1}{\mu} - 1\Big) H_n^j H_n^m \Big] ds \tag{2.63}$$

Here, ε and μ are the dielectric permittivity and magnetic permeability of the filling medium, respectively.

Therefore, to calculate the coupling coefficient by eqn. 2.63 it is sufficient to know the modal fields on the border of the filling medium in uniform waveguides and the function $\nu(s)$, describing the difference between the nonuniform waveguide at the given cross section and the uniform waveguide. For a uniform waveguide $\nu \equiv 0$ and all the coupling coefficients are equal to zero.

Eqn. 2.63 will be used as a basis for the analysis in the next section. It is especially convenient to use eqn. 2.63 when ε and μ are close to unity because a substitution of the empty waveguide fields in that equation gives us immediately the first terms of the expansion of the coupling coefficient S_{jm} in a power series of $(\varepsilon - 1)$ and $(\mu - 1)$.

It is also obvious that eqn. 2.58 for the wavenumber derivative can be rewritten in a form similar to eqn. 2.63.

The continuity condition for the functions $\varepsilon(x, y, z)$ and $\mu(x, y, z)$ formulated at the beginning of this section, provided the convergence of the series eqns. 2.42 – 2.43 in all points of the cross section. Thanks to that condition the term by term differentiation of these series was possible, and we obtained the coupled-wave eqns. 2.46 and the S_{jm} values given by eqn. 2.47. Nevertheless, the following transformation of eqns. 2.47, leading to obtaining eqn. 2.63 for discontinuous ε and μ, is in no way related to this condition. Analogously, the coupled-wave eqns. 2.46 for the coefficients in eqn. 2.43 do not depend on the smoothness of the functions ε and μ and hence are valid at any stage when approaching the limit from continuous to discontinuous functions ε and μ. They are also valid in the limit of discontinuous functions, however eqns. 2.43 and 2.44 at such a limit could not converge on the surface discontinuities of ε and μ and become nonuniformly convergent. The situation considered here is similar to one usually encountered by deduction of the boundary condition for electromagnetic fields on the border separating two media. These conditions can be also established by first employing the Maxwell equations valid for continuous media in a thin intermediate layer; further, these conditions remain valid when one approaches the limit towards the discontinuous media distribution, hence the final result is also valid.

The same consideration allows us in the next section to continue the transformation of the expression for the coupling coefficients and in such a way we approach the limit for an infinite value of $|\varepsilon|$ in eqn. 2.63.

Above, we performed the limit evaluation in eqn. 2.57 for discontinuous functions ε and μ. This limit could be also evaluated with eqn. 2.47, but then we would obtain expressions essentially more complicated than eqn. 2.63 for the coupling coefficients calculation. First, they would include a boundary integral similar to eqn. 2.63. This integral results from integrating over the thin intermediate layer where the field gradients are very large and the derivatives contained in eqn. 2.47 include functions similar to the delta functions as well as the derivatives in eqn. 2.57. Secondly, integrals similar to eqn. 2.47 would be retained in those expressions where the integration is performed over the whole cross section since their integrands as well as their counterparts in eqn. 2.47 are different from zero in the whole cross section. When we take the limit for $|\varepsilon| \rightarrow \infty$ in the next Section, both types of integrals would be retained in the expression for the S_{jm} if we employed this calculation method and the integral over the cross section would contain z derivatives of the transversal eigenfunctions. The resulting expressions would be essentially much more complicated than eqns. 2.94 or 2.97 derived in the next Section, hence we do not show them here. Let us only note that by employing eqns. 2.103 of the next section, the identity of these formulae and eqns. 2.97 for the coupling coefficients could be demonstrated by means of algebraic methods.

2.2.5 *Dielectric permittivity and magnetic permeability tensors*

In this subsection we will give a generalisation of the formal mathematical apparatus developed above for the case of a waveguide filled by a medium with tensorial parameters, following Barsukov (1959).

Let us suppose that the medium is characterised by its Hermite tensors of the dielectric permittivity and magnetic permeability

$$\mathrm{E} = \begin{pmatrix} \varepsilon_1 & -i\varepsilon_2 & 0 \\ i\varepsilon_2 & \varepsilon_1 & 0 \\ 0 & 0 & \varepsilon_3 \end{pmatrix}, \quad \mathrm{M} = \begin{pmatrix} \mu_1 & -i\mu_2 & 0 \\ i\mu_2 & \mu_1 & 0 \\ 0 & 0 & \mu_3 \end{pmatrix} \tag{2.64}$$

The orthogonality condition for the waveguide fields will be represented under the form of an integral

$$\int \left(E_y^m H_x^{j*} - E_x^m H_y^{j*} + H_x^m E_y^{j*} - H_y^m E_x^{j*} \right) dS \tag{2.65}$$

equal to zero for $j \neq m$. Obviously, the condition in eqn. I.9 for the case of scalar ε and μ could also be substituted by the condition in eqn. 2.65. However, eqn. I.9 is simpler. We normalise the fields in such a way that eqn. 2.65 is equal to $2kh_m$ for $j = m$.

All the considerations supporting the deduction of the coupled-wave eqns. 2.46 remain valid for this case as well. Then we obtain instead of eqn. 2.47 the following expression for the coupling coefficients

$$S_{jm} = \frac{1}{2kh_j} \int \left\{ \left[\vec{E}^{m\,*} \times \vec{H}^{j\,\prime} \right]_z + \left[\vec{E}^{j\,\prime} \times \vec{H}^{m\,*} \right]_z \right\} dS \tag{2.66}$$

By means of the same manipulations realised in the last subsection we can rewrite the expression giving S_{jm} as a generalisation of eqn. 2.57

$$S_{jm} = \frac{1}{2h_j\left(h_j - h_m\right)} \int \left\{ \vec{E}^{m\,*} \frac{\partial \mathrm{E}}{\partial z} \vec{E}^j + \vec{H}^{m\,*} \frac{\partial \mathrm{M}}{\partial z} \vec{H}^j \right\} dS \tag{2.67}$$

The tensor derivative is determined, as usual, by a new tensor with elements equal to the derivatives of the elements of the given tensor.

If the tensor components E and M are piecewise constant functions, then eqn. 2.67 can also be reduced to a boundary contour integral. The expressions for the intermediate layer fields will satisfy relationships which are a little more complicated than eqn. 2.62

$$E_s^m(n) = E_s^m,\; E_z^m(n) = E_z^m,\; E_n^m(n) = \frac{1}{\varepsilon_1(n)} E_n^m - i\frac{\varepsilon_2(n)}{\varepsilon_1(n)} E_s^m \tag{2.68}$$

where the symbols of eqn. 2.62 have been retained. Here we will not write down the equivalent formula for the $\hat{\vec{H}}^m$ components.

Integrating eqn. 2.67 by n over the intermediate layer, we obtain the expression for the coupling coefficient containing only the fields on the border of the filling insertion inside the waveguide

$$S_{jm} = \frac{1}{2h_j(h_m - h_j)} \oint \nu(s) \begin{bmatrix} \left(\dfrac{\varepsilon_1^2 - \varepsilon_2^2}{\varepsilon_1} - 1\right) E_s^j E_s^{m\,*} \\ + \dfrac{\varepsilon_1 - 1}{\varepsilon_1} E_n^j E_n^{m\,*} + i\dfrac{\varepsilon_2}{\varepsilon_1}\left(E_s^j E_n^{m\,*} - E_n^j E_s^{m\,*}\right) \\ +(\varepsilon_3 - 1) E_z^j E_z^{m\,*} + \left(\dfrac{\mu_1^2 - \mu_2^2}{\mu_1} - 1\right) H_s^j H_s^{m\,*} \\ + \dfrac{\mu_1 - 1}{\mu_1} H_n^j H_n^{m\,*} + i\dfrac{\mu_2}{\mu_1}\left(H_s^j H_n^{m\,*} - H_n^j H_s^{m\,*}\right) \\ +(\mu_3 - 1) H_z^j H_z^{m\,*} \end{bmatrix} ds \tag{2.69}$$

The results of this subsection can be used, for example, when the electromagnetic wave transformation phenomena in different ferrite inclusions are studied.

2.2.6 *Case of weak coupling*

Now let us consider applications of the coupled-wave differential eqns. 2.46 for the amplitudes $P_m(z)$. As will be demonstrated in the next Section, the same coupled-wave equations also describe the electromagnetic field in a variable cross-section waveguide. Due to this fact, all the results obtained later are applicable to such waveguides. Their coupling coefficients are also expressed by means of boundary contour integrals containing the inclination angle between the generatrices and the z-axis.

First, let us study the case of weak coupling, i.e. the case when all the coupling coefficients are small. This happens when the waveguide parameters vary slowly along the direction of propagation, i.e. ν_0, the magnitude of the mean value of the generatrice tangent, is small or in general when ε and μ have small variations for distances of the order of the linear dimensions of the cross section. A special case when the coupling coefficients have large values despite slowly varying parameters, because of the small value of the wavenumber in the denominator, will be studied in Sections 3.1 and 3.2.

As in the previous section we will start transforming the coupled-wave eqns. 2.46 in such a way that their matrix does not contain diagonal elements. We will introduce again the reduced amplitudes $p_j(z)$ determined now by the identity

$$P_j(z)=\sqrt{\frac{h_j(0)}{h_j(z)}}\,p_j(z)\exp\left(-i\gamma_j\right),\quad \gamma_j(z)=\int_0^z h_j dz \tag{2.70}$$

Substituting eqn. 2.70 into the coupled-wave eqns. 2.46 and taking into account the relationship in eqn. 2.50 between S_{jj} and h_j', we will obtain the following coupled-wave equations for the reduced amplitudes

$$p_j'(z)=\sqrt{\frac{h_v(0)}{h_j(0)}}S_{jv}\sqrt{\frac{h_j(z)}{h_v(z)}}\left(1-\delta_{jv}\right)p_v\exp\left(-i\left(\gamma_v-\gamma_j\right)\right) \tag{2.71}$$

According to the conditions in eqn. 2.48 and to the factors introduced in eqn. 2.70, the boundary conditions for $p_j(z)$ have the following form

$$\begin{aligned} &p_m(0)=1,\ \ p_j(0)=0 \text{ for } j>0, j\neq m\\ &p_j(L)=0\ \ \text{ for } j<0 \end{aligned} \tag{2.72}$$

Repeating the considerations described before in eqn. 2.34 we will retain the incident wave amplitude only in the right-hand side of eqn. 2.71. Then $p_m'=0$, i.e.

$$p_m(z) = 1, \quad P_m(z) = \sqrt{\frac{h_m(0)}{h_m(z)}} \exp(-i\gamma_m z) \tag{2.73}$$

As follows from the comparison with eqn. I.10 for the energy carried by the mode, the incident mode energy does not vary along the waveguide in the first order of magnitude.

Substituting eqn. 2.73 into eqn. 2.71 we obtain for $j \neq m$ the equation

$$p_j' = \sqrt{\frac{h_m(0)}{h_j(0)}} S_{jm} \sqrt{\frac{h_j(z)}{h_m(z)}} \exp\left(-i(\gamma_m - \gamma_j)\right) \tag{2.74}$$

From eqn. 2.74 and the boundary conditions in eqn. 2.72 we determine the amplitudes of the scattered parasitic modes

$$p_j = \pm \sqrt{\frac{h_m(0)}{h_j(0)}} \int_0^L \left[S_{jm} \sqrt{\frac{h_j}{h_m}} \exp\left(-i(\gamma_m - \gamma_j)\right) \right] dz \tag{2.75}$$

The upper sign in eqn. 2.75 gives us $p_j(L)$ for $j > 0$ and $j \neq m$ and the lower sign gives $p_j(0)$ for $j < 0$. The square of the absolute value of the integral in eqn. 2.75 gives us the ratio of the energy of the scattered parasitic mode and the incident mode energy. The physical meaning of the various elements in eqn. 2.75 has been discussed in Section 1.2 where another method has been employed.

Eqn. 2.75 together with the various expressions encountered above for the coupling coefficients reduces the given analysis problem for the waveguide with slowly varying parameters to an integral evaluation.

2.2.7 *Advantage of reduced amplitudes*

The approximated solutions given by eqns. 2.73 and 2.75 can also be deduced from the rigorous coupled-wave eqns. 2.71 in a more formal manner. To do this we will expand the reduced amplitudes p_j in a power series of the small parameter of this problem

$$p_j = p_j^{(0)} + \nu_0 p_j^{(1)} + \ldots \tag{2.76}$$

The form of the coupled-wave eqns. 2.71 allows us to separate in a trivial manner the terms with a different order of smallness. Since all the coupling coefficients are of the order of ν_0, $S_{jm} = \nu_0 \overline{S}_{jm}$, where $\overline{S}_{jm}$ does not contain any small parameter. Substituting eqn. 2.76 into eqns. 2.71 we obtain

$$p_j^{(0)\prime} = 0, \quad p_j^{(1)\prime} = \sqrt{\frac{h_v(0)}{h_j(0)}} \overline{S}_{jv} \sqrt{\frac{h_j(z)}{h_v(z)}} (1 - \delta_{jv}) p_v^{(0)} \exp\left(-i(\gamma_v - \gamma_j)\right) \tag{2.77}$$

The first of these equations gives us $p_j^{(0)} = 0$ for $j \neq m$ and $p_m^{(0)} = 1$. From the second equation it follows that $p_m^{(1)} = 0$ and the formula for $p_j^{(1)}$ is identical to eqn. 2.75.

However, this consideration could be wrong. The studied problem contains in addition to the small parameter ν_0 also a large parameter, i.e. the length L of the waveguide section. According to its definition, the total variation of the waveguide properties is proportional to $\nu_0 L$ and hence cannot be considered small. Therefore, it may happen that eqn. 2.75 for $p_j^{(1)}(z)$ giving small values at the beginning of the waveguide section will no longer be small for increasing z. Indeed, this happens for a degeneracy when the wavenumber of a mode is nearly equal to the wavenumber of the incident mode. In such cases, the exponential factor in the integrand of eqn. 2.75 would not be a rapidly oscillating function of z and the integral in eqn. 2.75 would no longer be small. The calculation method employed for these cases will be considered in Section 2.4.1 for the example of a twisted waveguide. We have encountered similar conditions in the theory of bent waveguides with a small curvature radius. Therefore, the condition to apply these approximations of weak coupling coefficients is not the smallness of the parameter ν_0 only or another similar parameter but the smallness of all the amplitudes given by eqn. 2.75. More exactly, all the expressions like eqn. 2.75 have to be small compared to the unity when we substitute any z value of the considered interval instead of the upper limit L.

Particularly, this discussion explains why it is more convenient to analyse the reduced amplitudes $p_j(z)$ and why it is not possible to perform the same expansion directly from the coupled-wave eqns. 2.46 for $P_j(z)$. The right-hand side of the equation for P_j' contains the term proportional to P_j. When substituting instead of the P_j amplitude its series expansion similar to eqn. 2.76, this term would become of the second order of smallness (of ν_0) for $j \neq m$. Nevertheless, by integration over a long section this term could give a contribution in $P_j(z)$ of the first order of ν_0, as it can be easily verified. Therefore, to obtain a correct result we would have to combine the terms of different orders of magnitude by integration over a long interval. Employing the variables p_j instead of P_j permits us to elaborate, with the help of the expansion in powers of ν_0, a solution diverging essentially more slowly along the waveguide. Physically, this is explained by the solution $p_m(z) = 1$, satisfying the energy conservation law, as we have noted above.

There is some analogy between the transformation from $P_j(z)$ to the reduced amplitudes $p_j(z)$ and the well known manipulations employed in the theory of wave propagation in inhomogeneous media (see, for example, paragraph 16 in Brekhovskikh (1980) for the deduction of geometrical optics equations).

2.2.8 Three special solutions

In Section 2.4 we will apply the results of this section to the computation of a set of engineering problems. Here, we will underline only three special cases in which the formulae for the scattered mode amplitudes have a more simple form.

The first case occurs when the coupling coefficient S_{jm} is of the same order of magnitude along the whole nonuniform waveguide section whereas at the ends of the section, at $z = 0$ and $z = L$, it becomes equal to zero step-wise. This can happen, for example, when the tangent of the inclination angle of the generatrice $\nu(s,z)$ in eqn. 2.63 has small variations in the interval $0 < z < L$. Then it is possible to integrate the integral of eqn. 2.75 by parts and to omit the integral member[1]. Proceeding in this manner we are neglecting values of the order of square of the mean value of $\nu^2(s, z)$ and of the order of the derivative of ν with respect to z. In such a way we yield an analytical expression for p_j in the form of a two-term sum analogous to eqn. 2.35

$$p_j = \mp i\left(\frac{S_{jm}}{h_m - h_j}\right)_{z=0} \pm i\sqrt{\frac{h_m(0)h_j(L)}{h_j(0)h_m(L)}}\left(\frac{S_{jm}}{h_m - h_j}\right)_{z=L} \times \exp\left(-i\left[\gamma_m(L) - \gamma_j(L)\right]\right) \qquad (2.78)$$

In this formula the upper sign is for the forward waves whereas the lower sign is for the backward waves.

As in the case of a bent waveguide, the forward parasitic wave amplitudes are larger than those of the backward waves, since the oscillating factor of the forward waves in the integrand of eqn. 2.75 is a less rapidly varying function than in the case of backward waves. Nevertheless, in straight waveguides this amplitude difference is smaller since the coupling coefficients S_{jm}, as we will see later, have the same order of magnitude for backward and forward waves in a more general case, whereas the coupling coefficients F_{jm} in a bend are essentially larger for the forward waves than for the backward waves.

The integration by parts performed for the derivation of eqn. 2.78 emphasised the coupling coefficient values in the tilt points of the generatrice. The compensation of parasitic modes excited along the nonuniform waveguide section is less accomplished at these points than at all other cross sections.

The second case first studied in Isaenko (1959) realising an approximate integration in eqn. 2.75 is connected with the character of the variations of the rapidly oscillating exponential factor in the equation. The derivative of the exponential factor in the integrand of eqn. 2.75 is equal to $i(h_j - h_m)$. If the wavenumber of the excited forward parasitic wave $h_j(z)$ is equal to the wavenumber of the fundamental wave $h_m(z)$ at a waveguide cross section, i.e.

$$h_j(z) = h_m(z) \qquad (2.79)$$

[1] If eqn. 2.78's value is equal to zero (at any frequency) it is not correct to calculate p_j in eqn. 2.75 since the remaining integral member will be of the same order of magnitude (second order of ν_0) as the third term of the sum in eqn. 2.76 neglected in eqn. 2.75. This observation is also valid for the third case which is mentioned at the end of this subsection.

then the corresponding value of z gives us a stationary phase point of the integral in eqn. 2.75. The exponential factor varies more slowly in the neighbourhood of this point than in other areas of the nonuniform waveguide section, the compensation is also weaker and the whole integral is approximately equal to the integral calculated over this neighbourhood, whereas the values of the functions before the exponential factor can be simply extracted out of the integral symbol.

The integral of eqn. 2.75 for forward waves can be rewritten in the form

$$p_j = \int_0^L \Phi(z) \exp(-i\sigma\psi(z))\, dz \tag{2.80}$$

where Φ is equal to the product of all the factors before the exponential factor in eqn. 2.75 whereas the function ψ is of the order of the unity everywhere and is determined by

$$\psi = \frac{\gamma_m(z) - \gamma_j(z)}{\gamma_m(L) - \gamma_j(L)} \tag{2.81}$$

At the same time, a large parameter of the problem σ can be emphasised, it is equal to the difference of the phase shifts of two considered modes along the whole nonuniform waveguide section.

$$\sigma = \gamma_m(L) - \gamma_j(L) \tag{2.82}$$

If the root of equation $\psi' = 0$ exists on the interval $0 < z < L$, i.e. there exists a root of eqn. 2.79, then the integral in eqn. 2.80 can be calculated by the stationary phase method employing the condition $|\sigma| >> 1$. The largest term in the expansion of eqn. 2.80 by the inverse powers of σ is of the order of $\sigma^{-1/2}$ and is equal to

$$p_j = \exp\left(-i\left[\gamma_m(z_s) - \gamma_j(z_s) - \pi/4\right]\right)\Phi(z_s)\sqrt{\frac{2\pi}{h_m'(z_s) - h_j'(z_s)}} \tag{2.83}$$

where $z = z_s$ is the root of eqn. 2.79. The neglected terms of the sum are of the order of $\sigma^{-3/2}$.

Eqn. 2.83 contains a small term $\Phi(z_s)$, nevertheless, the denominator of the same expression contains also a small value, i.e. the root of the derivative by z. Because of this, if there are stationary phase points on a nonuniform waveguide section, i.e. points where eqn. 2.79 is satisfied, the corresponding parasitic mode amplitude is relatively large and proportional to the square root of the generatrice inclination angle, i.e. of the small parameter of the problem. The amplitudes of the modes for which eqn. 2.79 has no roots are smaller and they contain the first power of the inclination angle. Eqn. 2.83 was employed in Isaenko (1959) to determine the amplitudes of the parasitic modes excited in a special complicated waveguide taper.

The third case, when it is also possible to calculate the integrals of eqn. 2.75 is from some points of view opposite to the first case. In this case the functions describing the nonuniform waveguide surface are continuous together with all their derivatives. This case was considered by Pokrovskii *et al.* (1959*a*) where integrals similar to eqn. 2.75 were analysed. As is usual in similar problems of quantum mechanics, these integrals are solved by the method of steepest gradient in the complex variable plane; they have exponentially small values of the order of $\exp(-C/\nu_0)$, where the parameter C depends on the form of the eqn. 2.75. Application of this calculation trick involves certain difficulties since in the real world the metallic surface cannot be described by an analytical function. Possibly, this calculation technique can be important in the theory of natural waveguides, for example, atmospheric waveguides.

2.2.9 *Electromagnetic fields in waveguides with inhomogeneous filling medium*

We have supposed for all the manipulations of this section the fields $\hat{\vec{E}}^m(x, y)$ and $\hat{\vec{H}}^m(x, y)$ as well as the wavenumbers h_m of the modes of uniform waveguides with inhomogeneous filling medium, i.e. waveguides where $\varepsilon(x, y)$ and $\mu(x, y)$ are well known. However, in many cases the determination of these fields poses an independent and complex problem by itself. Here we will describe briefly a method to solve this problem with any desirable accuracy.

Here (and later in subsection 4.4.2) we apply a technique different from the rest of this publication. The fields in a filled waveguide can be expanded in a series of fields of a non-filled reference waveguide. Such an expansion was first proposed by Schelkunoff (1952) and applied by Morgan (1957) but the mathematical apparatus developed there is very complicated; Brodskiy (1958) used the same expansion in an essentially more effective way. The method described later allows us to obtain results practically equivalent to Brodskiy, (1958), but in a more direct and simple way.

Let us introduce the symbols $\hat{\vec{E}}^{0m}$ and $\hat{\vec{H}}^{0m}$ as well as h_m^0 to represent the empty uniform waveguide mode fields and wavenumbers. These values satisfy eqns. I.4 with $\varepsilon = 1$ and $\mu = 1$. We will expand the transversal components of the fields of the filled waveguide in the same waveguide section in the series

$$E_x = P_v E_x^{0v},\ E_y = P_v E_y^{0v},\ H_x = P_v H_x^{0v},\ H_y = P_v H_y^{0v} \tag{2.84a}$$

where $P_j = P_j(z)$. Substituting these expansions into two of the six Maxwell equations we obtain the following expansions for the longitudinal components

$$E_z = \frac{1}{\varepsilon} P_v E_z^{0v}, \quad H_z = \frac{1}{\mu} P_v H_z^{0v} \tag{2.84b}$$

which differ from eqn. 2.84*a* by the supplementary factors $1/\varepsilon$ and $1/\mu$. The presence of such factors demonstrates, over other considerations, that strictly speaking the complete field in a filled waveguide cannot be expanded in series of the empty waveguide fields. Nevertheless, as follows from eqns. 2.43 – 2.44, there exists such an expansion in a series of the fields of the reference waveguide with exactly the same cross section.

Substituting eqn. 2.84 into the remaining four Maxwell equations we will obtain four equations similar to eqn. 2.45, one of which has the following form

$$P_v' E_x^{0v} = P_v \left\{ -ik\mu H_y^{0v} + \frac{\partial}{\partial x}\left(\frac{1}{\varepsilon} E_z^{0v}\right) \right\} \tag{2.85}$$

and we do not write down here the other three. Later, by means of the orthogonality condition for the fields of the empty waveguide, we will obtain for $P_j(z)$ the ordinary coupled-wave differential equations with constant coefficients

$$P_j' = M_{jv} P_v \tag{2.86}$$

The coefficients M_{jv} characterise the coupling between the modes of the empty waveguide caused by the difference of ε and μ from unity. After simple transformations, employing in particular the Stokes theorem and the boundary condition for $\vec{E}$, it follows from eqn. 2.85

$$M_{jm} = \frac{i}{2h_j^0} \int \left\{ \begin{array}{l} \frac{1}{\varepsilon} E_z^{0j} E_z^{0m} + \frac{1}{\mu} H_z^{0j} H_z^{0m} \\ +\varepsilon\left(E_x^{0j} E_x^{0m} + E_y^{0j} E_y^{0m}\right) \\ +\mu\left(H_x^{0j} H_x^{0m} + H_y^{0j} H_y^{0m}\right) \end{array} \right\} dS \tag{2.87}$$

Eqns. 2.86 – 2.87 have the same meaning as eqns. 19–20 of Brodskiy (1958). The determination of the coefficients M_{jm} with this method does not require, as in Schelkunoff (1952) and Morgan (1957), the solution of an infinite number of coupled-wave equations.

The mode fields in the waveguide with $\varepsilon = \varepsilon(x, y)$ and $\mu = \mu(x, y)$ are derived from the requirements that solutions of eqn. 2.86 satisfy the condition

$$P_j(z) = P_j(0) \exp(-ihz) \tag{2.88}$$

The substitution of eqn. 2.88 into eqn. 2.86 gives us, as in the case of bent waveguides, an infinite number of homogeneous algebraic coupled-wave equations for $P_j(0)$

$$(h\delta_{j\nu} - iM_{j\nu})P_{\nu}(0) = 0 \qquad (2.89)$$

The mode wavenumbers h are, consequently, the roots of the following equation (compare with eqn. 2.39)

$$\det\left|M_{jm} + ih\delta_{jm}\right| = 0 \qquad (2.90)$$

Eqns. 2.89 and 2.90 are similar to eqns. 2.38 and 2.39. Having solved eqn. 2.90 we can find for each root $h = h_j$ from eqn. 2.89 and from the normalisation condition in eqn. I.8 the amplitudes $P_j(0)$ corresponding to the given mode.

If $\varepsilon = 1$ and $\mu = 1$, then $M_{jm} = -ih_j^0\delta_{jm}$ and the matrix of the coupled-wave eqns. 2.89 will be diagonal and all the wavenumbers h_j^0 will be the roots of eqn. 2.90. If $\varepsilon = \text{const} \neq 1$ and $\mu = \text{const} \neq 1$ then only those coefficients M_{jm} with $j=\pm m$ will be not equal to zero. The determinant of eqn. 2.90 is decomposed in the product of second-order determinants and therefore the wavenumbers are determined from quadratic equations. It is easy to verify that these wavenumbers are equal to $\pm\left\{\left(h_j^0\right)^2 + k^2(\varepsilon\mu - 1)\right\}^{1/2}$ as follows from elementary considerations.

The efficiency of the method described above for the determination of the mode fields depends on the order of magnitude of the nondiagonal elements, M_{jm}, and on the degree of their decreasing with remoteness from the diagonal. For example, if $|\varepsilon(x,y) - 1|$ and $|\mu(x,y) - 1|$ have small values, then all the nondiagonal coefficients M_{jm} will be also small and the matrix in eqns. 2.89 will adopt a quasidiagonal form. Then the solution of eqn. 2.90 in the largest order of magnitude with respect to $|\varepsilon - 1|$ and $|\mu - 1|$ can be obtained in an analytical form.

In conclusion, let us note the following. The expansion 2.84 and the coupled-wave eqns. 2.86 describe also the field in a nonuniform waveguide of constant cross section, i.e. for the filling parameters $\varepsilon(x, y, z)$ and $\mu(x, y, z)$. In this sense they have the same meaning as for the coupled-wave eqns. 2.46. Applying the coupled-wave eqns. 2.86 to nonuniform waveguides where the coefficients M_{jm} would be now z dependent, the variables $P_j(z)$ must not satisfy the condition of eqn. 2.88 but some other boundary conditions. If the nonuniformly filled waveguide section is connected at both ends to empty waveguides, the boundary conditions will have the form of eqn. 2.48. In the more general case, the boundary conditions for eqn. 2.86 will be determined in their turn from algebraic coupled-wave equations derived from the continuity requirement for the fields at the ends of the nonuniform waveguide section.

In subsection 4.4.2 we will describe the method for the numerical solution of the problem of a nonreflecting protective layer. This method will employ exactly the same mathematical apparatus as that elaborated above.

The question of which set of coupled-wave equations, either eqns. 2.46 or eqns. 2.86 is more convenient for a given engineering problem has to be answered depending on the field structure in the studied nonuniform waveguide. The electromagnetic field in perfectly matched broadband devices is very similar at any cross section to the field in a uniform waveguide of the same cross section. Because of this, analysing those devices is more comfortable employing the coupled-wave eqns. 2.46, since the coupling coefficients will be small there and, as a rule, all the amplitudes P_j except one or two will also be small. All the considerations made in Section 2.1 would be very complicated if the coupled-wave eqns. 2.86 were employed; the field would be described by a large number of strongly coupled P_j amplitudes. It would also be difficult to employ those coupled-wave equations for general manipulations in the next section. As a general rule, probably it is more convenient to employ it only for mode determination in nonuniformly filled waveguides.

Nevertheless, in the last subsection of Section 5.3, we will study a device where the field is quite similar to the empty uniform waveguide field; for analysing this bend, expansions similar to eqn. 2.84 lead to the pursued goal faster than expansions similar to eqns. 2.43 – 2.44.

2.2.10 Particular case of nonlinear filling medium

A nontrivial generalisation of the cross-section method is its application to a uniform waveguide containing a nonlinear medium, i.e. a medium where $\varepsilon = \varepsilon(\vec{E})$, Golovanov (1991). Waves generated by monocromatic sources are incident on the waveguide section containing this filling medium (for example, a nonlinear substrate microstrip line) and it is required to find the transmitted and reflected mode amplitudes.

The fields satisfying the nonstationary Maxwell equations in a waveguide filled partially by a nonlinear medium should be expanded in a Fourier series of all the combinatory frequencies. Each field contained in this series satisfies two coupled-wave equations with structures similar to those of Maxwell equations for a monocromatic oscillation but including an additional term caused by the nonlinearity and by coupling all these equations. These fields are expanded in a series of the mode fields of the linearly filled waveguide corresponding to the given frequency and with the same structure as the series of the cross-section method. The coefficients of these expansions depend on z and satisfy ordinary differential coupled-wave equations similar to eqn. 2.46. Therefore, the initial problem is reduced to the boundary value problem for coupled-wave equations as in eqn. 2.46 and also to nonlinear algebraic coupled-wave equations. All these equations must be solved together. Several numerical computational algorithms have been developed to provide stable solutions with respect to computation errors.

2.2.11 *Analogy to waveguides with varying wall impedance*

As shown in subsection 1.2.4, a waveguide with varying wall impedance along z can be analysed by the same method as that used for waveguides with variable filling medium. In such a case we will use as the reference waveguide a uniform waveguide where the impedance is the same function of s (the same scalar value or the same matrix) as in the nonuniform waveguide at the given z. The electromagnetic field in the nonuniform waveguide can be represented by the same series eqns. 2.43 and 2.44 whereas the coefficients $P_j(z)$ satisfy the same coupled-wave eqn. 2.46 and the boundary conditions of eqns. 2.48. Eqn. 2.47 is valid for the coupling coefficients. Applying uncomplicated manipulations described by Shevchenko (1962), the coefficients can be transformed to the form of eqns. 1.38 and they will contain the impedance derivatives instead of the field derivatives. These manipulations are a combination of the methods employed in Sections 1.2.4 and 2.2.8.

Let us write the Maxwell equations in x, y and ζ co-ordinates for the mode number m and taking their z derivative,

$$\mathrm{rot}\hat{\vec{E}}^{m\prime} + ik\hat{\vec{H}}^{m\prime} = \vec{0}, \quad \mathrm{rot}\hat{\vec{H}}^{m\prime} - ik\hat{\vec{E}}^{m\prime} = \vec{0} \tag{2.91}$$

By combination of these equations with Maxwell equations for the mode number $-j$, ($j \neq$ m) we obtain

$$\mathrm{div}\left\{\left[\hat{\vec{E}}^{m\prime} \times \hat{\vec{H}}^{-j}\right] - \left[\hat{\vec{E}}^{-j} \times \hat{\vec{H}}^{m\prime}\right]\right\} = 0 \tag{2.92}$$

Let us integrate this equation following the scheme given for eqn. 1.35. First, let us study the integral over the cross section. Substituting eqn. 2.55 and the similar expression for $\vec{E}^{m\prime}$ in this integral we can eliminate the second terms (containing $\hat{\vec{H}}^{m}$ and $\hat{\vec{E}}^{m}$) because the integral of the difference of their products by $\hat{\vec{E}}^{-j}$ and $\hat{\vec{H}}^{-j}$ is identically equal to zero, according to eqn. 1.17. Substituting in the cross-section integral the first terms (i.e. $\hat{\vec{E}}^{m\prime}$ and $\hat{\vec{H}}^{m\prime}$) and interchanging $E^{-j}_{x,y}$ by $-E^{j}_{x,y}$ and $H^{-j}_{x,y}$ by $H^{j}_{x,y}$ following eqn. I.6 we will obtain the same integral as represented in eqn. 2.47. In this way, the first term of the sum in eqn. 1.35 (the integral over the cross section) will be equal to $2kh_jS_{jm}$.

The integrand of the boundary contour integral contains the difference $\left[\hat{\vec{E}}^{m\prime} \times \hat{\vec{H}}^{-j}\right]_n - \left[\hat{\vec{E}}^{-j} \times \hat{\vec{H}}^{m\prime}\right]_n$. According to eqn. 1.32, $E^{m\prime}_s = w'H^m_z + wH^{m\prime}_z$ and $E^{m\prime}_z$ can be expressed similarly. The second terms of the sum (proportional to w) are annihilated by the term $-\left[\hat{\vec{E}}^{-j} \times \hat{\vec{H}}^{m\prime}\right]_n$ and the boundary contour integral will be the same as that in eqn. 1.38. Integration over z gives us the term $i/\left(h_m - h_j\right)$. For S_{jm} we will obtain the same formula (eqn. 1.38) as was obtained by the small perturbations method.

In the same manner, it is possible to obtain from eqn. 2.47 the formula for S_{jm} for the case of a tensorial impedance eqn. 1.39. In accordance with eqn. 1.42 the terms w_{sz}', w_{zs}' and w_{ss}' will be encountered in the integrand of the boundary contour integral.

2.2.12 *Special cases of mode degeneracy*

The case was studied above in subsection 2.2.8 where the phase velocities of the fundamental mode and of a parasitic mode coincide at any cross section of the nonuniform waveguide section. In this subsection we return to this problem and study two special situations where this coincidence of mode properties is more essential, i.e. when eqn. 2.79 is also satisfied together with the coincidence of other characteristics of both modes. Since those cases are not usual, here we will limit ourselves to a brief description of the result only.

The first of the two cases is an obvious generalisation of the situation described by eqn. 2.79. Let us suppose that not only does the function $h_j(z) - h_m(z)$ become equal to zero at $z = z_s$, but its first derivative is also equal to zero, i.e. simultaneously with eqn. 2.79 the equation $h_j'(z) - h_m'(z) = 0$ is satisfied. Obviously, in such a case, eqn. 2.83 is no longer valid. Now the parasitic mode amplitude p_j is not of the order of magnitude of $\nu_0^{1/2}$ but of $\nu_0^{1/3}$ (where ν_0 is the mean inclination angle), i.e. it will be larger than for the case where only the wavenumbers are identical. In general, if the function $h_j(z) - h_m(z)$ and its first n derivatives are zero at $z = z_s$ then $p_j \sim \nu_0^{1/(n+2)}$, Dikarev (1980) (let us note that the author of this paper employs the term 'turning point' with respect to the roots of eqn. 2.79 and not with respect to the roots of equations $h_j(z) = 0$ or $h_m(z) = 0$).

This result can be illustrated by the following nonrigorous consideration based on eqn. 2.75. In the neighbourhood of $z = z_s$ and if eqn. 2.79 is satisfied, then the expansion of the phase of the integrand has the form

$$\begin{aligned}\gamma_m(z) - \gamma_j(z) &= (\gamma_m - \gamma_j)\big|_{z=z_s} \\ &+ \frac{1}{2}(h_m' - h_j')\Big|_{z=z_s}(z - z_s)^2 \\ &+ \frac{1}{6}(h_m'' - h_j'')\Big|_{z=z_s}(z - z_s)^3 + \ldots\end{aligned} \tag{2.93}$$

Let the first n derivatives of the difference between the functions $h_m(z)$ and $h_j(z)$ be equal to zero whereas the difference of their $(n + 1)$ derivatives is different from zero. Let us retain in eqn. 2.93 only the first nonvanishing term, i.e. the term proportional to $(z - z_s)^{n+2}$. Its coefficient will be of the order of $k^{n+2}\ \nu_0^{n+1}$. As a result of an integration over z, the large term $\nu_0^{-(n+1)/(n+2)}$ will appear.

The term S_{jm} before the exponent will be of the order of ν_0 and therefore the amplitude p_j will be of the of order of $\nu_0^{1/(n+2)}$, i.e. the same result as follows from rigorous analysis. The larger n is, i.e. the closer to each other the functions $h_m(z)$ and $h_j(z)$ in the neighbourhood of the cross section $z = z_s$ are, the longer the section is where both modes will have nearly the same phase and therefore the parasitic mode carries away more energy.

The second case of deep mode degeneracy is related to the fact that in inhomogeneously filled uniform waveguides ($\varepsilon = \varepsilon(x, y)$), for certain ε distribution along the cross section, there are solutions of Maxwell equations still in the form of eqn. I.3, nevertheless, their wavenumbers are complex when there are no losses either in the walls or in the filling medium. For example, such a situation can occur in a circular waveguide with a circular dielectric rod (this example is studied in detail by Veselov *et al.*, 1983). These modes are always excited in pairs with $\mathrm{Re}(h_1) = -\mathrm{Re}(h_2)$ and $\mathrm{Im}(h_1) = \mathrm{Im}(h_2)$ and in reality they form a standing wave field near the excitation source. If there is some absorption in the filling medium or in the walls (and only under this condition (Shevchenko, 1996)) not only the wavenumbers of two pairs of modes can coincide but their field structures can be identical as well. In such a waveguide the complete set of modes contains not only eigenmodes, i.e. modes similar to eqn. I.3 but also the so-called associated modes. The fields of those associated modes depend on z in a more complex manner than do the eigenmode fields. These fields are orthogonal (in the sense of eqn. 1.17) to all the eigenmode fields and because of this it is necessary to add the corresponding summing terms in eqns. 2.42 or 2.43. The theory of these uniform waveguides is described in great detail in Shevchenko (1969).

If the reference waveguide at the cross section $z = z_s$ of a nonuniform waveguide is a waveguide of this type, its complete set of modes contains associated modes. For $z \neq z_s$ the set of modes of the reference waveguide will already be complete, but when z approaches z_s two of these modes are nearly identical. For $z \rightarrow z_s$ this pair becomes an eigenmode and an associated mode.

The existence of associated modes at only one cross section requires the corresponding supplementary complement for the coupled-wave differential eqns. 2.46. Nevertheless, it seems that the theory of nonuniform waveguides containing cross sections with the properties mentioned above has not been developed up to now. Let us mention only the work of Bichutskaya and Novikov (1979) where results for a plane nonuniform waveguide with an impedance wall have been obtained. As can be expected from the physical considerations, the coincidence of not only the wavenumbers but also of the fields of the fundamental and parasitic modes at a cross section produces a very strong coupling. The parasitic mode amplitude will not be small for an arbitrarily smooth nonuniformity (i.e. despite a small value of ν_0), therefore the parasitic and the fundamental modes will be of the same order of magnitude and the amplitude of the fundamental mode will be changed by a finite value after propagation through this cross section. It seems as if nobody has considered up to now the practical applications of this phenomenon.

2.3 Calculations on variable cross-section waveguides by the cross-section method

2.3.1 Derivation of the coupling coefficients

A direct application of the cross-section method to variable cross-section waveguides encounters some difficulties since the fields $\vec{E}, \vec{H}$ and $\hat{\vec{E}}^m, \hat{\vec{H}}^m$ in the series such as eqn. 2.42 satisfy different boundary conditions. Therefore, on the cross-section boundary contours, these series do not converge to the functions which they must represent. As is well known, taking the derivative of these series causes some difficulties.

From the physical point of view the reason for the difficulty is the following: the field representation as a series expansion cannot be applied in the neighbourhood of a metal, i.e. exactly at the place where the boundary conditions should be applied.

To avoid a manipulation with nonuniformly convergent series we will proceed following a scheme based on the results of Section 2.2.

Let us compare the given variable cross-section waveguide (see Figure 2.4) with a constant cross-section reference waveguide (see Figure 2.5) filled by a medium with constant ε and μ. The filling is placed in such a way that the area without filling corresponds to the given waveguide with variable cross section. In other words, we will inscribe the given waveguide into another waveguide with a constant and larger cross section and we will fill the space between both waveguides by a medium with $\varepsilon = \text{const} \neq 1$ and $\mu = \text{const} \neq 1$.

Using the results obtained in the last section for the larger waveguide with non-uniform filling we can apply the series expansions of eqns. 2.43 and 2.44 as well as the coupled-wave eqns. 2.46 for the coupling coefficients. For the discontinuous filling medium given in the larger waveguide, the coupling coefficients are expressed by eqn. 2.63. These results are valid for any values of ε and μ of the filling medium.

Now we increase the dielectric permittivity ε of the filling medium and assume that it is complex. All the equations mentioned above are still valid, moreover, they are valid also when the limit $|\varepsilon| \to \infty$ is reached. At this limit the

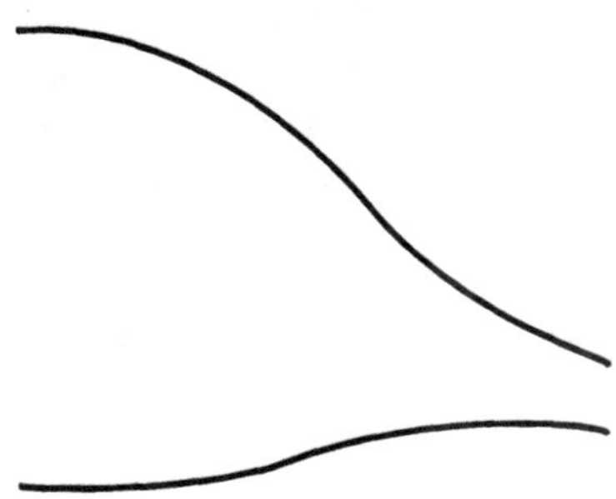

Figure 2.4 Waveguide with varying cross section

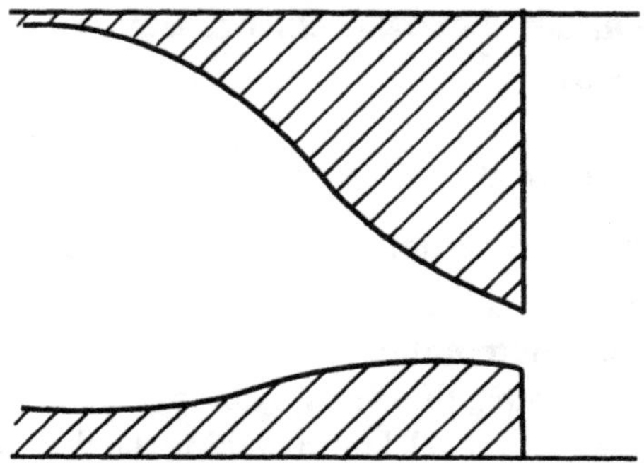

Figure 2.5 Auxiliary waveguide with constant cross section and appropriate filling medium

reference waveguide (Figure 2.5) is identical to the variable cross-section waveguide (Figure 2.4). Therefore, all the eqns. 2.43, 2.44, 2.46, 2.48, 2.50 and 2.51 derived for the variable filling medium waveguides are applicable directly to the case of variable cross-section waveguides as well. Moreover, all the equations and results obtained in subsections 2.2.7 to 2.2.9 are also applicable.

An indetermination of the kind $0 \cdot \infty$ will appear in eqn. 2.63 for the coupling coefficients, because the components E_s^m and E_z^m on the boundary contour, i.e. on the medium border, tend to zero for $|\varepsilon| \to \infty$. To obtain an analytical expression for the coupling coefficient in the variable cross-section waveguide we realise the continuous limit calculations in eqn. 2.63.

When $|\varepsilon| \to \infty$ the field components E_s^m, E_z^m and H_n^m tend to zero. Simultaneously, the tangential components of the electric and magnetic fields satisfy the following relations, according to the Leontovich boundary condition

$$\begin{aligned} E_s^m &= \sqrt{\frac{\mu}{\varepsilon}} H_z^m, \quad E_z^m = -\sqrt{\frac{\mu}{\varepsilon}} H_s^m \\ E_s^j &= \sqrt{\frac{\mu}{\varepsilon}} H_z^j, \quad E_z^j = -\sqrt{\frac{\mu}{\varepsilon}} H_s^j \end{aligned} \tag{2.94}$$

Let us remember that eqns. 2.94 correspond to fields in the uniform reference waveguides, the normal to their surface is given by $\vec{n}$.

Having substituted eqn. 2.94 into eqn. 2.63 we will obtain an expression under the integral symbol, where it is easy to calculate the limit for $|\varepsilon| \to \infty$. In such a way the following formula for the coupling coefficient in the variable cross-section waveguide is obtained

$$S_{jm} = \frac{1}{2h_j(h_j - h_m)} \oint \nu(s)\left(E_n^j E_n^m + H_z^j H_z^m - H_s^j H_s^m\right) ds \tag{2.95}$$

As can be expected, the magnetic permeability of the filling medium disappears in this resulting expression. Let us note that this fact could be employed to deduce eqn. 2.95 in the shortest way if one manipulates with scalar expressions instead of the vector ones. Then one should assume in the variable filling medium waveguide (Figure 2.5) introduced at the beginning of this section $\mu(x,y,z) = \frac{1}{\varepsilon(x,y,z)}$. Therefore, the product $\varepsilon \cdot \mu$ has a constant value in the variable filling medium reference waveguide. It is easy to demonstrate that the fields in such waveguides can be expressed by two scalar eigenfunctions ψ and φ with formulae similar to eqn. I.14 and, moreover, the separation of *TE* and *TM* modes is also possible as in empty waveguides. The functions ψ and φ satisfy self-conjugated equations similar to eqn. I.11 generating also a complete set of eigenfunctions, etc.

Eqn. 2.95 expresses the coupling coefficient through currents and charges excited by the propagating waves with superscripts j and m on the walls of the uniform reference waveguide. For nonuniform waveguides of variable cross section analysed in this section, the reference waveguide is an empty waveguide having its cross section identical to the non-uniform waveguide cross section at the given z.

Performing a double limit calculation, similar to that performed for the case of discontinuous ε filling and for $|\varepsilon| \to \infty$ in eqn. 2.58, it is also easy to determine the following contour integral expression for the derivative of the wavenumber

$$h'_m = \frac{1}{2h_m} \oint \nu(s)\left[(E_n^m)^2 + (H_z^m)^2 - (H_s^m)^2\right] ds \tag{2.96}$$

Eqns. 2.95 and 2.96 coincide with eqns. 1.53 and 1.57, obtained by more elementary methods in Section 1.3. On the other hand, applying eqns. 2.46 and 2.48 to a small nonuniformity, it is easy to obtain eqns. 1.50 and 1.51.

It is also easy to obtain from eqns. 2.95 and 2.96 the relation between $S_{-m,m}$ and h_m'

$$S_{-m,m} = -\frac{h_m'}{2h_m} - \frac{1}{2h_m^2} \oint \nu(s)\,(H_s^m)^2\, ds \tag{2.97}$$

this is simply another way to write eqn. 1.54.

For an empty waveguide, the electromagnetic fields can be expressed by means of transversal eigenfunctions. Substituting eqn. I.14 into eqn. 2.95 we obtain the coupling coefficients' formulae; these formulae will be employed in Chapter 4 for practical calculations. Eqn. 2.98*a* corresponds to the coupling of two *TE* waves, eqn. 2.98*b* to two *TM* waves and eqn. 2.98*c* is for the case that the mth *TE* wave is coupled to the jth *TM* wave. According to the reciprocity relation of eqn. 2.51, we do not need to analyse separately the fourth possible combination

$$S_{jm} = \frac{1}{2h_j\left(h_j - h_m\right)}\oint \nu(s)\left[\alpha_j^2\alpha_m^2\psi^j\psi^m + \left(h_jh_m - k^2\right)\frac{\partial\psi^j}{\partial s}\frac{\partial\psi^m}{\partial s}\right]ds \quad (2.98a)$$

$$S_{jm} = \frac{k^2 - h_jh_m}{2h_j\left(h_j - h_m\right)}\oint \nu(s)\frac{\partial\varphi^j}{\partial n}\frac{\partial\varphi^m}{\partial n}ds \quad (2.98b)$$

$$S_{jm} = -\frac{k}{2h_j}\oint \nu(s)\frac{\partial\varphi^j}{\partial n}\frac{\partial\psi^m}{\partial s}ds \quad (2.98c)$$

In these equations, $\vec{n}$ is the normal vector directed towards the metal, $\vec{s}$ has been chosen in such a way that $\vec{n}, \vec{s}$, and $\vec{z}$ form a right-hand side system of co-ordinate axes.

According to eqn. 2.96, the derivative of the wavenumber for the *TE* and *TM* waves, respectively, is equal to

$$h'_m = \frac{\alpha_m^2}{2h_m}\oint \nu(s)\left[\alpha_m^2\left(\psi^m\right)^2 - \left(\frac{\partial\psi^m}{\partial s}\right)^2\right]ds \quad (2.99a)$$

$$h'_m = \frac{\alpha_m^2}{2h_m}\oint \nu(s)\left(\frac{\partial\varphi^m}{\partial n}\right)^2 ds \quad (2.99b)$$

Eqn. 2.97 can be also written in a form similar to eqn. 1.54.

2.3.2 Coupled-wave equations

Coupled-wave eqns. 2.46 and eqns. 2.95 are fundamental for many applications, hence we will describe here another method of their deduction (Kovalev 1986). Such a method requires neither the limit evaluation performed in subsection 2.3.1 nor term by term differentiation.

The main idea of this method consists in that the amplitudes $\vec{E}^m$ and $\vec{H}^m$ appearing in eqn. I.3 should be considered as independent vectors. Obviously, they do not satisfy the Maxwell equations. Their z dependence is determined by the z dependence of the reference waveguides (compare with a similar discussion near eqns. 2.54 and 2.55, but for consideration with respect to $\hat{\vec{E}}^m$ and $\hat{\vec{H}}^m$). To compare $\vec{E}^m$ and $\vec{H}^m$ with the $\vec{E}$ and $\vec{H}$ fields in the nonuniform waveguide, one should obtain the expressions for $\mathrm{rot}\vec{E}^m$ and $\mathrm{rot}\vec{H}^m$. These expressions can be deduced from eqns. I.4 for $\varepsilon = \mu = 1$. One should add to their right-hand side and left-hand side the corresponding derivatives with respect to z. Therefore, the first line of eqns. I.4 will have the following form

$$\frac{\partial E_z^m}{\partial y} - \frac{\partial E_y^m}{\partial z} = -ikH_x^m - ih_m E_y^m - \frac{\partial E_y^m}{\partial z} \tag{2.100}$$

The remaining five formulae are elaborated in the same way, so we do not present them here. Eqns. 2.100 and the remaining five equations have the following vector form

$$\begin{aligned} \mathrm{rot}\vec{E}^m &= -ik\vec{H}^m + ih_m\left[\hat{z}_0 \times \vec{E}^m\right] + \left[\hat{z}_0 \times \frac{\partial \vec{E}^m}{\partial z}\right] \\ \mathrm{rot}\vec{H}^m &= ik\vec{E}^m + ih_m\left[\hat{z}_0 \times \vec{H}^m\right] + \left[\hat{z}_0 \times \frac{\partial \vec{H}^m}{\partial z}\right] \end{aligned} \tag{2.101}$$

where $\hat{z}_0$ is the unity vector along the z-axis.

The fields in a nonuniform waveguide satisfy eqns. I.1. Transforming eqns. 2.101 and I.1 in a way similar to the deduction of Lorentz lemma, i.e. multiplying eqn. 2.101 by $\vec{H}$ and $\vec{E}$ whereas eqn. I.1 is multiplied by $-\vec{H}^m$ and $-\vec{E}^m$, and then adding the four equations, we obtain

$$\begin{aligned} &div\left\{\left[\vec{H}^m \times \vec{E}^m\right] - \left[\vec{H} \times \vec{E}^m\right]\right\} \\ &= ih_m\left\{\left[\vec{H}^m \times \vec{E}\right] - \left[\vec{H} \times \vec{E}^m\right]\right\}\cdot\hat{z}_0 + \left\{\left[\frac{\partial \vec{H}^m}{\partial z} \times \vec{E}\right] - \left[\vec{H} \times \frac{\partial \vec{E}^m}{\partial z}\right]\right\}\cdot\hat{z}_0 \end{aligned} \tag{2.102}$$

Let us integrate this equation over the volume between the cross sections z=const and z+Δz=const. Normal components of the vectors included under the divergence symbol will form the integrand in the integrals over these cross sections and over the corresponding waveguide walls. The integral over waveguide walls contains the $\vec{E}$ and $\vec{E}^m$ tangential components to this surface. The term containing $\vec{E}$ disappears because of eqn. I.2. In the term containing $\vec{E}^m$ the term with E_s^m also disappears, whereas only the term containing E_t^m remains.

Here, *s* and *t* are two directions tangential to the metal and their corresponding unity vectors are placed in the cross-section plane for (*s*) and orthogonal to it for (*t*). The condition $E_s^m = 0$ is a consequence of the coincidence of the *s* directions in the nonuniform and reference waveguides.

The component E_t^m on the nonuniform waveguide surface is different from zero and equal to $E_n^m \nu$, where E_n^m has the same sense as in eqn. 2.95, i.e. it corresponds to the reference waveguide whereas ν is the tangent of the same angle as in eqn. 2.59, i.e. the same parameter characterises the differences of both waveguides as in eqn. 2.63, hence it is the same parameter as in eqn. 2.95.

Now taking the limit for $\Delta z \to 0$ we obtain

$$\frac{d}{dz}\int\left\{\left[\vec{H}^{m}\times\vec{E}\right]-\left[\vec{H}\times\vec{E}^{m}\right]\right\}\cdot\vec{z}_0\,dS-\oint \nu H_s E_n^m\,ds$$
$$=ih_m\int\left\{\left[\vec{H}^{m}\times\vec{E}\right]-\left[\vec{H}\times\vec{E}^{m}\right]\right\}\cdot\vec{z}_0\,dS+\int\left\{\left[\frac{\partial\vec{H}^{m}}{\partial z}\times\vec{E}\right]-\left[\vec{H}\times\frac{\partial\vec{E}^{m}}{\partial z}\right]\right\}\cdot\vec{z}_0\,dS \tag{2.103}$$

These expressions contain only the transversal components of the $\vec{E}$ and $\vec{H}$ fields. If we expand them in the series

$$E_{x,y}=P_v E^{v}_{x,y},\quad H_{x,y}=P_v H^{v}_{x,y} \tag{2.104}$$

and take into account the orthogonality conditions I.9, then we have on the left-hand side of eqn. 2.103 dP_m/dz, whereas on the right-hand side we have a linear combination of the P_n coefficients. The coefficients S_{mn} in such a combination will be expressed by very complex formulae containing both contour integrals and integrals over the cross section. Realising several formal manipulations similar to those between eqns. 2.47 and 2.57, it is possible to transform these coefficients to the form of eqn. 2.95.

Waveguides with variable filling media can also be studied by this method. Such a technique is relatively complicated, but its unquestionable advantage is its logical simplicity.

Here, we have used the trick of taking the limit $|\varepsilon|\to\infty$ for detailed deduction of the main formulae since such a deduction is more evident from a physical point of view and in such a way the analogy between different kinds of nonuniformity is emphasised.

2.3.3 Coupling coefficients far away from cut off

The coupling coefficients depend on the frequency because k and h participate in eqn. 2.98 and this dependence is different for forward waves ($j>0$) or for backward waves ($j<0$). Nevertheless, the coupling coefficients of eqn. 2.98 do not depend in practice on the frequency at very high frequencies. We will call frequencies high when the phase velocities of their considered waves are near to the speed of light, i.e. $\hbar_m\approx 1$ and $\hbar_j\approx 1$. Using the identity

$$\frac{h_m h_j-k^2}{h_j-h_m}=\frac{h_m\alpha_j^2+h_j\alpha_m^2}{\alpha_j^2-\alpha_m^2} \tag{2.105}$$

which follows from the definition of h_m in eqn. I.13, it is easy to write the expressions for S_{jm} when $\hbar_m\approx 1$ and $\hbar_j\approx 1$ corresponding to the three formulae of eqns. 2.98

$$S_{jm}=\frac{1}{2\left(\alpha_m^2-\alpha_j^2\right)}\oint \nu(s)\left[\alpha_j^2\alpha_m^2\psi^j\psi^m-(\alpha_j^2+\alpha_m^2)\frac{\partial\psi^j}{\partial s}\frac{\partial\psi^m}{\partial s}\right]ds \tag{2.106a}$$

$$S_{jm} = \frac{\alpha_m^2 \pm \alpha_j^2}{2\left(\alpha_m^2 - \alpha_j^2\right)} \oint \nu(s) \frac{\partial \varphi^m}{\partial n} \frac{\partial \varphi^j}{\partial n} ds \qquad (2.106b)$$

$$S_{jm} = \mp \frac{1}{2} \oint \nu(s) \frac{\partial \varphi^j}{\partial n} \frac{\partial \psi^m}{\partial s} ds \qquad (2.106c)$$

These expressions do not contain frequencies. The two last eqns. 2.106*b* and 2.106*c* are valid for both forward and backward waves; the upper sign corresponds to the forward waves and the lower sign to the backward waves. Eqn. 2.106*a* is written for forward waves. For backward waves we need to omit the first term and to substitute for the ratio $\frac{\left(\alpha_m^2 + \alpha_j^2\right)}{\left(\alpha_m^2 - \alpha_j^2\right)}$ by unity.

If one of the *TE* waves is circular symmetric, i.e. if $\frac{\partial \psi}{\partial s} \equiv 0$, there will be the first term only in eqn. 2.98*a*. According to eqn. 2.106*a*, the coupling coefficient then becomes zero for the backward *TE* waves. However, we should take into account the fact that eqns. 2.106 represent the largest term only (it is independent of the frequency) in the S_{jm} expansion in a series of negative powers of the frequency. For circular symmetric backward *TE* waves the next term will be the largest and therefore we should calculate S_{jm} by the following formula

$$S_{jm} = \frac{\alpha_j^2 \alpha_m^2}{4k^2} \oint \nu(s) \psi^m \psi^j \, ds \qquad (2.107)$$

If we do not take into account this exceptional case, the coupling coefficients tend to different finite values as the frequency increases for the forward and backward waves, whereas their order of magnitude is about ν_0 / a (here, a is of the order of the linear dimensions of the waveguide cross section). For circular symmetric *TE* waves as well as for some special cases, the coupling coefficient to backward waves tends to zero at high frequencies as $\nu_0 / k^2 a^3$.

As follows from the frequency independence of S_{jm} at high frequencies (and far away from the resonance frequencies), the losses in a small step are also independent of the frequency. Confronting eqns. 1.50 and 1.51 for the amplitudes P_j of the waves scattered by the waveguide step with eqn. 2.95 for the coupling coefficient S_{jm}, P_j can be obtained from S_{jm} by substitution of $\nu(s)$ in the integrand by the waveguide step height $\delta(s)$. Hence, the amplitudes P_j at high frequencies do not depend on frequency; their magnitudes are of the order of δ/a. The sum of the relative power losses on the waveguide step also does not depend on the frequency and, according to eqn. I.10, it is equal to

$$\sum_j \left|P_j\right|^2 h_j / h_m \qquad (2.108)$$

where all the propagating waves have been taken into account in the sum. The number of terms in the sum in eqn. 2.108 increases as frequency increases, however S_{jm} usually decreases rapidly as the number j of the wave increases (see, for example, eqns. 4.44 and 4.45). The main part of the power is carried away by the waves having low numbers (more exactly, with their number j nearly equal to m), the series of eqn. 2.108 converges very well and the sum varies very little with the variation of the number of terms of the sum. At high frequencies, power losses of any wave in an arbitrary cross-section waveguide are of the order of δ^2/a^2.

2.3.4 Coupling of degenerate modes

If two eigenvalues of two different waves are identical at some cross section of a nonuniform waveguide, i.e. if $\alpha_m = \alpha_j$, then their wavenumbers are identical at this cross section as well, and the denominators of the coupling coefficients given by eqns. 2.98*a* and 2.98*b* are equal to zero. Simultaneously, the nominators of these formulae disappear and the expressions are equal to an indetermination of the form $^0/_0$. Such a kind of cross section occurs, for example, in a smooth transition which converts the rectangular waveguide TE_{10} mode into the circular waveguide TE_{01} mode. Usually, in practical problems, eqns. 2.98 lead to analytical expressions without indeterminations or their indeterminations are solved by elementary techniques. Nevertheless, we give here the general transformation of the formula for S_{jm} to a form applicable at those special cross sections.

To deduce the equivalent formula of eqn. 2.98*a* we will employ two auxiliary identities connecting two integrals of transversal eigenfunctions of TE waves; one of the integrals is over the cross section and the other one is over the boundary contour of this cross section

$$\begin{aligned}&\oint \nu \frac{\partial \psi^m}{\partial s}\frac{\partial \psi^j}{\partial s}\,ds + \alpha_j^2 \int \psi^{m\prime}\,\psi^j\,dS + \alpha_m^2 \int \psi^m \psi^{j\prime}\,dS = 0\\ &\oint \nu \psi^m \psi^j\,ds + \int \psi^{m\prime}\,\psi^j\,dS + \int \psi^m \psi^{j\prime}\,dS = 0\end{aligned} \tag{2.109a}$$

It is very easy to obtain these identities, taking the z derivative of the orthogonality conditions

$$\int \nabla\psi^m\,\nabla\psi^j\,dS = 0,\ \int \psi^m \psi^j\,dS = 0 \tag{2.110}$$

further transforming the result according to the Green formula and applying the differential equation and the boundary condition for the ψ functions. Substituting the contour integrals into eqn. 2.98*a* following eqn. 2.109*a*, we obtain the following expression for the coupling coefficients between two TE waves

$$S_{jm} = \frac{1}{2h_j}\left\{\alpha_m^2 h_j \int \psi^m \psi^{j'}\, dS - \alpha_j^2 h_m \int \psi^{m'} \psi^j\, dS\right\} \tag{2.111a}$$

which does not contain indeterminations for $\alpha_m - \alpha_j \to 0$.

The similar transformation of eqns. 2.98*b* is based on the auxiliary identities

$$\oint \nu \frac{\partial \varphi^m}{\partial n}\frac{\partial \varphi^j}{\partial n}\, ds + \alpha_m^2 \int \varphi^{m'} \varphi^j\, dS + \alpha_j^2 \int \varphi^m \varphi^{j'}\, dS = 0$$

and

$$\int \varphi^{m'} \varphi^j\, dS + \int \varphi^m \varphi^{j'}\, dS = 0 \tag{2.109b}$$

which are obtained from eqn. I.11 in an analogous procedure as eqn. 2.109*a*. This transformation gives us the following expression for the coupling coefficients of two *TM* waves

$$S_{jm} = -\frac{1}{2h_j}\left(h_m \alpha_j^2 + h_j \alpha_m^2\right)\int \varphi^{m'} \varphi^j\, dS \tag{2.111b}$$

Eqns. 2.111 prove that the coupling coefficients can become infinite only if $|h_j| \to 0$. But for realising practical calculations, these formulae are substantially less comfortable than eqns. 2.98 since they contain the *z* derivatives of the transverse eigenfunctions.

Now we employ eqns. 2.111 to interrelate the coefficients S_{jm} and F_{jm}. The waveguide deformation considered above as a tilt (see Figure 2.6) can be considered also as a special case of cross-section variation where the waveguide boundary contour is shifted along the *x*-axis direction by a distance proportional to the *z* co-ordinate.

Such a waveguide belongs to the family of varying cross-section waveguides with a tilt of their generatrices. The amplitudes of scattered waves can be determined by means of eqn. 2.78 where, obviously, we have to retain the first term of the sum only. On the other hand, according to eqn. 1.6 these amplitudes are proportional to F_{jm}. Then the coupling coefficients F_{jm} can be expressed by means of the coupling coefficients S_{jm} calculated for the case of a special deformation shown in Figure 2.6.

Comparing eqns. 2.78 and 1.5 we obtain (for $j \neq m$)

$$F_{jm} = \frac{-i}{h_m - h_j}\frac{S_{jm}}{\Delta\vartheta} \tag{2.112}$$

Now let us calculate S_{jm} from eqns. 2.111. To do this we will introduce a co-ordinate system $\bar{x}, \bar{y}$ firmly connected with the waveguide boundary contour. It is obvious that

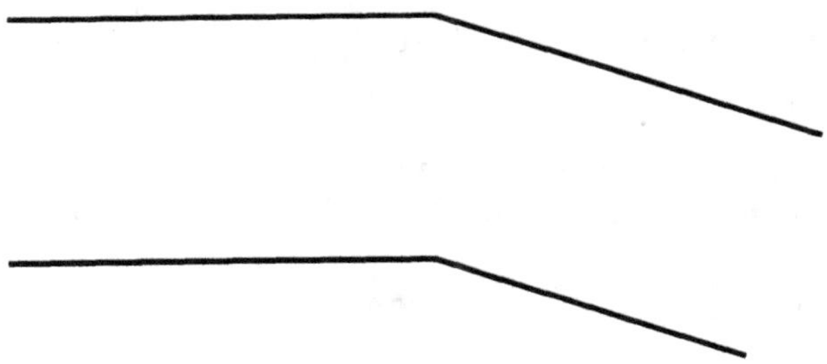

Figure 2.6 Waveguide tilt considered as a variation of waveguide cross section

$$\bar{x} = x - z \cdot \Delta\vartheta, \quad \bar{y} = y \tag{2.113}$$

Let us substitute the z derivatives in eqn. 2.111 by the $\bar{x}$ derivatives

$$\frac{\partial}{\partial z} = \frac{\partial}{\partial \bar{x}} \cdot \frac{\partial \bar{x}}{\partial z} = -\Delta\vartheta \frac{\partial}{\partial \bar{x}} \tag{2.114}$$

Let us start from eqn. 2.111*a*. It may be rewritten under the following form

$$S_{jm} = \frac{-\Delta\vartheta}{2h_j} \left\{ h_j \alpha_m^2 \int \psi^m \frac{\partial \psi^j}{\partial \bar{x}} dS - h_m \alpha_j^2 \int \psi^j \frac{\partial \psi^m}{\partial \bar{x}} dS \right\} \tag{2.115}$$

The integrals over the cross section in this formula are identical to the corresponding integrals determined in eqn. 2.18 and the coupling coefficient S_{jm} for this special case is equal to

$$S_{jm} = \frac{-\Delta\vartheta}{2h_j} \left\{ h_j \alpha_m^2 K^{mj} - h_m \alpha_j^2 K^{jm} \right\} \tag{2.116}$$

Substituting this result into eqn. 2.112 we find an expression for the coupling coefficient F_{jm} identical to eqn. 2.20.

According to eqns. 2.111*b* and 2.114 we obtain for two *TM* waves

$$S_{jm} = \frac{\Delta\vartheta}{2h_j} \left(h_m \alpha_j^2 + h_j \alpha_m^2 \right) \int \varphi^j \frac{\partial \varphi^m}{\partial \bar{x}} dS \tag{2.117}$$

Substituting this result into eqn. 2.112 and using the symbols of eqn. 2.18 we obtain an expression for F_{jm} identically equal to eqn. 2.21.

In a similar manner, eqn. 2.22 can be verified for the coupling coefficient between *TE* and *TM* waves.

The calculation realised above is useful not only for verifying eqns. 2.20 – 2.21, but it also allows one to explain why F_{jm} is larger for forward waves than for backward waves. From this point of view a tilt is not an elementary local non-uniformity but it forms an extended long nonuniformity where the interference phenomena between the elementary parasitic waves excited at different points along the waveguide are essential. However, as was mentioned above for these nonuniformities, the forward wave amplitudes are larger than those for the backward waves. Because of this, the difference in wavenumbers appears in the denominator of eqn. 2.112 and, hence, in the denominator of eqns. 2.20 – 2.21.

2.4 Solution of general nonuniform waveguide problem using the cross-section method

The most general kind of nonuniformity can be described as: the waveguide axis changes its direction and simultaneously the properties of the waveguide filling medium as well as the shape or location of its cross section are changed. From the point of view of the theory of small perturbations developed in the previous chapter, a general elementary nonuniformity is a superposition of the three kinds of elementary nonuniformity considered in Chapter 1. As we show later, the coupling coefficients are added in some sense. However, one should be careful: the dimensions of the coefficients F_{jm} and S_{jm} are different, i.e. F_{jm} is calculated per angular unit and S_{jm} per length unit.

2.4.1 Straight waveguide with varying cross section and inhomogeneous filling medium

The problem of a straight waveguide with varying cross section and varying filling medium (i.e. its ε and μ) can be analysed in the simplest way, since, according to the two previous sections, the coupling coefficients caused by both nonuniformities have not only the same geometrical nature but a common analytical expression as well. To derive the formula for S_{jm} in this case, let us inscribe a waveguide with varying cross section in a larger waveguide with constant cross section, as was done in Section 2.3, and the space between the waveguide surfaces is filled with a medium described by the parameters ε_0 and μ_0. The coupling coefficient in this auxiliary waveguide is given by eqn. 2.57. However, in contrast to the similar integral considered in Section 2.3, the integrand is different from zero not only within a narrow transition layer near the border of the medium with the parameters ε_0 and μ_0, but within the whole cross section of the initial waveguide. Later, taking the limit by $|\varepsilon_0| \to \infty$, we yield the coupling coefficient as a two-term sum of the integral of eqn. 2.57 over the cross section of the initial nonuniform waveguide and an integral like eqn. 2.95 obtained by integration over a transition layer.

This contour integral differs from eqn. 2.95 by the factor ε in front of its first term and by the factor μ ahead of its other two terms, since eqn. 2.95 is valid only if $\varepsilon=1$ and $\mu=1$ inside a waveguide. This difference is caused by additional factors ε and μ in eqns. 2.62 related to $E_n^m(n)$ and $H_n^m(n)$, and ε and μ are different from unity in eqn. 2.61 on the border of the region, where integration over n is performed. Hence, eqn. 2.63 has the following form

$$S_{jm} = \frac{1}{2h_j(h_j - h_m)} \oint \nu(s) \left\{ (\varepsilon_0 - \varepsilon)(E_s^j E_s^m - E_z^j E_z^m) - \left(\frac{\varepsilon^2}{\varepsilon_0} - \varepsilon \right) E_n^j E_n^m \right.$$
$$\left. + (\mu_0 - \mu)(H_s^j H_s^m - H_z^j H_z^m) - \left(\frac{\mu^2}{\mu_0} - \mu \right) H_n^j H_n^m \right\} ds \qquad (2.118)$$

Here, the fields are related to the same side of the border, where the permittivity and permeability are equal to ε and μ, respectively. Calculating the limit for $|\varepsilon_0| \to \infty$, one yields a formula which differs from eqn. 2.95 by the mentioned additional factors in the integrand.

In such a way, the coupling coefficient contains two terms caused by a cross-section variation and by a variation of the filling medium, respectively. Since a nonuniform filling is usually a dielectric insertion in a waveguide with a sharp border, it is worth applying eqn. 2.63 to the part of the coupling coefficient dependent on the nonuniform filling. To simplify the formula, we assume $\varepsilon \neq 1$ and $\mu = 1$ in the dielectric body and $\varepsilon = 1$ and $\mu = 1$ between the dielectric border and the metallic waveguide walls. Then the total coupling coefficient is equal to

$$S_{jm} = \frac{1}{2h_j(h_j - h_m)} \left\{ \oint \nu(s)(E_n^j E_n^m + H_z^j H_z^m - H_s^j H_s^m) ds \right.$$
$$\left. + \oint \nu(s) \left[(\varepsilon - 1)(E_s^j E_s^m - E_z^j E_z^m) - \left(\frac{1}{\varepsilon} - 1 \right) E_n^j E_n^m \right] ds \right\} \qquad (2.119)$$

Here, the functions $\nu(s)$ in the two integrals are different. In the first integral $\nu(s)$ determines the slope of the generatrix for the metallic walls, whereas in the second integral $\nu(s)$ is the slope of the generatrix for the dielectric insertion. The fields are related to the side of the border of the dielectric insertion, where $\varepsilon = 1$ and $\mu = 1$; note that the E_n components are different on the two sides. A more general formula can be easily derived if the magnetic permeability of the insertion is not equal to one and in the filling medium the space between the insertion and waveguide walls $\varepsilon \neq 1$ and $\mu \neq 1$. Eqn. 2.119 is the mathematical basis for calculating the compensating dielectric lenses in straight waveguides.

2.4.2 *Bent waveguide with inhomogeneous filling medium*

The coupling coefficient for the nonuniformity considered in the previous subsection, i.e. for a varying cross section and varying filling, could be immediately written in a general form as a single formula as in eqn. 2.47. Then the surface integral in eqn. 2.47 must be understood as including the contour integral over the contours of discontinuity, where the integrand functions have singularities. This form is inconvenient for real calculations, but it is useful to employ it for the following general analysis of bent waveguides.

Let us consider a bent waveguide with an inhomogeneous filling. Repeating the discussion of the last subsection, first let us limit ourselves to a constant cross-section waveguide with inhomogeneous filling. We start from a circular arc bend. The formal manipulations of Section 2.1 need to be repeated. However, one should take into account that now the fields $\vec{E}^j$ and $\vec{H}^j$ depend on the angle ϑ, since the functions $\varepsilon(x,y)$ and $\mu(x,y)$ in uniform reference waveguides depend on the location of the cross section ϑ = const.

Substituting eqns. 2.3 and 2.6 into the four Maxwell eqns. 2.7, we yield more complicated formulae instead of eqns. 2.8

$$\begin{aligned}
P_\nu' E_x^\nu &= -P_\nu E_z^\nu - iP_\nu h_\nu E_x^\nu r - P_\nu E_x^{\nu\prime} \\
P_\nu' H_x^\nu &= -P_\nu H_z^\nu - iP_\nu h_\nu H_x^\nu r - P_\nu H_x^{\nu\prime} \\
P_\nu' E_y^\nu &= -iP_\nu h_\nu E_y^\nu r - P_\nu E_y^{\nu\prime} \\
P_\nu' H_y^\nu &= -iP_\nu h_\nu H_y^\nu r - P_\nu H_y^{\nu\prime}
\end{aligned} \tag{2.120}$$

Here and everywhere in this section, as well as in Section 2.1, the prime symbol denotes the derivative with respect to the angle ϑ. In addition, employing the orthogonality condition to single out the derivative $P_j{}'$, we obtain

$$P_j' = \left(F_{j\nu} + T_{j\nu}\right)P_\nu \tag{2.121}$$

Here, F_{jm} denotes the same coefficient as in eqn. 1.7, and T_{jm} denotes the following values

$$T_{jm} = \frac{1}{2kh_j}\int\left(E_x^j H_y^{m\prime} - E_y^j H_x^{m\prime} + E_x^{m\prime} H_y^j - E_y^{m\prime} H_x^j\right)dS \tag{2.122}$$

Unlike in eqn. 2.47, the integrand in eqn. 2.122 contains the derivatives with respect to ϑ.

Eqn. 2.121 is valid for all values of j giving the desired coupled-wave differential equations describing the field in a nonuniform waveguide of a general kind. The boundary conditions for the coupled-wave eqns. 2.121 are identical to those of eqn. 2.10.

Using the same method as for the derivation of eqn. 2.14, it can easily be shown that the following condition for the matrix in eqns. 2.121 follows from the reciprocity principle

$$h_j\left(F_{-j,-m} + T_{-j,-m}\right) + h_m\left(F_{mj} + T_{mj}\right) = 0 \tag{2.123}$$

Of course, this relationship could be derived simply from eqns. 2.14 and 2.10. These relationships are valid for each of the two terms in the total coupling coefficient in eqn. 2.121.

The coefficient F_{jm} can be again represented in its symmetrical form of eqn. 2.16. However, the subsequent transformation to the form of eqns. 2.20 and 2.22 is impossible, since for our problem the fields cannot usually be expressed through transversal eigenfunctions in uniform reference waveguides, where the dielectric filling is inhomogeneous within the cross section.

We do not transform the coefficient T_{jm} of eqn. 2.122 into a form similar to eqn. 2.57, despite it giving us a convenient expression for the coupling coefficient in a waveguide with sharp filling boundaries and varying cross section. The general form of these formulae would require a lot of space in this book. Let us limit ourselves to a transformation applicable under the condition that the radius of curvature is much greater than any of the linear dimensions of the cross section. Taking into account that $\partial / \partial\vartheta = r\partial / \partial z$, one can put a mean value of r outside of the integral for T_{jm}. Then one obtains

$$T_{jm} = rS_{jm} \tag{2.124}$$

where S_{jm} is given in eqn. 2.47 and can be represented in the form of eqn. 2.57. Now we can avoid the restriction of a constant cross section imposed at the beginning of this section. A further transformation to a discontinuous distribution of ε and μ, to waveguides with varying cross section and to a general case of a waveguide with both varying cross section and varying filling medium can be performed in the same way as above. For S_{jm} in eqn. 2.124 it leads to either eqn. 2.63 or eqn. 2.119.

Further, let us introduce the variables $p_j(\vartheta)$, so that the matrix for the coupled-wave equations does not contain diagonal elements. Again, let us limit ourselves to the conditions by which eqn. 2.124 is valid. According to the definition of eqn. 2.25 of the curvature radius of the bend and the relationships of eqn. 2.50 and 2.124, the diagonal elements of the matrix for the coupled-wave eqns. 2.121 have the following form

$$F_{jj} = -irh_j, \quad T_{jj} = -rh_j' / 2h_j \tag{2.125}$$

For the sake of simplicity, let the radii of curvature (eqn. 2.25) be equal for all modes, here and everywhere later. Taking into account eqn. 2.125, we conclude that, as in eqn. 2.70, the variables $p_j(\vartheta)$ must be determined by the equations

$$P_j(\vartheta) = \sqrt{\frac{h_j(0)}{h_j(\vartheta)}} p_j(\vartheta) \exp(-i\gamma_j), \quad \gamma_j = r\int_0^{\vartheta} h_j d\vartheta \tag{2.126}$$

The coupled-wave equations for these variables are similar to eqns. 2.71

$$p_j{}'(\vartheta) = \sqrt{\frac{h_\nu(0)}{h_j(0)}}(F_{j\nu} + rS_{j\nu})\sqrt{\frac{h_j(l)}{h_\nu(l)}}p_\nu(1-\delta_{j\nu})\exp(-i(\gamma_\nu - \gamma_j)) \qquad (2.127)$$

Now one can easily consider a bend with a varying radius of curvature r. Similar to Section 2.1, one should introduce the arc length, l, measured along the bent waveguide as the independent variable instead of ϑ. The coupled-wave eqns. 2.127 with this variable have the following form

$$\frac{\partial p_j(l)}{\partial l} = \sqrt{\frac{h_\nu(0)}{h_j(0)}}\left(\frac{F_{j\nu}}{r} + S_{j\nu}\right)\sqrt{\frac{h_j(l)}{h_\nu(l)}}(1-\delta_{j\nu})\exp(-i(\gamma_\nu - \gamma_j)) \quad (2.128)$$

and the value of γ_j is equal to the output phase after the length l

$$\gamma_j = \int_0^l h_j dl \qquad (2.129)$$

The coupled-wave eqns. 2.128 are the desired generalisation of the coupled-wave eqns. 2.32 for a bent waveguide with constant cross section and eqns. 2.71 for a straight nonuniform waveguide. Eqns. 2.128 would be identical to eqns. 2.32 for $h_j(l) \equiv h_j(0)$ and $S_{jm}(l) \equiv 0$ and they would be identical to eqns. 2.71 for $r \to \infty$. According to eqn. 2.10 and to the normalisation accepted in eqn. 2.126, the boundary condition for the coupled-wave eqns. 2.128 is identical to that of eqn. 2.72.

The coupled-wave eqns. 2.128 and 2.72, together with the various expressions for the coupling coefficients presented above, describe the field for the most general case of a waveguide nonuniformity. Their application to the problem of a weak coupling leads to exactly the same results as for the coupled-wave eqns. 2.71 describing the field in a straight waveguide. If the incident mode amplitude is much larger than the amplitudes of other modes along a whole nonuniform section, i.e. the approximation of eqn. 2.73 is applicable, then eqn. 2.75 gives the amplitudes of all parasitic modes. However, the coupling coefficient S_{jm} in eqn. 2.75 has to be substituted by the following expression

$$S_{jm} \to S_{jm} + \frac{F_{jm}}{r} \qquad (2.130)$$

If these conditions are valid, for which the parasitic mode amplitudes can be expressed by the two-term formula of eqn. 2.78, then a similar formula is valid for the considered general case if the substitution of eqn. 2.130 is performed in eqn. 2.78. In other words, the additional field perturbation caused by a bend may be formally considered by the same technique as for nonuniformities not changing the direction of the axis. One should only consider a bend as introducing an additional coupling described by the coupling coefficient F_{jm}/r.

Strictly speaking, this additivity of the results relating to different kinds of nonuniformity is valid only for the conditions for which eqn. 2.124 is valid. For bends with a small curvature radius, eqn. 2.124 is not valid, and the coefficient T_{jm}, as we have already mentioned, could also be transformed to a form similar to eqn. 2.124 with a definite value of r, but then T_{jm}/r would be different from eqn. 2.47. In other words, a bend not only adds the second term in eqn. 2.130, it changes the first term as well. So, for example, the coupling coefficient caused by a cross-section variation is slightly different in a bent waveguide from that in a straight one. We do not present the corresponding formulae here. Partly because they take a lot of space, but mainly because the mostly interesting results are usually related to a weak coupling or in any case for large radii of curvature, when eqn. 2.124 and therefore eqn. 2.130 are applicable.

Of course, the coupling coefficients are additive for superposition of small nonuniformities. This is simply the consequence of the general principle of additivity of small perturbations. According to eqn. 1.6 and 1.27, the amplitudes of parasitic modes excited along a nonuniform section of the length Δz, where a tilt simultaneously occurs with an angle $\Delta\vartheta$ and a change of properties described by the matrix S_{jm}, are equal to

$$F_{jm}\Delta\vartheta + S_{jm}\Delta z \tag{2.131}$$

Eqn. 2.130 is the most important result of this section, as well as the conditions of its applicability. This formula, together with eqn 2.119, can form a basis for mathematical analysis of devices employing mutual compensation of various kinds of nonuniformity.

2.4.3 *Example of a resonant antenna with semitransparent aperture*

The computation method for the fields in a waveguide with varying cross section and varying wall impedance has been applied (see Voytovich *et al.*, 1993) to analyse a resonant antenna formed by a planar resonator with semitransparent wall. This is used as a radiating aperture of the antenna. To achieve a narrow far-field radiation pattern the antenna has to be long, i.e. the ratio of its length ($\gg \lambda$) to its height ($\sim \lambda/2$) has to be large. The gain factor of the antenna must be not high, otherwise the antenna would have a narrow operating frequency band. A whole set of eigenoscillations can be excited simultaneously in such a resonator instead of a single eigenoscillation. Then the aperture field does not have a simple sinusoidal form anymore and therefore the antenna far-field radiation pattern has high level sidelobes and can become a multibeam pattern.

To avoid this, it has been proposed to taper the aperture height and its transparency (i.e. practically, its impedance) along its length. It is more comfortable to consider this kind of antenna not as an open resonator but as a short-circuited waveguide with varying height and impedance. Then one needs to solve an excitation problem for such a waveguide with a given source. Unlike the field of a high quality resonator, the electromagnetic field in this waveguide depends strongly on the position along the waveguide and on the properties of the exciting transmission line.

As has been demonstrated by numerical simulation, the main phenomena in a single-mode planar waveguide is not the coupling to the parasitic modes but the local reflections of the fundamental mode on the nonuniformities. This result can be explained by the large differences between the wavenumbers of different modes. The parasitic modes and their coupling to the fundamental mode are essential only near the output of the exciting transmission line. It is simpler to describe mutual coupling of two waves of the same mode propagating in opposite directions not by the first-order differential equations for P_m and P_{-m} but by the following second-order differential equation $Q_m'' + h_m^2 Q_m = 0$ for Q_n given in eqn. 2.2. The wavenumber $h_m(z)$ of the corresponding reference waveguide can be determined from the simple transcendent equation containing the waveguide height and its impedance.

The main difficulty encountered when following the above proposed programme is the instability of the computation process. The direct application of this technique based on the solution of a Cauchy problem causes an unacceptable accumulation of errors in some cases. Special computation algorithms have been developed recently to avoid these difficulties.

Both direct analysis problems and optimisation problems have been solved for the case of a resonant antenna with semitransparent aperture, i.e. the tapering of the antenna height and its transparency have been determined, and the far-field radiation pattern has been optimised by a given criterion. This method has also been successfully used to solve the problem of a leaky wave antenna formed by a single-mode waveguide with the semitransparent broad wall.

***

Coupled-wave differential equations for mode amplitudes have been derived for basic kinds of nonuniform waveguide as well as for the general kind of nonuniform waveguide. Various expressions of coupling coefficients caused by different nonuniformities have been obtained and their additivity has been proven. The expressions for parasitic mode amplitudes have been derived for the case of weak coupling in the form of an integral over the nonuniform section. In this case, waveguide properties vary slowly along the waveguide. In some cases these amplitudes can be expressed in an explicit form depending only on the parameters relating to the ends of the nonuniform waveguide section.

Chapter 3

Special cases: cut-off cross sections and resonance frequencies

Generally, if waveguide properties vary slowly, coupling coefficients have small values and if there is no mode degeneracy, only one equation for each parasitic mode remains from the coupled-wave eqns. 2.9 and 2.46. Their solutions have the form of eqns. 2.35, 2.75 and 2.37, respectively. However, if a wavenumber h_j is small or equal to zero, the values of S_{jm} and F_{jm}/r are not small even for very small ν_0 and a/r, despite slowly varying parameters.

The so-called cut-off cross section can exist in a straight waveguide with varying h_j along its axis. There, $h_j = 0$ at a given frequency and $|h_j|$ is small. For the case of bent waveguides with constant cross section, $|h_j|$ can be small only in a narrow frequency band near the so-called resonance frequency, i.e. $|h_j| = 0$. In the first three sections of this chapter we study the special conditions which are valid when critical cross sections exist or the frequency is near the resonance.

3.1 Mode reflection and transmission in presence of a cut-off cross section

3.1.1 Formulation of coupled-wave equations using specific variables

Only in this subsection, the indices i and j denote TE modes, the indices m and n TM modes and the indices t and s can denote both TE and TM modes.

Let us indicate by '~' the values relating to a cut-off cross section, and let the eigenvalue at the cross section $z = \tilde{z}$ be equal to k for some mode index t, i.e. $\alpha_t = k$. Hence, $h_t(\tilde{z}) = 0$.

The coupling coefficients $S_{t,s}$ are then infinite for all or nearly for all s at $z = \tilde{z}$, and the $S_{t,s}$ have large values near $z = \tilde{z}$. Some coefficients in the coupled-wave eqns. 2.46 for P_t and P_{-t} are infinite at $z = \tilde{z}$ and, hence, the solution of eqn. 2.75 is not valid anymore.

The reflection of the incident mode at a cut-off cross section means a strong coupling between forward and backward waves. Near the cut-off cross section the decomposition of the total field into forward and backward waves does not correspond to the physical nature of the phenomena. It is, hence, inconvenient for our calculations.

The amplitudes P_t and P_{-t} can become infinite at some cross section. However, only the total fields $\vec{E}$ and $\vec{H}$ and not single terms of eqns. 2.43 and 2.44 must have finite values. In fact, eqns. 2.43 and 2.44 are the decomposition of the total fields into fields of the waves of both directions. Let us return to the variables Q_t and R_t related to P_t and P_{-t} by eqns. 2.2. These variables satisfy to the following coupled-wave equations

$$Q_t' + ih_t R_t = \sum_{s=1}^{\infty} Q_s (S_{ts} - S_{-ts}) \tag{3.1a}$$

$$R_t' + ih_t Q_t = \sum_{s=1}^{\infty} R_s (S_{ts} + S_{-ts}) \tag{3.1b}$$

where $t = 1,2, \ldots$ eqns. 3.1 can easily be derived from the coupled-wave eqns. 2.46 for the variables P_t with $t = \pm 1, \pm 2, \ldots$

We analyse the coupled-wave eqns. 3.1 for empty waveguides. There the total field can be decomposed into *TE* and *TM* modes. The coefficients $(S_{t,s} \pm S_{-t,s})$ in eqns. 3.1 have different forms for these two kinds of modes.

As follows from eqns. 2.98, I.5 and I.15

$$S_{j,n} - S_{-j,n} = 0, \; S_{n,j} + S_{-n,j} = 0 \tag{3.2a}$$

and the coefficients

$(S_{j,i} - S_{-j,i})$

and (3.2*b*)

$(S_{n,m} + S_{-n,m})$

as well as their derivatives in respect to z have finite values at the cut-off cross sections. The coefficients

$(S_{j,i} + S_{-j,i})$

and

$(S_{j,n} + S_{-j,n})$

as well as (3.2*c*)

$(S_{n,m} - S_{-n,m})$

and

$(S_{n,j} - S_{-n,j})$

have infinite values described by $1/h_j$ or $1/h_n$ at $h_j \to 0$ or at $h_n \to 0$, respectively. The products of these coefficients and h_j or h_n as well as their derivatives have finite values at the cut-off cross sections.

For the case of a cut-off cross section, according to eqn. I.14, $H_x^j = 0$ and $H_y^j = 0$ for TE modes. As follows from the series of eqns. 2.42, the coefficient R_j for TE modes can have finite values everywhere, but the coefficient Q_j for TM modes must have finite values everywhere. Hence, the amplitudes P_j and P_{-j} can have an infinite value at the cut-off cross section, but only in such a way that their difference has a finite value. Similarly, the amplitudes P_n and P_{-n} can become infinite, but their sum R_n must have a finite value.

So in eqns. 3.1 the coefficients of eqns. 3.2*c* as well as the variables Q_n and R_j have infinite values at the cut-off cross sections. Hence, these equations cannot be employed for analytical and numerical analysis.

We introduce in this subsection new variables instead of those with infinite values and we derive coupled-wave equations suitable for a numerical analysis. We employ a different technique in another subsection of this section, and in the next section as well, which allows us to apply the method usually used for an approximate analytical study of transmission lines with return points.

Following Tsimring and Pavelyev (1982), let us introduce new variables instead of Q_n and R_j

$$\overline{R}_j = h_j R_j \tag{3.3a}$$

$$\overline{Q}_n = h_n Q_n \tag{3.3b}$$

Unlike R_j and Q_n, these variables have finite values at cut-off cross sections. Besides this, to simplify the coefficients in the derived coupled-wave equations, we introduce the variables $\overline{R}_n = kR_n$ instead of R_n and $\overline{Q}_j = kQ_j$ instead of Q_j. Multiplying the coupled-wave eqns. 3.1 written for R_j' and Q_n' by h_j and by h_n, respectively and substituting the new variables, we yield four coupled-wave equations for $\overline{Q}_j$, $\overline{R}_j$, $\overline{Q}_n$ and $\overline{R}_n$

$$\begin{aligned}
&\overline{Q}_j' + ik\overline{R}_j = \sum_i \left(S_{ji} - S_{-ji}\right)\overline{Q}_i \\
&\overline{R}_j' + i\frac{h_j^2}{k}\overline{Q}_j = \sum_i \left(S_{ji} + S_{-ji}\right)\frac{h_j}{h_i}\overline{R}_i + \frac{h_j'}{h_j}\overline{R}_j + \sum_n \left(S_{jn} + S_{-jn}\right)\frac{h_j}{k}R_n \\
&\overline{R}_n' + ik\overline{Q}_n = \sum_m (S_{nm} + S_{-nm})\overline{R}_m \\
&\overline{Q}_n' + i\frac{h_n^2}{k}\overline{R}_n = \sum_m (S_{nm} - S_{-nm})\frac{h_n}{h_m}\overline{Q}_m + \frac{h_n'}{h_n}\overline{Q}_n + \sum_j \left(S_{nj} - S_{-nj}\right)\frac{h_n}{k}\overline{Q}_j
\end{aligned} \tag{3.4}$$

The coefficients at the right-hand side of these equations not only have finite values at the cut-off cross sections, as is evident, they also do not depend on frequency at all. Employing eqns. 2.99, 2.100 and 2.51, these coefficients can be expressed through the transversal eigenfunctions ψ^j and φ^n. The corresponding cumbersome formulae will not be given here since they can be derived by heavy, but elementary algebraic calculations.

The boundary conditions for solving these coupled-wave equations follow from eqn. 2.48. If the *TE* mode $j = 1$ is incident on a nonuniform waveguide section from the left-hand side (i.e. from $z = 0$) and there are no incident waves from the right-hand side ($z = L$), then at $z = 0$ those conditions for the variables $\overline{Q}_j, \overline{R}_j, \overline{Q}_n$ and $\overline{R}_n$ have the following forms

$$\overline{R}_1 + \frac{h_1}{k}\overline{Q}_1 = 2h_1, \quad \overline{R}_j + \frac{h_j}{k}\overline{Q}_j = 0 \text{ for } j > 1$$
$$\overline{Q}_n + \frac{h_n}{k}\overline{R}_n = 0 \text{ for } n > 0 \tag{3.5a}$$

and at $z = L$

$$\overline{R}_j - \frac{h_j}{k}\overline{Q}_j = 0 \text{ for } j < 0, \quad \overline{Q}_n - \frac{h_n}{k}\overline{R}_n = 0 \text{ for } n < 0 \tag{3.5b}$$

The boundary conditions have a similar form if a *TM* mode is incident on the nonuniform waveguide section. The coupled-wave eqns. 3.4 and 3.5 are equivalent to the coupled-wave eqns. 2.46 and 2.48, they represent their specification for empty waveguides.

Eqns. 3.4 have been used in the papers of Malygin *et al.* (1983), Zapevalov *et al.* (1984) and Pavelyev *et al.* (1987) to determine the complex eigenfrequencies of a system containing several coupled open cavities formed by nonuniform circular waveguide sections. The boundary condition of eqn. 3.5*a* is then substituted by a homogeneous condition, i.e. the value of $2h_1$ is substituted by zero.

3.1.2 Transformation of coupled-wave equations into second-order differential equations

The second method of avoiding infinite values in eqns. 3.1 is to transform them into second-order equations. First, let us differentiate eqn. 3.1*a* with respect to *z*. Taking into account eqn. 2.50 and using eqn. 3.1*b*, we yield

$$Q_j'' + h_j^2 Q_j = -i\sum_{m\neq j} R_m\left\{h_j\left(S_{jm} + S_{-jm}\right) + h_m\left(S_{jm} - S_{-jm}\right)\right\}$$
$$+\sum_m Q_m\left(S_{jm} - S_{-jm}\right)' + \sum_m\sum_q Q_q\left(S_{jm} - S_{-jm}\right)\left(S_{mq} - S_{-mq}\right) \tag{3.6}$$

We apply this equation to *TE* modes, i.e. we assume that the mode j is of *TE* type. The variable R_j does not take part in eqn. 3.6. Because of this, as well as according to eqns. 3.2 and the discussion thereafter, the right-hand side of eqn. 3.6 has finite values at the cut-off cross section as well as in its neighbourhood. Taking into account the terms of zero and first order of ν only, we obtain

$$Q_j'' + h_j^2 Q_j = G_j$$
$$G_j = -i\sum_{m \neq j} R_m \left\{h_j\left(S_{jm} + S_{-jm}\right) + h_m\left(S_{jm} - S_{-jm}\right)\right\} \tag{3.7a}$$

The terms of the order of ν^2 as well as of the order of the derivative ν' are neglected in this equation. It is important that the neglected terms do not have any singular points near the cut-off cross section. We apply eqn. 3.7*a* exactly to that neighbourhood.

In a similar way, differentiating eqn. 3.1*b* and using eqns. 3.1*a* and 2.50, we yield an equation similar to eqn. 3.6, where the expression $(R_j'' + h_j^2 R_j)$ is on the left-hand side and on the right-hand side are the sum of Q_m without the Q_j term and the sum of R_m. We employ this equation for *TM* modes, i.e. now we assume that the mode j is a *TM* mode. Taking into account the terms of zero and first order only, we obtain the following equation

$$R_j'' + h_j^2 R_j = G_j$$
$$G_j = -i\sum_{m \neq j} \left\{h_j\left(S_{jm} - S_{-jm}\right) + h_m\left(S_{jm} + S_{-jm}\right)\right\} Q_m \tag{3.7b}$$

In this approximation this equation is valid everywhere near the cut-off cross section. The analysis of waveguides with slowly varying cross section containing a cut-off cross section is based on eqns. 3.7. This analysis is performed later in this section and in Section 3.2.

We consider in the present section the field of an incident mode only, i.e. we assume that the wavenumber h_m of this mode is equal to zero at the cut-off cross section. First of all we are interested in the case of the cut-off cross section located near the beginning of a narrow waveguide.

Parasitic modes excited by the incident mode could be found later by the method described in the previous chapter, while substituting the solution determined in this section instead of eqn. 2.73 in formulae like eqn. 2.71. The case of the cut-off cross section for a parasitic mode and not for an incident mode must be treated separately; it is considered in Section 3.2.

Let us consider a narrowed waveguide. The derivative of the wavenumber in such a waveguide is negative near the cut-off cross section, $\tilde{h}_m' < 0$. A wave is incident from the broad waveguide. For certainty, let it be a *TE* mode, so that the variable $Q_m(z)$ has finite values everywhere. Its field is determined from eqn. 3.7*a*

$$Q''_m + h_m^2 Q_m = 0 \qquad (3.8a)$$

and from eqn. 3.1*a*, where again the first term only on the right-hand side is relevant

$$Q'_m + ih_m R_m = Q_m (S_{mm} - S_{-mm}) \qquad (3.8b)$$

The neglected terms in eqns. 3.8*a* and 3.8*b* have small values of the second order for all z. Using the last equation, one can exclude R_m and express P_m and P_{-m} through Q_m and Q_m' as:

$$P_m = i\frac{Q'_m - [ih_m + (S_{mm} - S_{-mm})]Q_m}{2h_m}$$
$$P_{-m} = i\frac{Q'_m - [-ih_m + (S_{mm} - S_{-mm})]Q_m}{2h_m} \qquad (3.9)$$

Our problem is to determine the reflection coefficient of the incident mode. The boundary condition for $Q_m(z)$ at $z = L$, i.e. at the beginning of the narrow waveguide, is required for this purpose. This boundary condition follows from the boundary condition $P_{-m}(L) = 0$ and from the second of eqns. 3.9. One yields

$$Q'_m - [-ih_m + (S_{mm} - S_{-mm})]Q_m = 0, \text{ at } z = L \qquad (3.10)$$

3.1.3 WKB solution of coupled-wave equations

Let us solve eqns. 3.8*a* and 3.10. Eqn. 3.8*a* is a generalisation of the formulae of Section 2.2 in the case when h_m^2 has small values or is equal to zero. Standard methods like WKB, i.e. the methods of geometrical optics, can be applied to eqn. 3.8 in the region where $|h_m|$ is not small, more exactly, in the region, where $|h'_m / h_m^2| \ll 1$. The approximate solution there is a combination of the functions $\exp(\pm i\gamma_m)/\sqrt{h_m}$, which were employed by us in Section 2.2. This condition is not valid near the cut-off cross section, hence these function do not satisfy eqn. 3.8*a*. Because of this we have to derive another solution of eqn. 3.8*a* employing the Airy functions. Later, we need to provide a smooth transition between the solutions from different regions.

If h_m^2 is small it can be approximated by the first term of its expansion in a Taylor series around $(z - \tilde{z})$

$$h_m^2 = -k^3 A^2 (z - \tilde{z}) \qquad (3.11)$$

By its definition, the small positive and dimensionless variable A^2 is equal to

$$A^2 = -\frac{1}{k^3}\left(h_m^2\right)'\Big|_{z=\tilde{z}} \tag{3.12}$$

It is easily derived using eqn. 2.96 and has a small value of the order of $\tilde{\mathrm{v}}$. Let us introduce a new variable t instead of z

$$t = A^{2/3}k(z - \tilde{z}) \tag{3.13}$$

So one yields

$$h_m = -iA^{2/3}kt^{1/2} \tag{3.14}$$

By its definition, the sign of the square root here is chosen in such a way, that $t^{1/2} = i\sqrt{(-t)}$ for $t < 0$, i.e. for real h_m, at $(z < \tilde{z})$, and $t^{1/2}$ has positive values for $t>0$, i.e. $z > (\tilde{z})$. Using this variable eqn. 3.8*a* has the following form

$$\ddot{Q}_m - tQ_m = 0 \tag{3.15}$$

where the derivatives with respect to t are denoted by the dots.

As is known, the general solution of eqn. 3.15 is given by a linear combination of the Airy functions $u(t)$ and $\upsilon(t)$

$$Q_m = Mu + N\upsilon \tag{3.16}$$

The ratio of the coefficients in eqn. 3.16 is determined by substitution of this solution into the boundary condition of eqn. 3.10 for Q_m. Here it is assumed that the beginning of the narrow waveguide is situated not very far from the cut-off cross section, i.e. the approximation of eqn. 3.11 can be used there. Hence, the solution of eqn. 3.16 is valid there. However, as is shown in the second subsection of Section 3.2, the obtained solutions are valid in the simpler case too, when the cut-off cross section is situated far away from the beginning of the narrow waveguide.

This ratio is derived from eqns. 3.16 and 3.10

$$\frac{N}{M} = -\frac{\dot{u}(t_L) + \left[t_L^{1/2} + B\right]u(t_L)}{\dot{\mathrm{v}}(t_L) + \left[t_L^{1/2} + B\right]\mathrm{v}(t_L)}$$

with

$$B = -\frac{1}{k}(S_{mm} - S_{-mm})A^{-2/3} \tag{3.17}$$

Here, t_L denotes the value of the variable t (eqn. 3.13) at $z = L$. The value t_L is proportional to the difference between the cut-off frequency of the narrow waveguide and the operating frequency. It is natural to denote the cut-off frequency of the narrow waveguide as $\alpha_m(L)$. The following expression of t_L, mostly suitable for real calculations, is derived from eqn. 3.14

$$t_L = 2A^{-4/3}\frac{\alpha_m(L) - k}{k} \tag{3.18}$$

The coefficient in eqn. 3.18 is a large number, since $A^{-4/3} \approx \tilde{\nu}^{-2/3} >> 1$.

The solution of eqn. 3.16 is valid in the same region where eqn. 3.11 can be used, i.e. approximately for $\hbar_m^2 << 1$ *or* $|t| << A^{-4/3}$. This region is overlapped by the region where the condition $|h_m'/h_m^2| << 1$ is satisfied, and it can be written in the form $|t| >> 1$ (the value of t is positive near $z = 0$). Therefore, eqn. 3.8*a* must approach the solutions obtained in the approximation of geometrical optics for $t >> 1$. Taking this into account, we use eqn. 3.16 to establish a relationship between P_m *and* P_{-m} on the boundary of the region $|h_m'/h_m^2| << 1$. This relationship can be treated as a boundary condition equivalent to a cut-off cross section and a narrow waveguide behind it. Later, this equivalent boundary condition allows us to limit ourselves to a region far from the cut-off cross section. A similar idea was proposed by Brekhovskikh and Ivanov (1952) for another problem leading also to eqn. 3.8*a*.

If the cut-off cross section is situated at the end of a smoothed nonlinear taper, i.e. where $\nu = 0$, then $h_m'(\tilde{z}) = 0$ and eqn. 3.11 starts with a higher power of $(z - \tilde{z})$. The solution is then expressed not through the Airy functions but through other special functions, e.g. cylinder functions, whose order is in a definite way related to the order of the power of $(z - \tilde{z})$. The general calculation technique is also valid in this case. If $h_m'(\tilde{z})$ is not equal to zero, but nevertheless has a very small value, then two terms must be taken into account in eqn. 3.11.

Later, we assume that $h_m'(\tilde{z}) \neq 0$ and the consideration can be limited to eqn. 3.11. Hence, some of the results obtained in this section, namely the results relating to the case of the small or finite distance $|\tilde{z} - L|$, are valid only with an additional assumption: the junction of the nonuniform waveguide and the narrow waveguide is not a smoothed one.

3.1.4 Phase of reflection coefficients at a cut-off cross section

Let us start from the calculation of the reflection coefficient when the mode does not penetrate in the narrow waveguide (Figure 3.1)

The absolute value of the reflection coefficient is then equal to one, and its phase has to be found. We introduce the variable $\delta(z)$ defined by the following relationship

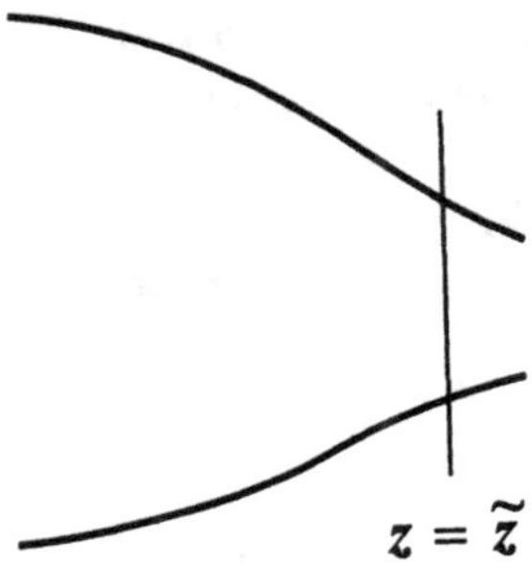

Figure 3.1 *Schematic of a cut-off cross section in a nonuniform waveguide (note that $z = L$ is located right of $z = \tilde{z}$)*

$$P_{-m}/P_m = \exp(i\delta) \tag{3.19}$$

The desired reflected phase $\delta(0)$ is the value of this variable at $z = 0$. It can easily be shown, that, according to eqns. 2.46, where only the terms containing P_m and P_{-m} must be taken into account, δ satisfies the nonlinear equation

$$\delta' - 2h_m - 2S_{-mm}\sin(\delta) = 0 \tag{3.20}$$

In the region where geometric optics is applicable, the third term in eqn. 3.20 is small and can be neglected if the condition $|h'_m/h^2_m| \ll 1$ is valid. The integral of this term over a large interval is small as well, since it contains the oscillating factor, and yields

$$\delta' - 2h_m = 0 \tag{3.21}$$

To determine $\delta(0)$ from this equation, a boundary condition for δ must be derived. This can be done, calculating the ratio P_{-m}/P_m near the boundary of the region given by $|h'_m/h^2_m| \ll 1$, following eqn. 3.9 and the general solution of eqn. 3.16. A simple calculation yields

$$\exp(i\delta) = \frac{\left[\dot{u} + \left(t^{1/2} + B\right)u\right] + \dfrac{N}{M}\left[\dot{\upsilon} + \left(t^{1/2} + B\right)\upsilon\right]}{\left[\dot{u} + \left(-t^{1/2} + B\right)u\right] + \dfrac{N}{M}\left[\dot{\upsilon} + \left(-t^{1/2} + B\right)\upsilon\right]} \tag{3.22}$$

Here the ratio N/M must be taken from eqn. 3.17.

To derive the equivalent boundary condition for eqn. 3.21, one should let $|t| \gg 1$ in eqn. 3.22. The variable t is negative in the region of our interest $z < \tilde{z}$, so asymptotic expressions of $u(t)$ and $\upsilon(t)$ must be used for large negative values of t. Performing the corresponding manipulations, we find the asymptotic value of the function δ, i.e. the largest terms in the expansion of δ from eqn. 3.22 into a series of inverse powers of $|t|$

$$\delta = -2\alpha + \delta_0 \tag{3.23}$$

Here, δ_0 has a constant value, i.e. it does not depend on t and is defined by the equation

$$\exp(i\delta_0) = \frac{\frac{N}{M} - i}{\frac{N}{M} + i} \tag{3.24}$$

The variable α is from the asymptotic expressions of the Airy functions for large negative values of their argument. It is equal to

$$\alpha = \frac{2}{3}(-t)^{3/2} + \pi/4 \tag{3.25}$$

As can be easily proved, the function of eqn. 3.23 satisfies eqn. 3.21. The desired solution of this equation approaches the general solution eqn. 3.22 in its corresponding region and the solution of eqn. 3.23 near to the cut-off cross section. The sought for value δ(0) is the solution of eqn. 3.21 in $z = 0$ (i.e. in $t < 0$). This solution can be written, for all t, in the form

$$\delta(z) = 2\int_{\tilde{z}}^{z} h(z)dz + \delta(\tilde{z}) \tag{3.26a}$$

This function is equal to $-4/3(-t)^{3/2} + \delta(\tilde{z})$, according to eqn. 3.11 and 3.13. The constant term $\delta(\tilde{z})$ can be obtained if we say that for $t >> 1$ the solution has the form $\delta = \delta_0 - \pi/2 - 4/3(-t)^{3/2}$, see eqns. 3.23 and 3.25. Hence, $\delta(\tilde{z}) = \delta_0 - \pi/2$. The same formula for $\delta(\tilde{z})$ can be found if we make the assumption that the obtained solution for $t >> 1$ is valid at $t = 0$ too. The condition $\delta(\tilde{z}) = \delta_0 - \pi/2$ gives the desired boundary condition for eqn. 3.21. The solution of this equation must satisfy this condition to approach the solution of eqn. 3.16 in the region near the cut-off cross section.

So the solution of eqn. 3.20 in the region of concern is given by

$$\delta(z) = 2\int_{\tilde{z}}^{z} h_m(z)dz - \pi/2 + \delta_0 \tag{3.26b}$$

and the desired phase of the reflection coefficient is equal to

$$\delta(0) = -2\tilde{\gamma}_m - \pi/2 + \delta_0 \tag{3.27}$$

The physical meaning of the first term is obvious. It is the phase acquired during propagation to the cut-off cross section and back, calculated in the geometric optics approximation. It can be easily found for any given form of the dependence of the transversal eigenvalue on the cross-section location, i.e. for any function $\alpha_m(z)$. For example, $1/\alpha_m$ for a cone is the linear function of z and

$$\frac{1}{\alpha_m(z)}=\frac{1}{\alpha_m(0)}-\frac{1}{C}z, \quad C=\text{const}, \quad C>0 \tag{3.28}$$

Simple manipulations yield

$$\tilde{\gamma}_m=C(g-\arctan(g)), \quad g=\frac{h_m(0)}{\alpha_m(0)} \tag{3.29}$$

The value δ_0 in eqn. 3.27 depends on the distance between the cut-off cross section and the beginning of the narrow waveguide. According to eqn. 3.24 and 3.17, δ_0 depends on the frequency difference determining the parameter t_L in eqn. 3.18. It gives an amendment to the phase of the reflected wave caused by the narrow waveguide. Power does not propagate in the narrow waveguide. Nevertheless, the field penetrates to the region $z>\tilde{z}$, exponentially attenuated along the distance from the cut-off cross section. So, the structure of the waveguide in this region influences the phase of the reflected wave. However, the influence of the narrow waveguide is significant only for finite t_L. Then δ_0 is very small for $t_L >> 1$, and in practical terms it is small already for $t_L \sim 1$, and the phase of the reflected wave differs by a constant term $(-\pi/2)$ from the phase calculated in the geometric optics approximation. In such a case it seems that the field does not reach the beginning of the narrow waveguide.

Let us note the phase shift of the same value in the well known problem of a normal wave incident on the ionosphere as well as in several other problems, leading to the same eqns. 3.8*a* with the condition of eqn. 3.11.

According to eqns. 3.24 and 3.17, the exact form of the function $\delta_0(t_L)$ depends on the type of mode and on the waveguide properties. The value of $(-\delta_0)$ versus t_L is presented in Figure 3.2 for the $TE_{0,n}$ modes in a circular waveguide.

According to eqns. 2.97 and 2.98, $S_{-m,m}=S_{m,m}$ for these modes and, hence, there is no B term in eqn. 3.17. However, this term has some value only for very small t_L, consequently, the curve in Figure 3.2 represents correctly the dependence $(-\delta_0)$ versus t_L of all modes for $t_L \gtrsim 0.4 \div 0.5$. This behaviour we have confirmed numerically by using the mode matching and generalised scattering matrix method which shows good agreement with the results presented in Figure 3.2.

A similar technique is employed to determine the reflection phase for *TM* modes but we do not represent those derivations here. Let us just note that the

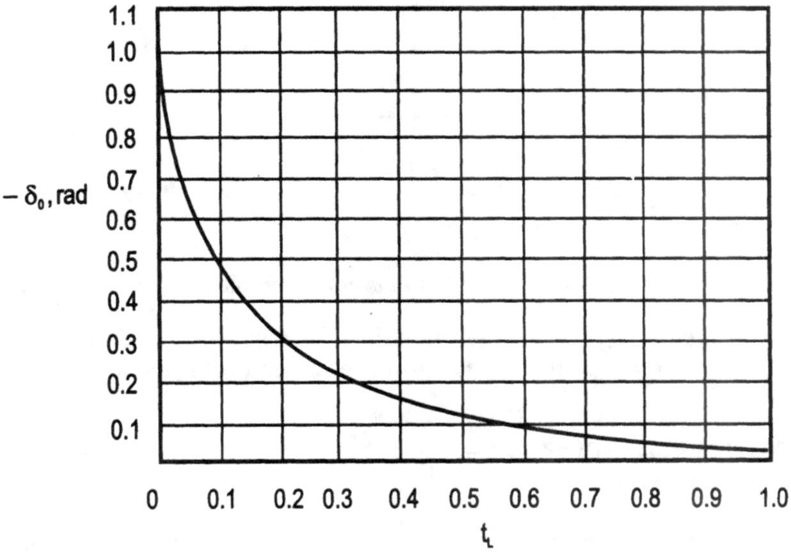

Figure 3.2 Amendment to the phase of the reflection coefficient for $TE_{0,n}$ modes, caused by a cut-off section in the circular waveguide

reflection phase of *TM* modes is equal to $(-2\tilde{\gamma}_m + \pi/2)$ for $t_L >> 1$. The same result is obtained in Section 3.2. There it is also shown that the reflection phase of *TE* modes is equal to $(-2\tilde{\gamma}_m - \pi/2)$ for $t_L >> 1$, as follows from eqn. 3.27. Moreover, this result is valid if the beginning of the narrow waveguide is so distant from the cut-off cross section that the expansion of eqn. 3.11 at $z = L$ employed above is not valid anymore.

3.1.5 Reflection coefficients close to cut-off

Now let us determine the reflection coefficient in the case when the mode penetrates the narrow waveguide. In this case h_m^2 is positive anywhere, i.e. $k>\alpha_m(L)$ and, strictly speaking, there is no cut-off cross section at all. However, the values of $(k - \alpha_m(L))$ and of the wavenumber are very small in the narrow waveguide, $\hbar_m << 1$, hence the coupling coefficients near the beginning of the narrow waveguide have large values. In such a case, the reflection coefficient can be nearly equal to one. The method developed above can be applied to this problem. One should only let the cut-off cross section to be situated outside the taper, i.e. outside the nonuniform waveguide as shown in Figure 3.3, so $\tilde{z} > L$.

Let us introduce the new variable $\rho(z) = P_{-m}/P_m \exp(-2i\gamma_m)$. As can be easily proved, it satisfies the Riccati equation

$$\rho' = -S_{-mm} \exp(-2i\gamma_m)\left(1 - \rho^2 \exp(4i\gamma_m)\right) \qquad (3.30)$$

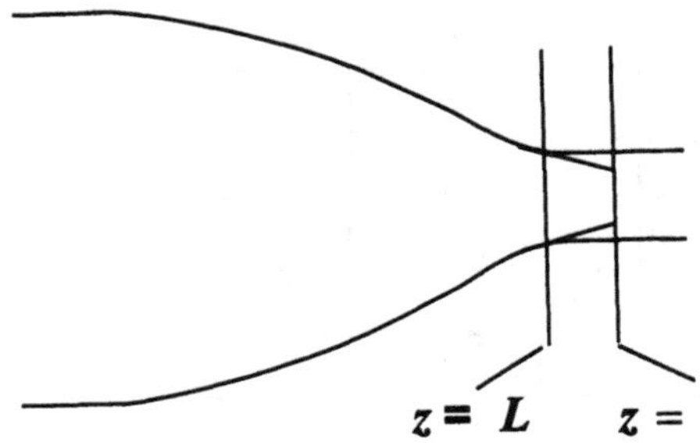

Figure 3.3 Schematic of a cut-off cross section in a narrow uniform waveguide

As we will prove later, if $|t_L|$ has a large value the last term in eqn. 3.30 can be neglected for all $|\rho| << 1$ and the reflection coefficient $\rho(0)$ is equal to

$$\rho(0) = -\int_0^L S_{-mm} \exp(-2i\gamma_m)dz \tag{3.31}$$

Of course, this solution is identical to eqn. 2.75 for $j = -m$. The later can be applied if a geometric optical approximation is valid along the whole taper.

Eqn. 3.31 gives for $\rho(0)$ a small value of the order of ν_0. However, only $\rho'(z)=0$ follows from eqn. 3.30 in the zero-order approximation. Hence, a constant term (generally of the zero order) can be added to $\rho(z)$. This term does not take part in eqn. 3.31. However, as we are going to prove now, this term is important for low and finite values of $|t_L|$. The value of this constant term must be determined in the same way as the equivalent boundary condition for eqn. 3.21 was determined in the previous subsection 3.1.4. One should write an explicit expression of $\rho(z)$ through the functions $u(t)$ and $v(t)$, similar to eqn. 3.22 for $\delta(z)$, and take a limit of this expression when t approaches large negative values. It gives the value of $\rho(z)$ in the common region of $|h'_m/h_m^2| << 1$ and eqn. 3.11. As can easily be proven, this value does not depend on z. It has the zero order of ν_0 and, according to the discussion above, it is valid in the whole of the region where $|h'_m/h_m^2| << 1$, including $z = 0$. Employing an obvious relation between $\rho(z)$ and $\delta(z)$ as well as eqns. 3.23 and 3.24, we yield for this constant term denoted by ρ_0 the following value

$$\rho_0 = \frac{\frac{N}{M} - i}{\frac{N}{M} + i} \exp(-2i(\gamma_m + \alpha_m)) \tag{3.32}$$

The ratio N/M has been given in eqn. 3.17 but, in contrast to the previous subsection, t_L has a negative value in this ratio.

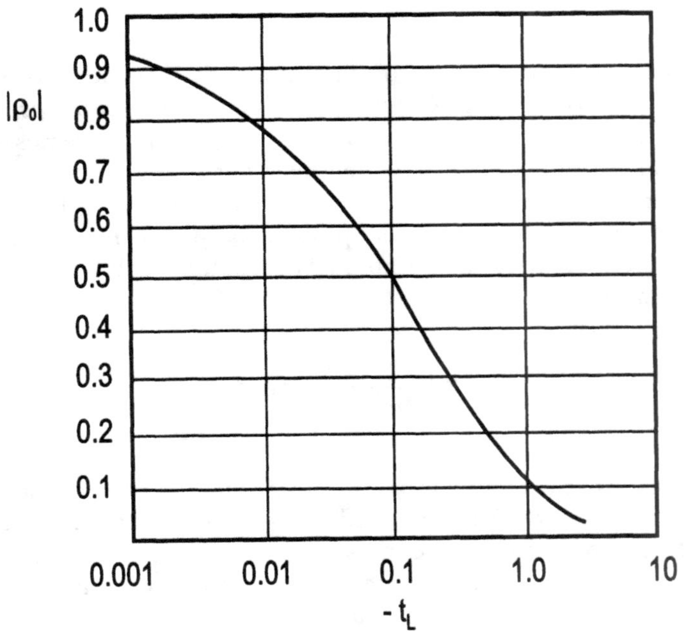

Figure 3.4 Reflection coefficient of TE_{01} mode in circular waveguide

According to eqn. 3.17, the dependence of ρ_0 on t_L differs for various waveguides. The function $|\rho_0|$ versus $(-t_L)$ is plotted in Figure 3.4 for the $TE_{0,1}$ mode in circular waveguides. Again, we have confirmed this analytical result numerically by using the mode matching and generalised scattering matrix method.

For TE_{0n} modes eqns. 3.17 are slightly simpler. Nevertheless, a general dependence of $|\rho_0|$ versus the frequency difference participating in eqn. 3.18 is the same for all modes. The reflection coefficient for a fixed frequency difference is smaller for a smoother shape of a taper near its narrow end. The region, where $|\rho_0|$ decreases from large to small values, is narrowed for decreasing $\nu(L)$. For example, if in a circular waveguide taper the angle between its generating line and the z-axis at its end is equal to 5°40' (i.e. $\nu(L) = -0.1$), $|\rho_0|$ of the TE_{01} mode varies from 1 at $t = 0$ to 0.5 at $t = -0.086$ for $[k - \alpha_m(L)]/k$ varying from 0 to 0.006, i.e. the operating frequency of the narrow waveguide increases by 0.6% from its cut-off frequency. If the angle is two times smaller, i.e. $\nu(L) = -0.05$, the same value of $|\rho_0| = 0.5$ is achieved for an operating frequency increment of 0.4% only, and the frequency increment by 0.6% causes in this case a decrease of $|\rho_0|$ to 0.4.

If $|t_L|$ increases, then for $(-t_L) >> 1$ the value of ρ_0 decreases and becomes small

$$\rho_0 = \frac{1}{8(-t_L)^{3/2}} \exp(-2i\gamma_m(L)) \tag{3.33}$$

This expression is already not of the zero order but of the first order of ν; however, it does not approach exactly to eqn. 3.31. To match these two expressions, one should find a second term in the expansion of ρ_0 as a Taylor series of ν. We do not present here those derivations, their main purpose is the determination of the next terms in the expansions of the solution of eqn. 3.22 and eqn. 3.30. We write down the result for a particular problem only, namely a circular conical taper and the TE_{01} mode. A more rigorous value of the reflection coefficient for this problem is equal to

$$\rho(0) = \rho_0 - \frac{i\nu}{4\mu_m}\left\{\frac{1+\rho_0^2}{g^3} - \rho_0\left(\frac{1}{3g^3} - \frac{1}{g} - \arctan(g)\right)\right\} \tag{3.34}$$

This expression already approaches eqn. 3.31, if that equation is written as a two-term formula like eqn. 2.78. Then it is sufficient to take into account only the first term in the last parentheses in eqn. 3.34.

Usually the first term of eqn. 3.34, i.e. the term ρ_0, is significantly larger than the term with $[-i\nu/(4\mu g^3)]$ despite the fact that both of these terms are of the same first order of ν at $|t_L| >> 1$. Hence, the simple formula $\rho(0) = \rho_0$ gives, practically, good results not only for finite values of $|\rho_0|$ but for its small values as well. According to eqn. 3.34, when $|\rho_0|$ has a small value it is proportional to $\nu(L)$.

Let us note in the conclusion of this Section that, as can be easily shown, eqn. 3.8*a* has a rigorous solution for a circular conical taper in the whole region $0 \le z \le L$, if the condition of eqn. 3.28 is valid

$$\alpha_m^{1/2} Z_p(Ck/\alpha_m) \tag{3.35}$$

where Z_p is an arbitrary cylinder function and $p^2 = C^2 + (1/4)$. However, since the order and the argument of the function Z_p are much larger than one, it is worth employing asymptotic expressions of Z_p for actual calculations. As is well known, Z_p can be expressed through Airy functions of the argument of eqn.3.36

$$2^{1/3}(p - Ck/\alpha_m)(Ck/\alpha_m)^{-1/3} \tag{3.36}$$

in the regions, where this argument has finite values. It is not difficult to show that the value of eqn. 3.36 is equal to the variable t of eqn. 3.13 everywhere, where eqn. 3.36 has finite values. Hence, also for a cone, where the rigorous solution of eqn. 3.8*a* does exist, it is worth writing down the solution in the region of finite t immediately in the form of Airy functions, and in the region of large $|t|$ it can be represented as a linear combination of the functions $h_m^{-1/2}\exp(\pm i\gamma_m)$. We apply the same technique in the next Section.

3.2 Parasitic mode excitation in presence of a cut-off cross section of the propagating mode

3.2.1 General considerations

Now there is a cut-off cross section of an excited parasitic mode at $z = \tilde{z}$, i.e. $h_j(\tilde{z}) = 0$, but the incident mode can propagate in the narrow waveguide. Let us determine the amplitude of the parasitic mode propagating in the wide waveguide.

The solutions of the following equation in the whole nonuniform waveguide will be required for this problem

$$Q_j'' + h_j^2 Q_j = 0 \tag{3.37}$$

Let $V(z)$ and $U(z)$ be two solutions of eqn. 3.37 normalised in such a way that their Wronskian is equal to one

$$U'V - V'U = 1 \tag{3.38}$$

We define them to be proportional to the Airy functions $u(t)$ and $\upsilon(t)$ in the neighbourhood of the cut-off cross section, where the expansion of eqn. 3.11 is valid. As follows from eqn. 3.38, the proportionality factor must be equal to $A^{-1/3}\,k^{-1/2}$. Hence we have for small and finite values of t

$$U(z) = \pi^{1/2}\,A^{-1/3}\,k^{-1/2}u(t) \tag{3.39a}$$

$$V(z) = \pi^{1/2}\,A^{-1/3}\,k^{-1/2}\upsilon(t) \tag{3.39b}$$

here, A and t are defined by eqns. 3.12 and 3.13 with the substitution of the subscript j instead of the subscript m.

In the region of large values of $|t|$ the functions $U(z)$ and $V(z)$ need to be defined in such a way that they analytically approach eqn. 3.39 for finite $|t|$ values. Substituting the asymptotic expansions of the Airy functions in eqn. 3.39 and expressing the variable t through h_j and $\gamma_j - \tilde{\gamma}_j$, we obtain the fact that the functions $U(z)$ and $V(z)$ have to be defined at $z < \tilde{z}$, i.e. for the real h_j variable, by the following conditions

$$U(z) = h_j^{-1/2}\cos\left(-\gamma_j + \tilde{\gamma}_j + \pi/4\right) \tag{3.40a}$$

$$V(z) = h_j^{-1/2}\sin\left(-\gamma_j + \tilde{\gamma}_j + \pi/4\right) \tag{3.40b}$$

Let us assume for $z > \tilde{z}$, i.e. when $h_j = -i|h_j|$

$$U(z) = (ih_j)^{-1/2} \exp(i(\gamma_j - \tilde{\gamma}_j)) \tag{3.41a}$$

$$V(z) = (ih_j)^{-1/2} \exp(-i(\gamma_j - \tilde{\gamma}_j)) \tag{3.41b}$$

The function $U(z)$ increases rapidly when one moves in the area behind the cut-off, i.e. when the difference $z - \tilde{z}$ increases, and the function $V(z)$ decreases rapidly in this area.

According to eqns. 3.40 and 3.41, far from the cut-off cross section these functions are linear combinations of the solutions $h_j^{-1/2} \exp(\pm i\gamma_j)$ and in the neighbourhood of the cut-off they approach the finite solutions of eqns. 3.39. The regions where U and V can be represented in the form of eqn. 3.39 are overlapping with the regions where eqns. 3.40 and 3.41 are valid.

We will assume immediately in this section that the cut-off cross section is situated so far from the beginning of the narrow waveguide that the parasitic mode field is practically equal to zero at the beginning of the narrow waveguide. As will be shown further, the final results do not contain the function $U(z)$ and the field is fully determined by the function $V(z)$.

3.2.2 *Phase of reflection coefficient at the cut-off cross section*

Repeating the discussion of the problem from the previous Section 3.1, we will again determine the phase of the mode reflected from the cut-off cross section. For a *TE* mode, the field is described by eqn. 3.18*a*. The field must not increase for $z > \tilde{z}$. This requirement substitutes the boundary condition at $z=L$, since we have assumed that the beginning of the narrow waveguide is situated far away from the cut-off cross section. Then the function CV is the solution of eqn. 3.8*a*, and the constant C can be determined from the condition $P_m(0) = 1$. In the neighbourhood of the beginning of the wide waveguide, i.e. near $z = 0$ the function $V(z)$ has the form of eqn. 3.40*b*, hence

$$Q_m = \frac{C}{2i}\left\{\exp\left(i\left(-\gamma_m + \tilde{\gamma}_m + \pi/4\right)\right) - \exp\left(-i\left(-\gamma_m + \tilde{\gamma}_m + \pi/4\right)\right)\right\} \tag{3.42}$$

This expression must be substituted in eqn. 3.9 and the term $(S_{mm}-S_{-mm})$ can be neglected there, since it is essential only in the neighbourhood of a cut-off cross section, i.e. for small $|h_m|$. In such a way one can show that the first term in eqn. 3.42 is equal to $P_m(z)$ and the second term is $P_{-m}(z)$, hence eqn. 3.42 corresponds to the representation of Q_m as the subtraction $Q_m = P_m(z) - P_{-m}(z)$. Now the coefficient C and the desired reflection coefficient can be easily derived

$$P_{-m}(0) = \exp\left(-2i\left(\tilde{\gamma}_m + \pi/4\right)\right) \tag{3.43}$$

This formula is identical to the result of subsection 3.1.4: $\delta(0) = -2\tilde{\gamma}_m - \frac{\pi}{2}$ for $t_L >> 1$. In such a way this result is generalised for the case of a large distance between the cut-off cross section and the beginning of the narrow waveguide.

The field of *TM* modes is described by an equation analogous to eqn. 3.7*b*

$$R_m'' + h_m^2 R_m = 0 \tag{3.44}$$

Its solution is, again, given by the function *CV*, and $R_m(z)$ has the same form as eqn. 3.42 far away from the cut-off cross section. However, R_m is equal to the sum and not to the difference of the eqn. 2.2 amplitudes, therefore the second term in eqn. 3.42 corresponds to $(-P_{-m}(z))$ and

$$P_{-m}(0) = \exp\left(-2i\left(\tilde{\gamma}_m - \pi/4\right)\right) \tag{3.45}$$

The phase of the *TM* mode differs from the phase of the *TE* mode by the sign of the constant term $\pi/2$ only, which is added to the geometrical phase shift $(-2\tilde{\gamma}_m)$.

Let us compare these results with the reflection from a metallic short circuit of the waveguide instead of the cut-off cross section. On such a short we would have $P_{-m} = P_m$. Taking this into account, we can hypothetically interpret eqns. 3.43 and 3.45 in the following way: the phase of the mode reflected from the cut-off cross section is equal to the phase of the mode reflected from a metallic mirror in the same waveguide, if this mirror is shifted by one eighth of the wavelength from the cut-off cross section. This shift must be in the incident mode direction for *TM* modes and in the opposite direction for *TE* modes. This interpretation is a hypothetical one since, comparing the $\pm\pi/4$ phase shifts in eqns. 3.43 and 3.45 with the geometrical shift of the mirror, one assumes that the phase velocity has a finite value and small variations. This is not the case in the neighbourhood of a cut-of cross section.

3.2.3 Symmetrical form of equivalent boundary conditions

Now we return to the main problem of this section and study the parasitic mode field. For the sake of definiteness, we assume that the considered parasitic mode with the number *j* is a *TE* mode, so its field is described by eqn. 3.7*a*. As in Section 3.1, we apply this equation to derive the boundary condition for the first-order coupled-wave differential eqns. 2.46 or eqns. 2.71. This boundary condition is equivalent to the presence of a cut-off cross section. Further, applying this boundary condition one could use the first-order differential eqns. 2.46 almost everywhere in the nonuniform waveguide. These equations are essentially simpler than the second-order eqns. 3.7 and, more important, selecting there the terms of the first order of ν_0, one does not face any difficulties near tilts of the waveguide contour line and, generally, in any areas of rapid $\nu(z)$ variations versus z.

According to eqn. 3.38, the solution of the inhomogeneous eqn. 3.7*a* can be written in the form

$$Q_j = U\int_L^z VG_j dz - V\int_C^z UG_j dz \tag{3.46}$$

The lower limit of the first integral was derived from the requirement that, unlike $U(z)$, $Q_j(z)$ does not increase behind the cut-off cross section, i.e. for $z \to L$. We retain an indefinite lower limit of the second integral.

According to eqn. 3.1*a*, far from the cut-off cross section Q_j and R_j are related in the highest order terms of ν_0 by the following equation

$$Q_j' + ih_j R_j = Q_m \left(S_{jm} - S_{-jm}\right) \tag{3.47}$$

One can exclude R_j from this equation and eqn. 2.2, so $P_j(z)$ and $P_{-j}(z)$ are expressed in such a way through $Q_j(z)$ and $Q_j'(z)$. We yield similar to eqn. 3.9

$$\begin{aligned} P_j(z) &= \frac{i}{2h_j}\left[Q_j' - ih_j Q_j - \left(S_{jm} - S_{-jm}\right)Q_m\right] \\ P_{-j}(z) &= \frac{i}{2h_j}\left[Q_j' + ih_j Q_j - \left(S_{jm} - S_{-jm}\right)Q_m\right] \end{aligned} \tag{3.48}$$

To obtain the desired equivalent boundary condition, let us substitute eqn. 3.47 into eqn. 3.48 and exclude the second integral in eqn. 3.46, i.e. let us derive an algebraic relationship between $P_j(z)$ and $P_{-j}(z)$ containing no arbitrary constant, C. The derived relationship between $P_j(z)$ and $P_{-j}(z)$ is valid everywhere; we apply it at points far from the cut-off cross section. The function $V(z)$ for these points is given by eqn. 3.40*b*. The expressions $ih_j V \pm V'$ appearing in this manipulation are equal to

$$ih_j V \pm V' = \mp h_j^{1/2} \exp\left(\pm i\left(\gamma_j - \tilde{\gamma}_j - \pi/4\right)\right) \tag{3.49}$$

In such a way one derives the following relationship between $P_j(z)$ and $P_{-j}(z)$, existing far away from the cut-off cross section

$$\begin{aligned} &h_j^{1/2} P_j(z)\exp\left(-i\left(\gamma_j - \tilde{\gamma}_j + \pi/4\right)\right) - h_j^{1/2} P_{-j}(z)\exp\left(i\left(\gamma_j - \tilde{\gamma}_j + \pi/4\right)\right) \\ &= \int_L^z VG_j dz - VQ_m\left(S_{jm} - S_{-jm}\right) \end{aligned} \tag{3.50}$$

The functions $Q_m(z)$ and $R_m(z)$ contained in eqn. 3.50 describe the fundamental mode field. It is enough to derive them in the highest order. The fundamental mode propagates in this order without distortions, $P_{-m}(z) \equiv 0$ and

$$R_m(z) = Q_m(z) = \sqrt{\frac{h_m(0)}{h_m(z)}} \exp(-i\gamma_m) \tag{3.51}$$

Let us rewrite the boundary condition of eqn. 3.50 in a more convenient, symmetrical form. We will employ the identity valid for any function $f(z)$ having no singular points on the interval from a given z till L

$$fV\exp(-i\gamma_m) = \int_L^z f(V' - ih_m V)\exp(-i\gamma_m)\,dz + \int_L^z f'V\exp(-i\gamma_m)\,dz \tag{3.52}$$

Here we have additionally assumed that $V(L)=0$. This identity can be easily proven by integration by parts. Applying eqn. 3.52 to the second term in eqn. 3.50 and neglecting the last term in eqn. 3.52, since it is smaller by its order of magnitude, the right-hand side of eqn. 3.50 can be rewritten as a single integral expression.

So one yields the desired symmetrical form of the equivalent boundary condition which is more convenient both for a physical analysis and for specific calculations

$$P_j(z)\exp\left(-i\left(\gamma_j - \tilde{\gamma}_j - \pi/4\right)\right) - P_{-j}(z)\exp\left(i\left(\gamma_j - \tilde{\gamma}_j - \pi/4\right)\right)$$

$$= \sqrt{\frac{h_m(0)}{h_j(z)}} \int_z^L \frac{\exp(-i\gamma_m)}{\sqrt{h_m}} \left\{S_{jm}\left(ih_j V + V'\right) + S_{-jm}\left(ih_j V - V'\right)\right\} dz \tag{3.53}$$

The integrand contains such combinations of the coupling coefficients which, as was shown in Section 3.1 (see text after eqn. 3.2*c*), have no singular points in the whole area of integration and particularly at the cut-off cross section. In this formula z can have an arbitrary value in the region where eqn. 3.40*b* is valid.

An equivalent boundary condition can be derived by that method for any other possible cases. If mode j is a *TM* mode, eqn. 3.7*b* must be employed near the cut-off cross section, and if mode m is incident from the narrow waveguide, the solution of eqn. 3.51 is modified in an obvious way, etc.

3.2.4 Parasitic mode amplitudes

The field of a parasitic mode can be derived from the first-order coupled-wave eqns. 2.46 and from a boundary condition in the form of eqn. 3.53 equivalent to the existing cut-off cross section. The second boundary condition has the form $P_j(0)=0$, it provides the absence of this kind of incident mode.

By the condition $P_j(0)=0$, the solution of these coupled-wave equations for an arbitrary z outside the cut-off area (and for $z < \tilde{z}$) is similar to eqn. 2.75

$$P_j(z) = \sqrt{\frac{h_m(0)}{h_j(z)}}\exp(-\gamma_j)\int_0^z S_{jm}\sqrt{\frac{h_j}{h_m}}\exp\left(-i(\gamma_m-\gamma_j)\right)dz$$
$$P_{-j}(z) = \sqrt{\frac{h_m(0)}{h_j(z)}}\exp(-\gamma_j)\int_0^z S_{-jm}\sqrt{\frac{h_j}{h_m}}\exp\left(-i(\gamma_m+\gamma_j)\right)dz + \sqrt{\frac{h_j(0)}{h_j(z)}}\exp\left(i\gamma_j\right)P_{-j}(0) \tag{3.54}$$

Substituting this solution into the boundary condition of eqn. 3.53, we derive $P_{-j}(0)$, the desired amplitude of a parasitic mode propagating into the wide waveguide

$$P_{-j}(0) = \sqrt{\frac{h_m(0)}{h_j(0)}}\left\{\exp\left(-2i\left(\tilde{\gamma}_j+\pi/4\right)\right)\int_0^z S_{jm}\sqrt{\frac{h_j}{h_m}}\exp\left(-i\left(\gamma_m-\gamma_j\right)\right)dz\right.$$
$$-\int_0^z S_{-jm}\sqrt{\frac{h_j}{h_m}}\exp\left(-i\left(\gamma_m+\gamma_j\right)\right)dz - \exp\left(-i\left(\tilde{\gamma}_j+\pi/4\right)\right) \tag{3.55}$$
$$\left.\times\int_z^L \frac{\exp\left(-i\gamma_m\right)}{\sqrt{h_m}}\left[S_{jm}\left(ih_jV+V'\right)+S_{-jm}\left(ih_jV-V'\right)\right]dz\right\}$$

In eqn. 3.55, z can also have any value but the point z cannot be near the cut-off cross section. Indeed, the factors S_{jm} and S_{-jm} in the last integral are identical to the corresponding factors in the first two integrals, hence the sum of all three integrals is an invariant in relation to z.

There is some conditional stipulation by representing $P_{-j}(0)$ in eqn. 3.55 as the sum of three terms. That stipulation is connected with an arbitrary choice of the point z. However, that representation has a quite definite physical meaning suggested by the results of Sections 1.2 and 1.3. The forward and backward propagating waves with index j are excited by the propagating mode m along each nonuniform waveguide section from $z = 0$ to the cross section z participating in eqn. 3.55. The superposition of all the parasitic modes propagating backward occurs at the cross section $z = 0$ and, according to eqns. 1.29 or 2.75, the result of that superposition is identical to the second term in eqn. 3.55. The parasitic modes propagating forward are incident on the cut-off cross section and reflected from it. They acquire the same additional phase factor, eqn. 3.43, as if this wave was incident from the outside of the nonuniform waveguide. In such a way, the first term appears in eqn. 3.55. Thereafter, the last term in eqn. 3.55 results from the parasitic wave field excited near the cut-off cross section. In this area, there is no sense in separating forward and backward propagating waves (compare with a similar comment in Brekhovskikh (1980) and subsection 4.3.7). The parasitic mode is excited there in a more complicated way as described in Chapter 2.

The amplitudes of parasitic waves propagating forwards are usually larger than the backward propagating wave amplitudes (see Section 2.2). Hence, if z in eqn. 3.55 is chosen to be not very distant from the cut-off cross section, the first term is the largest term and the third term must be taken into account in exceptional circumstances only. The proposed interpretation of eqn. 3.55 allows us to determine for other possible problems the two main terms in the field of the parasitic mode propagating into the wide waveguide without solving eqns. 3.7 and without deriving an equivalent boundary condition. For example, if the exciting wave is incident from the narrow waveguide situated at the left-hand side of the cut-off cross section, then the backward (and not forward) propagating waves have an additional phase factor, eqns. 3.43 or 3.45, corresponding to the parasitic wave reflection from the cut-off cross section.

3.2.5 Parasitic mode amplitudes in linear tapers to a cut-off cross section

Let us apply the derived results to waveguides where $\nu(s,z)$ is of the same order of magnitude for all z and is stepwise equal to zero at $z = -0$, i.e. to waveguides with a tilt of their contour line. To reduce the volume of manipulation we assume, as in Section 2.2, that those tilts are situated only at the ends of a nonuniform waveguide.

We have derived in subsection 2.2.8 eqn. 2.78 for the amplitudes of the modes scattered from these nonuniform waveguides. This formula is valid in the absence of a cut-off cross section.

As follows from the previous consideration, one should yield an expression similar to eqn. 2.78 for the parasitic mode field. Obviously, there will be no term corresponding to the narrow waveguide, but there will be terms corresponding both to the forward propagating waves reflected from the cut-off cross section and to the backward propagating waves.

For further manipulations, it is convenient to rewrite $P_{-j}(0)$ in the form of a single integral. Such an expression is derived by assuming $z = 0$ either in the solution of eqn. 3.55 or in the boundary condition of eqn. 3.53

$$P_{-j}(0) = -\sqrt{\frac{h_m(0)}{h_j(0)}}\exp\left(-i\left(\tilde{\gamma}_j + \pi/4\right)\right)$$
$$\times\int_0^L \frac{\exp(-i\gamma_m)}{\sqrt{h_m}}\left[S_{jm}\left(ih_jV + V'\right) + S_{-jm}\left(ih_jV - V'\right)\right]dz \qquad (3.56)$$

According to eqn. 3.49, the products $\left(ih_jV \pm V'\right)\exp\left(\mp i\gamma_j\right)$ do not contain any rapidly oscillating factors in the main part of the integration area in eqn. 3.56. Taking this into account, let us separate in the integral of eqn. 3.56 two terms containing S_{jm} and S_{-jm}, respectively. Further, we emphasise these products in both terms and integrate by parts. For example, the first term then has the following form

$$\int_0^L \frac{\exp(-i\gamma_m)}{\sqrt{h_m}} S_{jm}\left(ih_jV + V'\right) dz$$

$$= -i\left\{\frac{\left(ih_jV + V'\right)S_{jm}}{\sqrt{h_m}\left(h_m - h_j\right)}\right\}\Bigg|_{z=0} - i\int_0^L \exp\left(-i\left(\gamma_m - \gamma_j\right)\right)\left\{\frac{\exp(-i\gamma_j)\left(ih_jV + V'\right)S_{jm}}{\sqrt{h_m}\left(h_m - h_j\right)}\right\}' dz \quad (3.57)$$

Here we have assumed in addition $V(L) = 0$ and $V'(L) = 0$. Substituting $(ih_jV \pm V')$ at $z = 0$ by eqn. 3.49, we derive the formula for the desired amplitude of the backward propagating wave from eqn. 3.57 and from the second similar expression containing S_{-jm}:

$$P_{-j}(0) = -i\left\{\frac{S_{jm}}{\left(h_m - h_j\right)}\right\}_{z=0} \exp\left(-2i\left(\tilde{\gamma}_j + \pi/4\right)\right) + i\left\{\frac{S_{-jm}}{\left(h_m + h_j\right)}\right\}_{z=0} + I \quad (3.58)$$

Further, we will write down an expression for the integral term I derived from the second terms of the formulae similar to eqn. 3.57.

The first two terms in eqn. 3.58 have a simple physical meaning. They must be compared with eqn 2.78 which is valid in the absence of a cut-off cross section. The first term corresponds to the forward propagating waves reflected from the cut-off cross section and the second term to the backward propagating waves. The derived formula illustrates the interpretation of parasitic field excitation given in the previous subsection. This interpretation allows us to derive formulae similar to eqn. 3.58 employing eqn. 2.75 and the expressions of the reflection coefficients from the cut-off cross section.

From this point of view the integral I in eqn. 3.58 corresponds to the influence of the neighbourhood of the intermediate cut-off cross section. Numerous but elementary manipulations omitted here lead us to the following expression

$$I = i\sqrt{\frac{h_m(0)}{h_j(0)}} \exp\left(-i\left(\tilde{\gamma}_m + \pi/4\right)\right)$$

$$\times\int_0^L \exp(-i\gamma_m)\left\{\left[i\frac{h_m h_j\left(S_{jm} + S_{-jm}\right) + h_j^2\left(S_{jm} - S_{-jm}\right)}{\sqrt{h_m}\left(h_m^2 - h_j^2\right)}\right]' V + \left[\frac{h_j\left(S_{jm} + S_{-jm}\right) + h_m\left(S_{jm} - S_{-jm}\right)}{\sqrt{h_m}\left(h_m^2 - h_j^2\right)}\right]' V'\right\} ds \quad (3.59)$$

The equation $V'' + h_j^2 V = 0$ was employed in these manipulations. Besides this, to avoid the convergence problem of the integrals appearing in these manipulations, it is convenient to assume the free-space wavenumber k having a complex value. Obviously, k must have a real value in the resulting expression.

Analysing eqn. 3.59, it is important to take into account the fact that the coupling coefficients participate in the integrand only as combinations of eqn. 3.2. These combinations, apart from having finite values at the cut-off cross section, do not contain the first power of h_j. Therefore, the integrand functions in eqn. 3.59 have finite values everywhere including the cut-off cross section. The coefficients of V and V' in eqn. 3.59 are of the order either of v_0^2 or of v'.

Since, as follows from eqn. 3.39*b*, V and V' increase slightly near the cut-off cross-section, the integral I is possibly of somewhat lower order. However, because the first two terms in eqn. 3.58 are of the order of v_0, one can neglect the third term in eqn. 3.58 for all problems where very accurate computations are not needed.

So we have proved that the amplitude of a parasitic wave for waveguides with a contour line tilt depends only on values corresponding to the point of the tilt. This result obtained in subsection 2.2.8 for a waveguide without a cut-off cross section is also valid in the general case. As was mentioned above, this result does not contradict the nonlocal excitation of parasitic waves excited along the whole nonuniform waveguide.

The analytical expressions of the coupling coefficients in rectangular and circular waveguides are represented in Section 4.2.

3.3 Waveguide tilts operating near the cut-off frequency of the generated parasitic mode

3.3.1 General considerations — resonance phenomena

A second case which needs a special analysis is the resonance effect which appears in a bent waveguide with constant cross section when the operating frequency approaches the cut-off frequency of one of the parasitic modes coupled by the bend.

From the mathematical point of view this case presents a special situation due to the fact that the coupling coefficients F_{jm} tend to become infinite when $k \rightarrow \alpha_j$, $h_j \rightarrow 0$. This is the reason why the solutions of eqns. 1.6, 1.14 and 2.35 cannot be used.

First of all, we are interested in the physical meaning of that phenomenon; therefore, we will limit ourselves to the study of the effect caused by a tilt with a small angle ϑ_0, that means $\vartheta_0 \ll 1$. The resonance phenomena in this case are simpler than for a bend with a finite angle, moreover in some sense they are more intense. Furthermore, for this problem, the resonance phenomena may be studied completely in an analytical form.

The resonance phenomena caused by any small deformation and particularly that of a small angle tilt are similar in some sense to the forced waveguide excitation by a current at the cut-off frequency of the excited wave. It is known that for a given current the amplitude of the excited wave reaches very high values even for small excitation currents. Therefore, for studying this problem correctly it is necessary to take into account the inverse reaction of the excited wave on the exciting element. For the case of a waveguide tilt, the analogue of the exciting element is the incident wave. For example, in Figure 1.1 the field of the transmitted waves (propagating in the right-hand waveguide) can be considered as the field created by the currents of the incident wave propagating into the left-hand waveguide.

Therefore, in this case, the resonance phenomena will obviously cause first, the field of the incident wave to be strongly modified, more precisely, a high amplitude reflected wave will be generated.

In order to start our analysis let us assume that the waveguide walls are perfect conductors. As we will see later that this assumption does not cause any mathematical infinities and, therefore, that the resonance phenomena in waveguides can be analysed rigorously without considering ohmic losses. Nevertheless, despite the fact that taking losses into account does not cause qualitatively new phenomena, it is absolutely necessary to do so for accurate calculation of the amplitudes of the excited modes.

A similar problem was considered in Jouguet (1947a) where the amplitudes of the scattered TE_{n0} waves are calculated when a TE_{p0} wave is incident at the junction of two semiinfinite rectangular waveguides, one of which is straight and the other of which has a small H-plane curvature (see Figure 4.23, later). The resonance phenomenon was also analysed in the same paper; Jouguet studied only the case when the operating frequency was exactly equal to the cut-off frequency of the parasitic wave and did not take into account the finite conductivity of the waveguide walls.

3.3.2 Reflected and transmitted wave amplitudes

As in the first sections of this chapter, we will use the variables Q_j and R_j which are related to the amplitudes P_j and P_{-j} by eqns. 2.2. These variables satisfy the following coupled-wave equations

$$\begin{aligned}\frac{dQ_j}{d\vartheta} &= \sum_{m=1}^{\infty}\left(F_{jm} - F_{-jm}\right)R_m \\ \frac{dR_j}{d\vartheta} &= \sum_{m=1}^{\infty}\left(F_{jm} + F_{-jm}\right)Q_m\end{aligned} \qquad (3.60)$$

which may be easily derived from the coupled-wave eqns. 2.9.

As was shown in Section 2.1, far away from the resonance conditions the incident wave is almost not distorted and for each mode with number j, the system of coupled-wave eqns. 3.60 can be reduced to two equations for the variables R_j and Q_j. Solving these equations, one can find further $P_j(\vartheta)$ and $P_{-j}(\vartheta)$; the obtained formulae are obviously the same as eqn. 2.28. For the frequency band near the resonant frequency, it is also necessary to take into account the perturbation of the incident mode and to consider simultaneously the four coupled-wave equations

$$
\begin{aligned}
\frac{dQ_j}{d\vartheta} &= \left(F_{jm} - F_{-jm}\right)R_m + \left(F_{jj} - F_{-jj}\right)R_j \\
\frac{dQ_m}{d\vartheta} &= \left(F_{mm} - F_{-mm}\right)R_m + \left(F_{mj} - F_{-mj}\right)R_j \\
\frac{dR_j}{d\vartheta} &= \left(F_{jm} + F_{-jm}\right)Q_m + \left(F_{jj} - F_{-jj}\right)Q_j \\
\frac{dR_m}{d\vartheta} &= \left(F_{mm} + F_{-mm}\right)Q_m + \left(F_{mj} + F_{-mj}\right)Q_j
\end{aligned}
\tag{3.61}
$$

As in eqns. 3.7 the other waves can be neglected.

When $k \to \alpha_j$, $h_j \to 0$, the coupling coefficients F_{jm} tend to become infinite. It is easy to show, for instance using eqn. 2.16, that these coefficients have similar features to the S_{jm} coefficients: if j is a *TE* mode, then the difference $F_{jm} - F_{-jm}$ does not tend to become infinite for any value of m, and, if j is a *TM* mode, then the sum $F_{jm} + F_{-jm}$ has no singularities. Therefore, only one coefficient of the coupled-wave eqns. 3.61 tends to become infinite, when $h_j \to 0$; from this point of view eqns. 3.60 are more appropriate as starting equations than eqn. 2.9. All singularities in the coefficients of these equations, as was to be expected from the considerations given at the beginning of Section 3.1, will disappear if we introduce for *TE* modes the new variable $R_j h_j$ instead of R_j. This new variable, as well as Q_j, will not approach to an infinite value when $h_j \to 0$; for *TM* modes it is more convenient to introduce the variable $Q_j h_j$.

Let us limit ourselves to the study of a bend with a very small angle ϑ_0, i.e. we will assume that $\vartheta_0 << 1$. It should be noted that, for a finite angle bend, the usual solution method of coupled equations with constant coefficients can be used; the solution for each of the variables is the sum of four exponents such as $e^{p\vartheta}$. From the analysis of the characteristic equation for these coupled-wave equations it follows that all its roots p are finite when $h_j \to 0$.

Let us start from the limiting case when the frequency is exactly equal to the cut-off frequency, $h_j = 0$. For definity we will assume that the modes j and m are of *TE* type; all the results are qualitatively the same for any other combination. To reduce the notation let us assume that $K^{jj} = 0$ where the integrals K^{jm} are defined by eqn. 2.18; following Section 2.1, this assumption is usually fulfilled. The coupled-wave equations for the variables Q_m, R_m, Q_j and $h_j R_j$ can be solved by expansion into a power series of the small magnitude ϑ_0. For the highest order terms of these expansions the calculation gives us

$$
\begin{aligned}
R_m &= 2\left(1 - \frac{\vartheta}{\vartheta_0}\right) + O(\vartheta_0), Q_m = O(\vartheta_0) \\
R_j h_j &= O(\vartheta_0^2), Q_j = \frac{2i\hbar_m}{kK^{jm}} \cdot \frac{1}{\vartheta_0} + O(1)
\end{aligned}
\tag{3.62}
$$

According to the last equation, in the curved part of the waveguide the electromagnetic field of the parasitic mode has very high values of the order $1/\vartheta_0$.

According to eqn. 3.62, $P_{-m}(0) = 1$ and $P_m(\vartheta_0) = 0$ in the highest order terms with respect to ϑ_0, so that the incident wave is totally reflected. Thus, a bend with an arbitrary small angle causes total reflection of the incident wave when the operating frequency is exactly equal to the cut-off frequency of the excited parasitic wave. Later it will be shown that, in real situations, this effect is strongly weakened.

Calculating the next terms of the expansions of eqn. 3.62 for ϑ_0, one can obtain more accurate expressions for the reflection and transmission coefficients

$$\begin{aligned} |P_{-m}(0)| &= 1 - \frac{1}{18}|F_{mm}|^2 \vartheta_0^2 + O(\vartheta_0^3) \\ |P_m(\vartheta_0)| &= \frac{1}{3}|F_{mm}|\vartheta_0 + O(\vartheta_0^2) \end{aligned} \tag{3.63}$$

These formulae are valid only when the product $|F_{mm}|\vartheta_0$ is small, i.e. for bends with a short electrical length.

Far from the cut-off frequency the reflection coefficient is of the order of ϑ_0^2 according to the results derived in Section 2.1, and the transmission coefficient differs from 1 by terms of order ϑ_0^2. For studying the transformation of eqn. 3.63 to these conditions, it is necessary to solve the four coupled-wave eqns. 3.61 for R_m, Q_m, R_j and Q_j using the fact that the values of the two parameters ϑ_0 and $\hbar_j$ are simultaneously small. Expanding the variables in series of small order parameters, one has to assume first that $\vartheta_0/\hbar_j \ll 1$, then that $\vartheta_0/\hbar_j \cong 1$ and finally that $\vartheta_0/\hbar_j \gg 1$. For all these three cases we will obtain for Q_m and Q_j identical analytical expressions in the highest order terms.

The amplitudes of the scattered waves of our interest may be expressed only in terms of Q_m and Q_j using eqns. 2.2 and 2.10. Therefore, also for these amplitudes, one can obtain the following expressions valid for any relation between ϑ_0 and $\hbar_j$

$$P_{-m}(0) = \frac{\tau}{1+\tau}, \quad P_m(\vartheta_0) = \frac{1}{1+\tau} \tag{3.64a}$$

$$P_j(\vartheta_0) = -P_{-j}(0) = -i\frac{kK^{jm}\vartheta_0/2\hbar_j}{1+\tau} \tag{3.64b}$$

Here, τ indicates an important parameter in our entire analysis

$$\tau = \frac{k^2 (K^{jm})^2 \vartheta_0^2}{4\hbar_m \hbar_j} \tag{3.65}$$

which is proportional to the ratio $\vartheta_0^2 / \hbar_j$ of two small magnitudes.

The power of the reflected and transmitted wave of the propagating mode, respectively, divided by the power of the incident wave is equal to

$$|P_{-m}(0)|^2 = \frac{|\tau|^2}{|1+\tau|^2}, |P_m(\vartheta_0)|^2 = \frac{1}{|1+\tau|^2} \tag{3.66}$$

If the operating frequency is lower than that of the parasitic cut-off mode, i.e. if $\hbar_j^2 < 0$, the parasitic waves of course do not take away any power, but when $\hbar_j^2 > 0$, both parasitic waves take away equal amounts of power equal to

$$|P_j(\vartheta_0)|^2 \hbar_j / \hbar_m = |P_{-j}(0)|^2 \hbar_j / \hbar_m = \frac{\tau}{(1+\tau)^2} \tag{3.67}$$

It is easy to verify that the entire power taken away by the scattered waves is equal to the power of the incident wave both when the operating frequency is higher than the cut-off value $(\hbar_j^2 > 0, \ |\tau| = \tau)$ and when it is lower $(\hbar_j^2 < 0, \ |\tau| = -i\tau)$.

Eqn. 3.64 allow us to observe how the resonance phenomena appear as the operating frequency approaches the cut-off frequency, i.e. when the magnitude $|\hbar_\varphi|$ decreases and consequently the parameter τ increases. When $|\tau| << 1$, strictly speaking there is not yet any resonance; the main wave amplitude does not change in order to propagate through the bend and the parasitic wave amplitudes of eqn. 3.64*b* are calculated as before by eqn. 1.6 and 1.12, which according to eqn. 2.20 are identical to eqn. 3.64*b* when $\tau << 1$. If $\tau << 1$, approaching the cut-off frequency it appears only in the identity of the amplitudes of the forward and backward parasitic waves; these amplitudes can be of finite magnitude, even high, nevertheless, the energy carried away by the parasitic waves is always very small. In accordance with eqn. 3.64*a*, the influence of these waves on the incident wave is also small, mainly because the coupling coefficient F_{mj} has a finite value and the angle ϑ_0 is small.

The resonance phenomenon appears when $|\tau| \sim 1$. In such a case, the perturbation of the transmitted wave is essential and at the same time the energy taken away by the parasitic waves is finite. Then, the parasitic wave field will be very intense, namely of the order τ / ϑ_0. The maximal value is reached at perfect resonance, when $\hbar_j = 0$ and $|\tau| \to \infty$, that is to say, when following eqn. 3.64*b*, or the last formula in eqn. 3.62 which is in accordance with eqn. 3.64*b*, this field will be inversely proportional to ϑ_0; in this case the incident wave power is totally reflected. The parasitic waves carry the maximum of power away when $\tau = 1$; then 1/4 of the incident power is reflected and 1/4 is transmitted through the bend in the main wave and two times 1/4 of the incident power is scattered to the forward and backward parasitic waves, respectively.

3.3.3 Influence of finite conductivity of the waveguide wall

In real waveguides the powerful resonance condition $|\tau| \geq 1$ is, practically, never satisfied. Due to the fact that the walls of real waveguides have a finite conductivity, $\hbar_j$ is a complex number which never can be equal to 0 for any operating frequency.

The complete expression for the propagation constant in the waveguide with nonideal walls, a deduction of which is given in the next Section to avoid confusion here, shows that the minimum value of $|\hbar_j|$ is always reached at a frequency a little lower than the cut-off frequency; for *TE* modes it is given by

$$\left|\hbar_j\right|_{min} = k\left\{\frac{d}{2}\oint\left(\psi^j\right)^2 ds\right\}^{1/2} \qquad (3.68a)$$

where *d* is the skin layer depth. For *TM* modes

$$\left|\hbar_j\right|_{min} = \left\{\frac{d}{2}\oint\left(\frac{\partial\varphi^i}{\partial n}\right)^2 ds\right\}^{1/2} \qquad (3.68b)$$

In order to reduce the notation in eqns. 3.68, the relative magnetic permeability of the waveguide wall material is assumed to be $\mu=1$. The order of magnitude of $|\hbar_j|_{min}$ is $(d/a)^{1/2}$.

The entire mathematical method developed in subsection 3.3.2 is also valid when the conductivity is taken into account. Nevertheless, for a given angle ϑ_0 the parameter $|\tau|$ cannot be greater than

$$|\tau|_{max} = \frac{k^2\left(K^{jm}\right)^2\vartheta_0^2}{4\hbar_m\left|\hbar_j\right|_{min}} \qquad (3.69)$$

This magnitude is proportional to $\vartheta_0^2(a/d)^{1/2}$. The proportionality factor depends on the waveguide cross section and on the wavenumbers. For instance, if a TE_{01} mode (wave *m*) propagates into a circular waveguide bend and its frequency is equal to the cut-off frequency of the coupled mode TE_{12} (wave *j*), then substituting eqns. I.18 and I.20 into eqn. 2.18 and into the integral in eqn. 3.68*a* we derive

$$K^{jm} = -\frac{\sqrt{2}\mu_j\mu_m}{k\sqrt{\mu_j^2-1}}\frac{1}{\mu_j^2-\mu_m^2} \text{ and } \oint\left(\psi^j\right)^2 ds = \frac{2a}{\mu_j^2-1} \qquad (3.70)$$

Since at the TE_{12} mode cut-off $\hbar_m = 0.7$ and $\mu_j = ka$, we obtain from eqn. 3.69

$$|\tau|_{\max} = \frac{\mu_j}{1.4\sqrt{\mu_j^2 - 1}} \frac{\mu_m^2}{\left(\mu_j^2 - \mu_m^2\right)^2} \sqrt{\frac{a}{d}}\vartheta_0^2 = 0.057\sqrt{\frac{a}{d}}\vartheta_0^2 \tag{3.71}$$

For the TE_{1q} waves ($q > 0$), the numerical magnitude of the coefficient in eqn. 3.71 will be even smaller. Let us suppose that $a/d \approx 2.5\times10^4$, which corresponds to a≈10 mm and copper conductivity at $\lambda \approx 10$ mm (30 GHz), then

$$|\tau|_{\max} \approx 9\vartheta_0^2 \tag{3.72}$$

Because all the calculations in the previous subsection were done under the assumption that $\vartheta_0 << 1$, the parameter $|\tau|_{\max}$ in any case is much smaller than unity. The resonance cannot be very strong but it will be appreciable.

According to eqn. 3.64*a* the reflection coefficient $|P_{-m}(0)|_{\max}$ for small values of $|\tau|_{\max}$ is equal to $|\tau|_{\max}$, i.e. to the value determined by eqn. 3.69 or eqn. 3.72. For instance, even if $\vartheta_0 = 1/5$, it will reach only the value $|P_{-m}(0)|_{\max} = 0.36$. The amplitude of the parasitic waves according to eqn. 3.64*b* is greater than that of the reflected main wave, since it is proportional to the first power of ϑ_0 instead of ϑ_0^2 as is $P_{-m}(0)$. Insertion of eqns. 3.68*a* and 3.70 in 3.64*b* gives the maximum value

$$|P_{-j}(0)| = |P_j(\vartheta_0)| = \frac{1}{\sqrt{2}} \frac{\mu_m}{\mu_j^2 - \mu_m^2} \sqrt{\frac{a}{d}}\vartheta_0 = 0.2\sqrt{\frac{a}{d}}\vartheta_0 \tag{3.73}$$

For the assumed value of $a/d \approx 2.5\times10^4$ we get $|P_j(\vartheta_0)| \cong 32\vartheta_0$, which can be very large compared to 1 even for small angles.

The frequency dependence of the reflection coefficient and of the parasitic waves amplitudes is determined by the term $1/|\hbar_j|$ for small $|\tau|$. The smaller d/a is, the higher is the sharpness of the frequency dependence, nevertheless, as is usual in waveguides, the resonance maxima are relatively wide. In accordance with eqn. 3.86 the value of $|\hbar_j|$ increases by a factor of $\sqrt{2}$ which means that the fields of the parasitic waves and of the reflected wave decrease by the same factor when the frequency changes by the following amount

$$\frac{\Delta k}{k} = \frac{d}{4}k^2 \oint \left(\psi^j\right)^2 ds \tag{3.74}$$

(for the *TE* modes). For the TE_{12} mode we have $\Delta k/k \approx 0.5d/a$ as follows from eqn. 3.70. When $a/d = 2.5\times10^4$ then $\Delta k/k \approx 2\times10^{-5}$ which means a bandwidth of about 1.5 MHz for $\lambda \approx 8$ mm (37.5 GHz).

Finishing this section, let us note that, in practice, unintentional nonuniformities of the cross section at different locations along the waveguides, which are always occuring, also reduce the resonance effect at an operating frequency equal to the cut-off frequency of the constant waveguide cross section. To estimate this effect we can consider $|\hbar_j|^2_{min}$, which is proportional to d/a, according to eqn. 3.68, containing additionally another term equal to the relative root mean square value of the deviations of the linear cross-section dimensions. In other words, the unintentional cross-section variations are equivalent to some increase of the skin depth d from the point of view of their influence on the resonance effects.

3.4 Wavenumber in waveguide with nonperfectly conducting walls

3.4.1 Wavenumber and wall impedance

Let us derive in this section a formula for the wavenumber in a waveguide with the walls of high but finite conductivity as we have used in the previous section. Because the wavenumber $\hbar_j$ is very small in the considered operating frequency band, we cannot apply either the method used in Section 1.3 for a similar problem or the commonly used power-loss method which can be applied only for not very low and real values of $\hbar_j$.

Corresponding results for *TM* modes were first obtained by a not very rigorous method in the book of Kisun'ko (1949). The expression of the wavenumber in a circular waveguide was obtained by Alpert (1940).

In a waveguide with nonperfectly conducting walls, the ideal metal boundary condition of eqn. I.2 is substituted for by the Leontovich boundary condition. According to this, the tangential field components on the metallic surface are related by

$$E_s = wH_z, \qquad E_z = -wH_s \tag{3.75}$$

These relations are approximately valid up to the first order terms of $|w|$. Here, w is the so-called wave impedance of the walls, $w = \sqrt{\mu/\varepsilon}$ where μ and ε are the parameters of the wall material. The electric permittivity ε of a metal is a large, pure imaginary value, whereas the magnetic permeability μ can be considered to be a real value. Introducing the skin layer depth d, yields

$$w = \frac{(1+i)}{2}\mu kd \tag{3.76}$$

The absolute value of w is small. We will determine the wavenumber h by an expansion of the fields in a Taylor series of $|w|$. Let us consider only the case without mode degeneracy, i.e. we assume that each value of α^2 in a waveguide with perfectly conducting walls corresponds to only one *TE* or *TM* mode.

Since a uniform waveguide is considered in this section in which no mode conversion occurs, we can omit the subscript of the wavenumber, of the transversal eigenfunctions etc . The values corresponding to a waveguide with perfectly conducting walls we will indicate by the subscript 0.

Writing down the field components taking part in the boundary condition of eqn. 3.75 with the help of the transversal eigenfunctions ψ and φ according to eqn. I.14, we can obtain the coupling boundary conditions for the values of these functions on the cross-section boundary:

$$\begin{aligned} -ih\frac{\partial\varphi}{\partial s} + ik\frac{\partial\psi}{\partial n} &= w\alpha^2\psi \\ \alpha^2\varphi &= w(ik\frac{\partial\varphi}{\partial n} + ih\frac{\partial\psi}{\partial s}) \end{aligned} \tag{3.77}$$

In particular, from these conditions follows the fact that pure *TM* modes ($\psi\equiv0$) or pure *TE* modes ($\varphi\equiv0$) can propagate only in waveguides with perfectly conducting walls, i.e. if $w = 0$ or if the fields are circularly symmetric, i.e. $\partial/\partial s \equiv 0$. In the general case, the solution contains both functions ψ and φ, which means propagation of hybrid modes.

The eigenfunctions ψ and φ satisfy the same eqns. I.11. One should determine these functions and the eigenvalue α^2 from these equations and from the boundary conditions of eqn. 3.77. The wavenumber h is related to α by the condition of eqn. I.13. The eigenvalue can be written as an expansion in a Taylor series of w

$$\alpha^2 = \alpha_0^2 + w\alpha_1^2 + \ldots \tag{3.78}$$

The square of the wavenumber can be expanded by a similar series

$$h^2 = h_0^2 - w\alpha_1^2 + \ldots,\quad h_0^2 = k^2 - \alpha_0^2 \tag{3.79}$$

One has to determine the second term of this expansion, i.e. the coefficient α_1^2. However, the main term of this expansion h_0^2 is equal to zero at the cut-off frequency, i.e. if $k = \alpha_0$. Near the cut-off frequency the main term is comparable with the second term, and therefore, following from eqn. 3.79, an obvious expansion of the wavenumber

$$h = \sqrt{h_0^2 - w\alpha_1^2 + \ldots} = h_0 - w\frac{\alpha_1^2}{2h_0^2} + \ldots \tag{3.80}$$

is not applicable for all frequencies including the neighbourhood of the cut-off frequency. As follows from eqn. 3.77, both functions ψ and φ also cannot be expanded simultaneously in Taylor series of w applicable everywhere; the field components cannot be expanded either. Therefore, one cannot directly expand the variables from eqn. 3.77 in Taylor series of w.

3.4.2 Wavenumber near the cut-off frequency

Let us use an artificial procedure that allows us to apply the method of successive approximation to the system of eqns. I.11 and the boundary conditions of eqn. 3.77 and so to derive solutions which are also valid in the neighbourhood of the cut-off frequency. We introduce auxiliary functions in such a way that only the square of the wavenumber will participate in the boundary conditions. For the sake of simplicity, it is convenient to introduce different auxiliary functions for *TE* and *TM* modes.

Let us start with the *TE* modes. We multiply the second equation in eqns. 3.77 by h and introduce an auxiliary function $\Phi = h\varphi$. Let us expand the functions ψ and Φ into series, taking into account the fact that there are no *TM* modes with the index zero since there is no degeneracy, i.e. $\Phi_0=0$. The expansion of Φ begins with a term of the order w:

$$\psi = \psi_0 + w\psi_1 + \ldots, \quad \Phi = w\Phi_1 + \ldots \tag{3.81}$$

Now let us substitute eqns. 3.78, 3.79 and 3.81 into eqns. I.11 and into the boundary conditions of eqn. 3.77. Separating the different orders, we obtain for the zero power of w

$$\nabla^2\psi_0 + \alpha_0^2\psi_0 = 0 \tag{3.82a}$$

$$\left.\frac{\partial\psi_0}{\partial n}\right|_C = 0 \tag{3.82b}$$

and for the first power of w:

$$\nabla^2\psi_1 + k^2\psi_1 = -\alpha_1^2\psi_0 \tag{3.83a}$$

$$\begin{aligned} \left.\frac{\partial\psi_1}{\partial n}\right|_C &= -\frac{i\alpha_0^2}{k}\psi_0 + \left.\frac{1}{k}\frac{\partial\Phi_1}{\partial s}\right|_C \\ \phi_1|_C &= \left.\frac{ih_0^2}{\alpha_0^2}\frac{\partial\psi_0}{\partial s}\right|_C \end{aligned} \tag{3.83b}$$

Excluding from the equations above the Φ_1 function's boundary values, we derive a boundary condition for ψ_1 containing only the solution of equations of zero order

$$\left.\frac{\partial\psi_1}{\partial n}\right|_C = \frac{-i\alpha_0^2}{k}\psi_0 + \left.\frac{ih_0^2}{k\alpha_0^2}\frac{\partial^2\psi_0}{\partial s^2}\right|_C \tag{3.84}$$

The unknown α_1^2 value is determined by the following condition. Eqn. 3.83*a* with its boundary condition of eqn. 3.84, must have a finite solution if eqns. 3.82 have a nontrivial solution. To determine the α_1^2 value, we multiply the first eqn. 3.82*a* by ψ_1 and eqn. 3.83*a* by ψ_0, then we subtract the equations and integrate over the cross section, transforming further this integral using Green's formula. After substituting the boundary conditions of eqns. 3.82*b* and 3.84 into the contour integral, we perform an integration by parts and, taking into account the normalisation conditions of eqn. I.16, we obtain

$$\alpha_1^2 = \frac{i}{k}\left\{\alpha_0^4 \oint \psi_0^2 ds + h_0^2 \oint \left(\frac{\partial \psi_0}{\partial s}\right)^2 ds\right\} \qquad (3.85)$$

Having determined α_1^2, we now derive two terms of the h^2 expansion in the Taylor series of w, according to eqn. 3.79. Substituting the w value from eqn. 3.76, we obtain

$$h^2 = h_0^2 + (1-i)M + O\left(|w|^2\right) \qquad (3.86)$$

where M denotes the value

$$M = \frac{\mu d}{2}\left\{\alpha_0^4 \oint \psi_0^2 ds + h_0^2 \oint \left(\frac{\partial \psi_0}{\partial s}\right)^2 ds\right\} \qquad (3.87)$$

The minimal value of $|h|$ considered as a function of frequency is achieved at $h_0^2 = -M$ and is equal to

$$|h|_{\min} = \sqrt{M} \qquad (3.88)$$

where only the first term of M caused by a longitudinal current is relevant. In addition, one can let $\alpha = k$; eqn. 3.68*a* can be derived in such a way.

Far from the cut-off frequency, at $h_0^2 > 0, h_0^2 >> M$, eqn. 3.86 yields the attenuation coefficient in the form

$$h'' = M/2h_0 \qquad (3.89a)$$

If one substitutes the M value from eqn. 3.87, eqn. 3.89*a* is equal to the formula obtained by a usual power-loss method.

The phase velocity $\upsilon = c/\mathrm{Re}(h)$ is equal to

$$\upsilon = \upsilon_0\left(1 - M/2h_0^2\right) \qquad (3.89b)$$

The phase velocity is a little less than in the waveguide with perfectly conducting walls with the same cross section. From the point of view of its influence on the phase velocity, the introduction of a finite conductivity is equivalent to an enlargement of the cross-section linear sizes by $M/(2\alpha_0^2)$, which is proportional to the skin depth *d*.

If the frequency is below cut off, i.e. $h_0^2 < 0$, $|h_0|^2 >> M$, eqn. 3.86 yields

$$h = -i\left(|h_0| - \frac{M}{2|h_0|}\right) + \frac{M}{2|h_0|} \tag{3.90}$$

The second term of this equation is proportional to the power flow propagating in a waveguide with nonperfect walls at a frequency below cut off.

The transition from eqn. 3.89 to eqn. 3.90 is described by eqn. 3.86. The latter gives an explicit expression of Re(h) and Im(h) for arbitrary relation between h_0^2 and M. Re(h) and Im(h) can be determined from the equation system

$$\begin{aligned} &(\mathrm{Re}(h))^2 - (\mathrm{Im}(h))^2 = h_0^2 + M \\ &2(\mathrm{Re}(h))(\mathrm{Im}(h)) = M \end{aligned} \tag{3.91}$$

For example, at $k = \alpha_0$, i.e. at the cut-off frequency when $h_0 = 0$, one gets $h = \sqrt{M}$ (1.10-i0.46).

This means that for the TE_{12} mode at 28 GHz in a circular copper waveguide, h = 3.3(1.10 – i0.46) meter^{-1} in agreement with Alpert (1940)and also Schaffeld and Bayer (1956).

Generally, at frequencies far from cut off, the wavenumber h, according to eqn. 3.86, is different from h_0 by terms of the order $k|w|$, but in the direct neighbourhood of the cut-off frequency the difference between h and h_0 is of the order $k|w|^{1/2}$.

Here, we will not determine the perturbation of the transversal eigenfunctions and the fields when a transition from perfect to nonperfect conducting walls is performed. The perturbation can easily be determined using the equations system of 3.83. For all frequencies, ψ_1 is of the order w and φ is of the order $wh_0^2/(kh)$, therefore, for example, the relation E_z/H_z is of the order w far from the cut-off frequency. The field components connected with a TM mode are very small near the cut-off frequency, where $|\hbar_0|$ is small, particularly the E_z component.

For calculating in the same way two terms in the expansion of the wavenumber for TM modes, one should introduce the function $\Psi = h\psi$ instead of ψ and expand the functions φ and Ψ in Taylor series of w. The function φ_1 satisfies the system of equations

$$\begin{aligned} &\nabla^2\varphi_1 + \alpha_0^2\varphi_1 = -\alpha_1^2\varphi_0 \\ &\varphi_1|_C = \frac{ik}{\alpha_0^2}\frac{\partial\varphi_0}{\partial n} \end{aligned} \tag{3.92}$$

The eigenvalue α_1^2 is determined from the condition, that this system of equations has the solution

$$\alpha_1^2 = ik\oint\left(\frac{\partial\varphi_0}{\partial n}\right)ds \tag{3.93}$$

All eqns. 3.86, 3.88 – 3.91 derived above for *TE* modes are also valid for *TM* modes if we substitute instead of *M* given in eqn. 3.87 the following value

$$M = \frac{\mu k^2 d}{2}\oint\left(\frac{\partial\varphi_0}{\partial n}\right)^2 ds \tag{3.94}$$

The value of φ_1 is of the order of w for all frequencies; ψ is of the order $wh_0^2/(kh)$ and the component H_z vanishes at the cut-off frequency.

***

Methods to solve the coupled-wave differential equations for mode amplitudes are developed in this Chapter for two problems, where the weak coupling approximation is insufficient already for very small and smooth nonuniformities (a cut-off cross section and a resonant frequency may exist). The reflection coefficients are derived for an incident wave on a cut-off cross section, as well as the amplitude of a parasitic wave, scattered by a nonuniform waveguide in cut off at some cross section and the amplitudes of the waves, scattered by a tilt of a uniform cross-section waveguide at one frequency, which can be arbitrarily close to the cut-off frequency of an excited parasitic wave. The latter problem encouraged us to derive in Section 3.4 an expression for the complex wavenumber in a waveguide with nonperfect walls. The expression is valid for all frequencies, including in the neighbourhood of the cut-off frequency.

***

3.5 Step-wise variation of waveguide parameters

3.5.1 Step-wise variation of waveguide cross section

We shall consider two other cases of strong coupling in this section, namely the problems of a waveguide cross-section step and of a step-wise variation of the cross-section geometrical topology. Their formulation is diametric to the problems considered in subsections 2.2.6 to 2.2.8.

Let us assume that a nonuniform waveguide section has a short length l_0, (i.e. $kl_0 << 1$; $\nu >> 1$), and that the waveguide properties vary so rapidly along this section that the difference between the waveguide parameters on both ends of

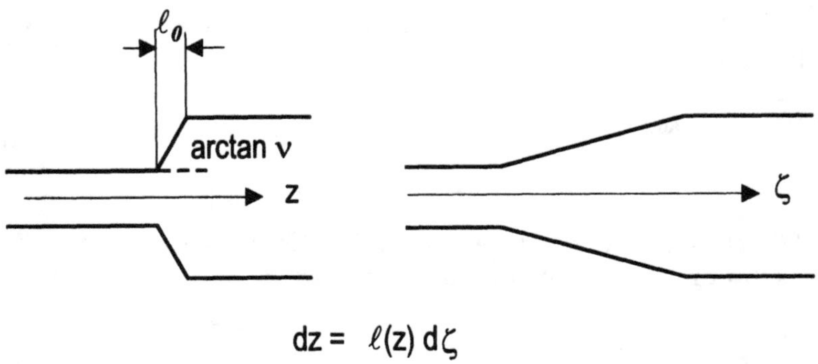

Figure 3.5 *Schematic showing the principle of rapidly varying co-ordinate,* ξ

this section is not small. In such a case, the function participating in eqn. 2.95 and in the formulae following thereafter, e.g. in eqns. 2.98 and eqns. 3.7 etc. has a large value proportional to $1/l_0$. All the coupling coefficients S_{jn} in eqn. 2.95 have large values as well. If the length l_0 approaches its zero limit but the waveguides differ from each other as before, this nonuniformity is transformed into a step-wise cross-section change. Nevertheless, the theory developed above is also valid for these nonuniformities.

All the coefficients S_{jn} in the coupled-wave eqns. 2.46 for P_j are of the order of $1/l_0$. The derivatives P_j' are of the same order. Let us substitute the co-ordinate z by a new rapidly varying co-ordinate ξ defined in such a way that it varies by a finite value along the section with the length l_0 as Figure 3.5 shows (Sporleder and Unger, 1979; Pavelyev *et al.*, 1987). We assume

$$dz = l(z)\, d\xi \tag{3.95}$$

where $l(z)$ is of the order of l_0. Let us define $l(z)$ by the following formula

$$l(z) = \frac{\int ds}{\int \frac{v(s)}{r(s)}\, ds} \tag{3.96}$$

where $r(s)$ is the distance between the point on the contour and any selected axis. Further, we multiply eqn. 2.46 by $l(z)$. The coefficients in eqn. 2.46 become the products $l(z)\, S_{jn}$. These values remain finite when we approach the limit $l_0 \to 0$. Clearly, $l(z) \to 0$ simultaneously, also, for all z. Now the coupled-wave eqns. 2.46 have the following form

$$\frac{dP_j}{d\xi} = \sum_n l(z) S_{jn} P_n \tag{3.97}$$

The second terms on the left-hand side of eqns. 2.46 have disappeared by this trick. Therefore, the nonuniform section has such a small length that we could neglect the phase shift along it.

The solution of this problem as well as of several other problems in this chapter is nonadiabatic, hence it is more convenient to employ the variables Q_j and R_j (j>0) instead of the variables P_j and to start from the coupled-wave eqns. 3.1 instead of eqns. 2.46. The second terms on the left-hand side of eqns. 3.1 also disappear by introducing the co-ordinate ξ instead of z and by approaching the limit $l_0 \to 0$. Furthermore, the coefficients of the sums there will be $l(z)(S_{jn} \mp S_{-jn})$. If j is a *TE* mode, then, according to the first formula in eqn. 3.2*a*, one yields the following formula from eqn. 3.1*a*

$$\frac{dQ_j}{d\xi} = \sum_i l(z)(S_{ji} - S_{-j,i})Q_i \tag{3.98a}$$

and the variables Q_i participating in this formula sum must also correspond only to *TE* modes. In the same way, if n is a *TM* mode, then, according to the second formula of eqn. 3.2*b* one yields the following formula from eqn. 3.1*b*

$$\frac{dR_n}{d\xi} = \sum_m l(z)(S_{nm} - S_{-n,m})R_m \tag{3.98b}$$

and the variables R_m participating in this formula sum must correspond to *TM* modes only. Therefore, the set of variables Q_j with the subscript j corresponding to *TE* modes satisfy the coupled-wave equations containing the variables of the same set only. The same is valid for the set of variables R_n with the subscript n corresponding to *TM* modes. Obviously, these variable sets are not independent, they are connected with each other and with other variables by boundary conditions. The conditions for the variables Q_j and R_n have approximately the same form as eqn. 3.8*a*. We do not write down here two other sets of coupled-wave equations derived from eqn. 3.1 by substitution of ξ instead of z.

For circular waveguides the factor $l(z)$ given by eqn. 3.96 is equal to ρ/ν and, since $\nu dz = d\rho$ one has $\xi = ln(\rho)$. The coefficients of the coupled-wave eqns. 3.98 are derived from the coefficients of eqn. 3.4 for a circular cross section by omitting the factor ν/ρ.

The physical meaning of this method describing the mode conversion by a cross-section step is a co-ordinate system enlargement along the z-axis according to the law of eqn. 3.95 (see Figure 3.5).

In such a way, the step is enlarged to a smooth nonuniformity where the cross-section method is applied and all the coupling coefficients are derived. Further, an 'inverse shrinkage' is performed, i.e. $l(z) \to 0$. Probably, to prove the validity of this procedure, one should prove that the result does not depend on the choice of the function $l(z)$ in eqn. 3.95, since there is not a single possible choice according to eqn. 3.96.

Instead of a rigorous proof of validity of the coupled-wave eqns. 3.98, a simple numerical experiment was performed (certainly, it is not a proof). A radial step in a circular waveguide was considered. The radius increases 1.32 times. The TE_{01} mode is incident from the narrow waveguide, its radius is equal to a, and only TE_{0n} modes are excited for this symmetry. Within the frequency band $7.2 < ka < 8.8$ the amplitudes of all propagating modes were calculated, i.e. of the backward propagating TE_{01} and TE_{02} waves and of the forward propagating TE_{01}, TE_{02} and TE_{03} waves. One evanescent mode TE_{04} together with those propagating modes was taken into account in the coupled-wave eqns. 3.98. The results were compared with those obtained by an independent method providing the field continuity on both sides of a step leads to an integral equation or to coupled-wave algebraic equations, i.e. to a more simple technique than that of the coupled-wave differential equations. The results are equal within the plot accuracy in the whole frequency band for all forward amplitudes. The very small backward amplitudes deviate by approximately 15%.

Yet it is not worth applying the cross-section method for analysis of a single cross-section step. However, if there are both smooth and step-wise nonuniformities in a waveguide, they can all be studied by the same technique — by the cross-section method developed above.

3.5.2 Step-wise variation of waveguide cross-section geometrical topology

In this subsection we apply our method to another problem with a step-wise nonuniformity. Let us consider a junction of coaxial and circular waveguides (Pavelyev and Starov, 1995). The external radius of the coaxial waveguide is equal to the circular waveguide radius and the inner coaxial radius is small, hence the nonuniformity mainly consists in the change of the cross-section geometrical topology. The *TEM* mode is incident from the coaxial. Only TM_{0n} modes are excited due to this symmetry.

For the sake of simplicity we have considered in the Introduction (Section I.3) only single-connected waveguide cross sections. The inner rod affects essentially the fields of almost all waveguide modes and enables *TEM*-mode propagation.

In the considered problem of a concentric coaxial waveguide, the transversal eigenfunctions of circular symmetric *TM* modes, similar to eqn. I.18, are equal to

$$\varphi^j(\rho) = M^j\left[J_0(\alpha_j\rho)N_0(\alpha_j b) - J_0(\alpha_j b)N_0(\alpha_j\rho)\right] \tag{3.99}$$

where a and b are the external and inner coaxial radii, respectively, and ρ and β are the radial and angular cylindrical co-ordinates. The boundary condition of eqn. I.12*b* is satisfied at $\rho = b$ (i.e. $\varphi^j(b) = 0$) for arbitrary α_j and the equation to determine α_j is the boundary condition at ρ = a: $\varphi^j(a) = 0$. The eigenvalues for $b << a$ are nearly equal to the roots of the equation $J_0(\alpha_j a) = 0$ and they weakly depend on b. The same statement is valid for the normalisation factor M^j determined according to eqn. I.17. However, the transversal field components

$$E^j_\rho(\rho) = -ih_j \frac{d\varphi^j}{d\rho} \quad \text{and} \quad H^j_\beta(\rho) = -ik\frac{d\varphi^j}{d\rho} \tag{3.100}$$

have large values of the order of $1/b$ on the inner rod surface at ρ= b.

Only two field components exist in the *TEM* mode, i.e. not only $H_z \equiv 0$, but also $E_z \equiv 0$ and besides this $h_j = k$. The fields are expressed through the transversal eigenfunction $\varphi^j(\rho)$ by the same formulae as eqn. 3.100. However, the boundary condition for this function is weaker than for eqn. I.12*b*, namely $\partial\varphi/\partial s = 0$. The eigenvalue of the problem containing eqn. I.11*b* and this boundary condition is $\alpha_j = 0$. The *TEM*-mode transversal eigenfunction is equal to

$$\varphi(\rho) = \frac{1}{\sqrt{2\pi\, ln\left(a/b\right)}}\, ln\, \rho \tag{3.101}$$

We have applied the normalisation condition in the form of eqn. I.16, the condition of eqn. I.17 is not equivalent to this condition and cannot be applied here. The E_ρ and H_β fields of the *TEM* mode also are of the order of $1/b$ on the inner rod.

The coupling coefficients in a coaxial waveguide with varying inner radius $b(z)$ are described by the same formulae as eqn. 2.98*b* with the integral over the circumference $\rho = b(z)$. In these formulae ν is the tangent of the angle between the inner rod contour line and z-axis, and if one of the modes participating is *TEM*, the corresponding values are $\alpha_j = 0$ and $h_j = k$. The coupling coefficient $S_{-m,m}$ of the *TEM* mode can be calculated by eqn. 2.97 with $h_m' = 0$.

As follows from the estimations of the fields at $\rho = b$ considered above, the coupling coefficients both between TM_{0n} modes and between them and the *TEM* mode are of the order of $1/b$, i.e. they have large values for small b. Contrary to the problem considered in the previous subsection, the S_{jm} have large values not because of the large ν values, but because the nonuniformity is situated in an area of high field magnitudes.

The transition from a coaxial to a hollow waveguide means the break of the current supporting the inner conductor. Hence, despite an arbitrary thin conductor, this break produces a nonsmall, i.e. finite, perturbation. Let us remark that if the incident wave was neither *TEM* nor *TM* but a TE_{0n} wave without longitudinal currents, then the break of the thin inner rod would cause only a small reflection and small mode conversion.

The same technique of defining a new rapidly varying co-ordinate ξ instead of z was employed as in the previous problem of the junction of two waveguides with different radii. The abrupt end of the rod was substituted by a slowly decreasing inner rod with $b = b(z)$, further, the limit of the step-wise ending was approached. The independent variable was defined as $\xi = lnb(z)$. The coupled-wave eqns. 2.46 were rewritten with this variable (more exactly, with the variable $1/\xi$) and the terms describing the phase shift were neglected as before. These equations were solved with the obvious additional boundary condition: the *TEM*-mode amplitude is equal to zero at the end of the inner conductor. Two ratios of external to inner radius were selected, namely $a/b = 15$ and $a/b = 100$. The absolute value of the *TEM*-mode reflection coefficient was computed in the frequency band from small kb values until $kb = 12$. The reflection coefficient is nearly equal to unity for small kb values and it decreases to a small value with increasing kb. However, due to reflection in TM_{0n} waves (n increasing with kb), this decrement is not monotonous. The decrement at smaller kb values is more rapid as at larger kb. The results were compared at several frequencies with the results of the independent computation method mentioned above and both results are absolutely identical.

In this section we studied nonuniformities such as a step, i.e. a discontinuous nonuniformity or in other words the interconnection of two different waveguides. Two such discontinuities are considered: the change of the cross-section boundary (the step) and the change of the topological connection (for example, the connection of a coaxial waveguide with a hollow cylindrical waveguide). To perform this analysis a method based on the replacement of the z co-ordinate by a rapidly varying co-ordinate has been applied. The physical meaning of this method consists in making a gradual variation where the abrupt cross-section change takes place. This method can be seen as an inverse shrinkage of the transition region.

* * *

The results of the last section are somehow unexpected. As was shown there, the cross-section method can be applied for calculation of nonsmall step-wise variations. First of all, this result is interesting from the methodical point of view. We have considered a waveguide with continuously varying parameters in Chapter 1 as the limit of the corresponding waveguide with a large number of small steps. In fact, there is some reciprocity, i.e. one derives also effective computational algorithms considering a step-wise variation as the limit of the continuous variation. Possibly, this reciprocity can be applied not only to the waveguide theory.

Chapter 4

Straight nonuniform waveguides

The mathematical technique developed in previous chapters is employed in this Chapter to solve some real problems of wave propagation in straight nonuniform waveguides. Examples illustrate different features of the cross-section method.

4.1 Twisted waveguides

4.1.1 Coupling coefficients in twisted waveguides

Let us consider a waveguide twist as a first example. A waveguide twist is a special case of a cross-section variation, where only the cross-section location is varied, but the cross-section shape as well as its size are constant. The cross-section contour C has the same shape anywhere along the waveguide, its location is defined by an angle Φ between a fixed direction on the cross-section contour and the x-axis of a fixed co-ordinate system x, y, z. The angle Φ depends on z, i.e.$\Phi = \Phi\,(z)$. We will start from the calculation of the coupling coefficients in twisted waveguides. Rectangular as well as elliptical waveguides will be considered, the latter having a cross section with a small eccentricity.

Deriving the function $\nu(s)$ contained in eqn. 2.98, we will not employ the complicated methods of differential geometry. An auxiliary technique based on the relation $\nu(s)\Delta z = \delta(s)$ is used instead. Here, $\delta(s)$ is the height of a step in an auxiliary stepped waveguide, with a limit which is the given nonuniform waveguide if each step's height is decreased. First, we have to find the distance $\delta(s)$ between the corresponding points of two cross sections twisted one to another by an angle $\Delta\Phi$ between them. Thereafter, we will divide $\delta(s)$ by Δz and find a limit for $\Delta z \to 0$. Obviously, $\delta(s)$ has to be determined only up to the terms of the first order of $\Delta\Phi$.

The co-ordinate axes denoted as $\bar{x}, \bar{y}$ are firmly connected with the cross section and twisted together with it when moving along the waveguide. For a rectangular waveguide these axes correspond to the sides of the rectangle, as

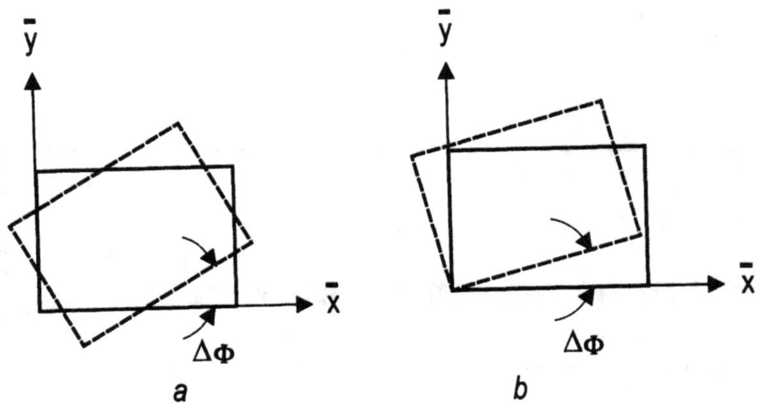

Figure 4.1 *Geometry of rectangular waveguide twists*
a) The central point is fixed
b) The vertex is fixed

shown in Figure 4.1*a*. The second cross section twisted by an angle $\Delta\Phi$ is also shown in the same figure.

For example, as follows from obvious geometric considerations, on the lower side at $\bar{y}=0$

$$\delta(\bar{x}) = \Delta\Phi\left(\frac{a}{2} - \bar{x}\right) + O\left[(\Delta\Phi)^2\right] \tag{4.1}$$

In eqn. 4.1 it was assumed that the central point of the cross section is fixed (Figure 4.1*a*), i.e. this point is unchanged by the twist. If the vertex $\bar{x}=0$, $\bar{y}=0$ was fixed (Figure 4.1*b*), instead of eqn. 4.1 one would yield

$$\delta(\bar{x}) = \Delta\Phi\cdot\bar{x} + O\left[(\Delta\Phi)^2\right] \tag{4.2}$$

In this case the coupling coefficients derived later would be different. For the sake of clarity we will analyse the twist corresponding to Figure 4.1a. Deriving the expressions for $\delta(s)$ corresponding to the three other sides (which are similar to eqn. 4.1) and rewriting to $\nu(s)$ we yield

$$\begin{aligned} &\bar{y}=0,\ \ \nu(s)=\left(\frac{a}{2}-\bar{x}\right)\Phi', \quad \bar{x}=a,\ \ \nu(s)=\left(\frac{b}{2}-\bar{y}\right)\Phi' \\ &\bar{y}=b,\ \ \nu(s)=\left(\bar{x}-\frac{a}{2}\right)\Phi', \quad \bar{x}=0,\ \ \nu(s)=\left(\bar{y}-\frac{b}{2}\right)\Phi' \end{aligned} \tag{4.3}$$

where $\Phi' = d\Phi/dz$.

In such a way, the calculation of the coefficients S_{jm} by eqn. 2.98 is reduced to simple integrals. It may be easily proven that twists in overmoded rectangular waveguides cause coupling between modes separated by an odd integer in both indices. We will limit ourselves only to the case when the incident mode m is the fundamental mode in a rectangular waveguide, i.e. the TE_{10} mode with the transversal eigenfunction of eqn. I.21.

$$\psi^m = N^m \cos\left(\frac{\pi x}{a}\right) \tag{4.4}$$

To design a twist for the propagating TE_{10} mode, it is usually sufficient to consider coupling to TE_{01} only, since the wavenumbers of the other coupled modes (e.g. TE_{21}, TM_{21}, TE_{03}) are far away. A twist does not couple the TE_{10} mode to the TE_{n0} modes. The TE_{01} mode has an orthogonal polarisation and its transversal eigenfunction is

$$\psi^j = N^j \cos\left(\frac{\pi y}{b}\right) \tag{4.5}$$

A simple derivation from eqns. 2.98a and 4.3–4.5 gives us

$$S_{jm} = -\frac{8}{\pi^2}\frac{h_j + h_m}{2h_j}\Phi' \tag{4.6}$$

Let us note that this formula can also be derived by calculating S_{jm} not from eqns. 2.98 but directly from eqn. 2.47. In this case, there will be no additional contour integral in eqn. 2.47 after the calculation of the limit discussed at the end of subsection 2.2.4. It may easily be shown that this additional contour integral containing only charges and longitudinal currents is equal to zero for the two considered modes TE_{10} and TE_{01}. This is caused by the fact that the charges as well as the longitudinal currents of these waves are excited on different parts of the cross-section contour, and these waves are hence coupled to each other by transversal currents only. For this same reason, in this case the second term of eqn. 2.98a disappears. Let us also note that, due to the same reason, the coupling coefficient of the waves propagating in opposite directions is in this case proportional to the difference of their wavenumbers as well as for the TE_{0q} waves in a circular waveguide, as follows from eqn. 4.6. This coupling coefficient is therefore significantly smaller than for the forward propagating waves.

The coupling coefficient to the backward TE_{10} wave, i.e. of the same kind as the incident one, can be derived in the simplest way using eqn. 2.97. Since the wavenumber does not vary along a twist, the term h'_m in eqn. 2.97 disappears. The integral term is equal to zero as well, since $\nu(s)$ has opposite signs on the opposite cross-section sides $\bar{y} = 0$ and $\bar{y} = b$ hence $S_{-mm} = 0$.

The reflection coefficient of the waveguide section twisted by a small angle $\Delta\Phi$ is therefore proportional to $(\Delta\Phi)^2$. It is proportional to $(\Phi')^2$ for a long twist with a slowly varying angle Φ. It can easily be proven that the twist does not couple the TE_{10} wave to the TE_{n0} waves. The coupling coefficient to other waves can also be derived from eqns. 2.98 and 4.3, see for example Doane (1988).

Considering elliptical waveguides, let us limit ourselves to a small eccentricity, so that they are nearly circular waveguides. In such a way we will be able to obtain all results in an elementary form without the introduction of elliptical functions.

If the eccentricity is small, the cross-section contour can be described in polar co-ordinates (ρ,β) (Figure 4.2) as

$$\rho = a + l\cos(2\beta) \tag{4.7}$$

where the positive value l is related to the eccentricity as $e^2 = 4l/a$; being $l << a$.

The equation of a twisted contour is derived by the substitution of the polar angle β by $(\beta - \Delta\Phi)$ in eqn. 4.7. Including up to the first order terms of $\Delta\Phi$, $\delta(s)$ is equal to the difference of the radius vectors corresponding to a same angle β for the twisted and the nontwisted contours, respectively. Determining $\nu(s)$, i.e. dividing $\delta(s)$ by Δz and assuming $\Delta z \to 0$ we obtain

$$\nu(s) = 2l\sin(2\beta)\Phi' \tag{4.8}$$

Let the incident mode be nearly equal to the TE_{11} mode in a circular waveguide and polarised along the major elliptical axis. If $l << a$, to determine the coupling coefficients one does not need to take into account the perturbation caused by the difference of our cross section from a circular cross section in the expressions of transversal eigenfunctions, and we can assume

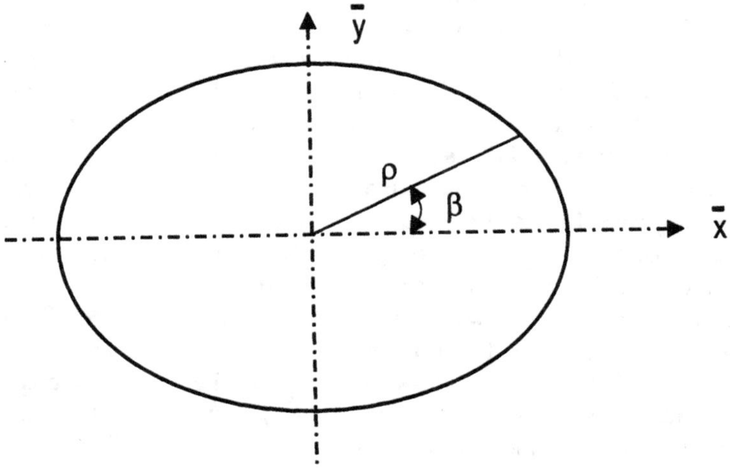

Figure 4.2 Cross section of an elliptical waveguide

$$\psi^m = N^m J_1(\alpha_m \rho)\sin(\beta) \tag{4.9}$$

where N^m is given by eqn. I.20. This approximated expression of the transversal eigenfunction can be employed everywhere including the points of the same contour, because the elliptical contour of eqn. 4.7 is not only nearly equal to the circular contour $\rho = a$, but also because the angles between them are small everywhere. However, the difference between the eigenvalues α_m in a slightly elliptical waveguide and the eigenvalues μ_m/a of the circular waveguide given by eqn. I.19 is significant.

The mode nearly equal to the TE_{11} mode with orthogonal polarisation has the transversal eigenfunction

$$\psi^j = N^j J_1(\alpha_j \rho)\sin(\beta) \tag{4.10}$$

The coupling coefficient between these two waves, derived from eqn. 2.98 and from the three last formulae, is equal to

$$S_{jm} = \frac{l}{a}\frac{\alpha_m}{\alpha_m - \alpha_j}\frac{\mu_m^2 + 1}{\mu_m^2 - 1}\Phi' \tag{4.11}$$

We will rewrite this equation substituting in it the eigenvalues difference for two TE_{11} modes of orthogonal polarisation in a slightly elliptical waveguide. This difference is proportional to l, it can easily be found using as an example eqn. 1.55. One only has to take into account during the calculation that $\delta(s)$ in eqn. 1.55 means the distance between two elliptical contours twisted by $\pi/2$ one to another, so that $\delta = -2l\cos(2\beta)$. An elementary derivation omitted here yields

$$\alpha_m - \alpha_j = \alpha_m \frac{\mu_m^2 + 1}{\mu_m^2 - 1}\frac{l}{a} \tag{4.12}$$

Therefore, the coupling coefficient between the two TE_{11} waves caused by the twist is simply equal to

$$S_{jm} = \Phi' \tag{4.13}$$

It can easily be proven using eqn. 4.8, that in the highest order of l/a a twist does not couple forward and backward waves in a slightly elliptical waveguide. The TE_{11} wave with the transversal eigenfunction of eqn. 4.9 is coupled to all TE_{1q} waves with orthogonal polarisation and to those TM_{1q} waves where $\varphi^j \sim \sin(\beta)$. However, for the TE_{11} wave with the transversal eigenfunction of eqn. 4.10 only, S_{jm} does not have a small value despite having the factor l in eqn. 4.8.

4.1.2 Parasitic wave amplitude

The coupling coefficients of eqns. 4.6 and 4.13 are proportional to the derivative of the angle Φ with respect to the length. If the variation of the angle either over the lengths a or λ is small, the coupling is small as well. The spurious wave amplitudes for a weak coupling are determined in eqn. 2.75. This formula is valid if the incident wave amplitude p_m does not change significantly along a twist transition, i.e. if the solution of eqn. 2.73 is valid. However, there can be some conditions when the incident wave amplitude changes significantly despite the very small coupling coefficients. In this case, the approximation of eqns. 2.73 and 2.75 cannot be applied. This may occur near mode degeneracy, more exactly when two wavenumbers of waves coupled by a waveguide deformation are nearly equal so that the phase difference σ between both waves acquired along the whole nonuniform section is small or finite. For a twist we have

$$\sigma = (h_m - h_j)L \tag{4.14}$$

where L is the twist length.

In this case, the parasitic wave amplitude has a finite value. From the physical point of view the reason is that the elementary parasitic waves excited at each particular section of the nonuniform waveguide section 1.2 superpose each other with a very small phase difference. In its turn, the interaction of this parasitic wave with the incident wave changes significantly the amplitude of the latter. We will study this effect on the example of a twisted waveguide. We will face again the same problem as in considering a bent circular waveguide.

Let us analyse the conditions required for a complete ($\sigma = 0$) or for a noncomplete degeneracy, beginning from the study of TE_{10} wave propagation along a twisted rectangular waveguide. The waves TE_{01} and TE_{10} are completely degenerate if the waveguide cross section is a square ($a = b$).

The field of these modes is described by the coupled-wave eqns. 2.71 for normalised amplitudes. To describe the conditions near the degeneracy we need to retain only two of these equations, related to forward waves m and j

$$\begin{aligned} p'_m &= -\frac{h_j}{h_m} S_{jm} \exp(-i(\gamma_j - \gamma_m))\, p_j \\ p'_j &= S_{jm} \exp(i(\gamma_j - \gamma_m))\, p_m \end{aligned} \tag{4.15}$$

The reciprocity conditions of eqn. 2.51 were taken into account when deriving these equations.

Since S_{jm} is proportional to Φ', it is convenient to use the twist angle Φ as the independent variable instead of z. After the substitution of eqn. 4.6 given for S_{jm}, eqns. 4.15 acquire the following form

$$\frac{dp_m}{d\Phi} = \sqrt{\frac{h_j}{h_m}}\, A\, p_j \exp(-i(\gamma_j - \gamma_m))$$
$$\frac{dp_j}{d\Phi} = -\sqrt{\frac{h_m}{h_j}}\, A\, p_m \exp(i(\gamma_j - \gamma_m)) \tag{4.16}$$

where A denotes the magnitude

$$A = \frac{8}{\pi^2} \frac{h_m + h_j}{2\sqrt{h_m h_j}} \tag{4.17}$$

According to eqn. 2.72, the boundary conditions for the coupled eqns. 4.16 are: $p_m(0) = 1$ and $p_j(0)=0$.

Since h_m and h_j do not depend on z, the exponential factors in eqns. 4.16 are proportional to z. The solution of the coupled-wave eqns. 4.16 depends on the twist type, i.e. on the function $\Phi(z)$. For the sake of clarity we will consider a uniform helical twist, in which Φ is proportional to z

$$\Phi = \theta \frac{z}{L} \tag{4.18}$$

Here, θ is the total twist angle. Then we can substitute the following expression in eqns. 4.16

$$\gamma_j - \gamma_m = q\Phi \tag{4.19}$$

where the parameter q, playing an essential role in further analysis, is equal to the ratio of the total phase difference σ and the total twist angle θ

$$q = \frac{\sigma}{\theta} \tag{4.20}$$

According to eqn. 4.14, the parameter q can be defined also as the ratio of the wavenumber difference $(h_m - h_j)$ to the twist angle per unit length θ/L. The first of these values describes local proximity to the mode degeneracy conditions (here local means not depending on the twist length), and the second of these values describes the speed of the local change of the waveguide properties.

Small values of $|q|$ mean that the conditions are close to mode degeneracy, and $q = 0$ corresponds to complete degeneracy, which takes place only in a square waveguide. If $|q|>>1$, there is practically no degeneracy at all, as we will see now. So, the influence of deformation is not determined by the twist angle per length unit θ/L and only the coupling coefficients are proportional to this value. These deformation effects are determined by the ratio between this value and the difference Δh, and this ratio is characterised by the parameter q.

The solution of the coupled-wave eqns. 4.16 for the normalised amplitudes p_m and p_j can be found in a simple way if eqn. 4.19 is valid. The values of P_m and P_j at the end of the twist, i.e. at $z = L$ and $\Phi = \theta$ are then determined by eqn. 2.70. These amplitudes are equal to

$$
\begin{aligned}
P_m &= (\cos(\kappa) + \frac{i\sigma}{2\kappa}\sin(\kappa))\exp(-i\frac{h_m + h_j}{2}L) \\
P_j &= -\frac{A\theta}{\kappa}\sin(\kappa)\exp(-i\frac{h_m + h_j}{2}L)
\end{aligned}
\tag{4.21}
$$

Here κ denotes the value

$$\kappa = \theta\sqrt{A^2 + q^2/4} \tag{4.22}$$

Let us first assume that $q = 0$ in the solution of eqn. 4.21. Then

$$P_m = \cos\left(\frac{8}{\pi^2}\right)\Phi, \quad P_j = -\sin\left(\frac{8}{\pi^2}\right)\Phi \tag{4.23}$$

The power of the TE_{10} wave is fully converted to the TE_{01} wave and *vice versa* as one moves along the twisted waveguide. The full conversion takes place each $\pi^3/16$ radians, i.e. after each 111°. Using eqns. 4.16, it could easily be shown that this result is valid for the square waveguide (i.e. for $\gamma_j - \gamma_m = 0$), and remains valid for any arbitrary function $\Phi(z)$.

The parasitic wave amplitude is always less than one for a finite $|q|$. Therefore, a part of the incident power is converted periodically. The maximal power of the TE_{01} wave is equal to

$$\frac{A^2\theta^2}{A^2\theta^2 + \frac{\sigma^2}{4}} \tag{4.24}$$

Finally, for $|q| >> 1$, i.e. when the total phase difference σ at the twist end is much greater than the total twist angle θ (Sveshnikov, (1958)), the parasitic wave amplitude is equal to

$$\left|P_j\right| = \left|\frac{2A}{q}\sin(\kappa)\right| \tag{4.25}$$

The same result can be derived in another way, using eqn. 2.75, which corresponds to the case of weak coupling. Therefore, when the difference of the wavenumbers is large enough (more exactly when $|q| >> 1$), the parasitic wave amplitude is inversely proportional to this difference.

The perturbation caused by a twist can be characterised not only by the amplitude of the parasitic wave with orthogonal polarisation, but by the variation of the wavenumber of an incident wave as well. The output phase of the fundamental TE_{10} wave can easily be found from eqn. 4.21. Dividing this phase by the section length, L, we obtain the so-called effective wavenumber of the incident wave in a twisted section. When $|q| >> 1$ and $|\sigma| >> 1$, i.e. far away from the mode degeneracy, this magnitude is equal to

$$h_m + A^2 \frac{\theta^2}{L^2} \cdot \frac{1}{h_m - h_j} \qquad (4.26)$$

Generally speaking, variations of the wavenumber proportional to $1/L^2$ are caused by coupling to all the remaining waves including the evanescent waves. However, as follows from eqn. 4.26, we can see that the most significant coupling is the coupling to those waves with wavenumbers which are closer to h_m and for which the coupling coefficient is not small. In almost all practical cases only coupling to the TE_{01} wave must be taken into account for calculating the effective wavenumber variation in a twist, if this variation is not very small.

Eqn. 4.26 can also be derived in another way. For this the so-called eigenmodes of a twisted waveguide must be considered, i.e. the modes propagating without distortion along such a waveguide. These eigenmodes are determined as those solutions of the coupled-wave eqns. 4.16 where $P_m(z)$ and $P_j(z)$ have the same dependence on z. The eigenmodes do not satisfy the usual boundary conditions and all eigenmodes are excited in a twist, when a TE_{10} wave is incident from a uniform waveguide.

It can be easily derived from eqns. 4.16 that the wavenumber of an eigenmode is equal to

$$h_m + \frac{h_m - h_j}{2}\left\{\sqrt{1 + 4A^2/q^2} - 1\right\} \qquad (4.27)$$

Since this formula was derived from the two coupled-wave eqns. 4.16, it is valid by the same assumption, i.e. if it is sufficient to take into account the coupling of only two waves.

If there is no coupling at all, i.e. if $|q| \to \infty$, the value in eqn. 4.27 approaches the wavenumber h_m of the TE_{10} mode and the corresponding eigenmode approaches this mode. When $|q| >> 1$ the difference between the eigenmodes and the modes excited along a nonuniform section disappears if a wave is incident on this section from a uniform waveguide. The wavenumber of eqn. 4.27 then approaches the effective wavenumber of eqn. 4.26.

If the waveguide cross section is nearly a square, i.e. $|a - b| << a$, but the length L is so long that $|\sigma| >> 1$, $|q| >> 1$ and eqn. 4.26 is valid, then a simple expression is obtained for the effective wavenumber[1]

---

[1] Formula 45 of Lewin (1955) approaches eqn. 4.28 when $|a - b| << a$. Wavenumbers of the eigenmodes were determined far from degeneracy. Eqn. 4.27 can be seen as the generalisation of Lewin's formula, valid for any q if the consideration can be limited to two-wave coupling only.

$$h_m\left(1+\frac{256}{\pi^4}\frac{a^2}{L^2}\frac{a}{a-b}\right) \tag{4.28}$$

4.1.3 Rectangular waveguide twist with nonconstant rate of twist

Fortunately, by using a nonconstant variation in Φ', the average mode conversion may be reduced substantially, as shown in Figure 4.3 (Doane, 1988). The linearly varying, or triangular, rate of twist variation is quite effective, especially over large bandwidths. This variation is described by

$$\Phi' = \begin{cases} 4\theta z/L^2 & 0 \le z \le L/2 \\ 4\theta(L-z)L^2 & \dfrac{L}{2} \le z \le L \end{cases} \tag{4.29}$$

The maximum rate of twist, at $z = L/2$, is double that in the constant rate of twist case. Other variations of Φ' such as cosine, cosine-squared, or hyperbolic secant also give lower average mode conversion than the constant Φ' twist, however the triangular variation appears to be the best (Doane, 1988).

4.1.4 Twisted elliptical waveguide

Now let us consider a twisted slightly elliptical waveguide and the coupling of two TE_{11} waves polarised along the major and minor elliptical axes, respec-

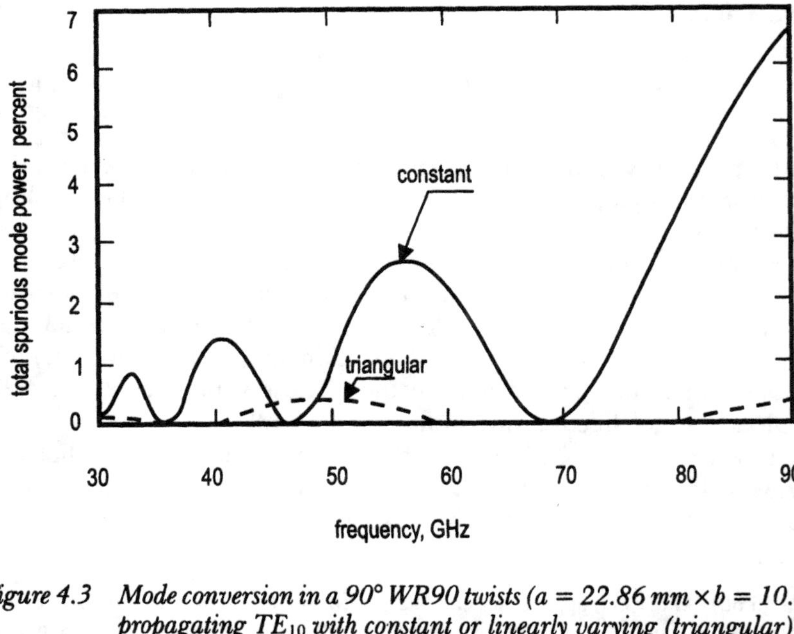

Figure 4.3 Mode conversion in a 90° WR90 twists (a = 22.86 mm × b = 10.16 mm) propagating TE_{10} with constant or linearly varying (triangular) rates of twist and L = 457 mm (Doane, 1988)

tively. Let us again limit ourselves to a uniform helical twist with a twist angle Φ proportional to z (eqn. 4.18). The amplitudes of both waves in such a waveguide are described by the same coupled eqns. 4.21, where, according to eqn. 4.13, $A = -1$ must be substituted.

Two TE_{11} waves, orthogonally polarised and having a phase difference, form an elliptically polarised wave at the end of the twisted section. It is convenient to describe the output field by the so-called polarisation ellipse. Its components are determined by the usual optical technique using oscillation amplitudes and phases for the two orthogonal directions given in eqns. 4.21. The axes of this polarisation ellipse are

$$|E_{\min}| = \sqrt{\frac{1-T}{2}}, \quad |E_{\max}| = \sqrt{\frac{1+T}{2}} \tag{4.30}$$

where

$$T = \sqrt{1 - \frac{\theta^2 \sigma^2}{\kappa^4} \sin^4(\kappa)}$$

The major axis rotation angle θ_0 of the polarisation ellipse with regard to the incident mode polarisation is determined from the equation

$$\tan(2(\theta_0 - \theta)) = -\frac{\kappa \sin(2\kappa)}{\theta\left(\cos(2\kappa) + \frac{q^2}{4}\right)} \tag{4.31}$$

Another parameter describing the perturbation caused by a twist is the cross polarisation of the output field, i.e. the component of the output field perpendicular to the polarisation of the incident wave. If the factor $\exp[-i(h_m + h_j)L/2]$ is omitted, the cross polarisation is equal to

$$B = -\frac{\theta}{\kappa} \sin(\kappa)\cos(\theta) + \sin(\theta)\cos(\kappa) + i\frac{\sigma}{2\kappa}\sin(\kappa)\sin(\theta) \tag{4.32}$$

The last four formulae give a full description of the output field after a nonuniform section, which allows us to analyse the perturbation caused by the twist.

The values of $|E_{\min}|$ against the total twist angle are represented by solid lines in Figure 4.4 for several σ values, and those of $|E_{\max}|$ are shown by dashed lines. The corresponding $|B|$ values are plotted in Figure 4.5.

If $|\sigma|$ is small, i.e. $|\sigma| << 1$, the rotation angle θ_0 of the major axis of the polarisation ellipse is small as well, being of the order of σ^2, $|E_{\max}| \cong 1$ and $|E_{\min}|$ is, in practice equal to $|B|$. The magnitude of B is then equal to

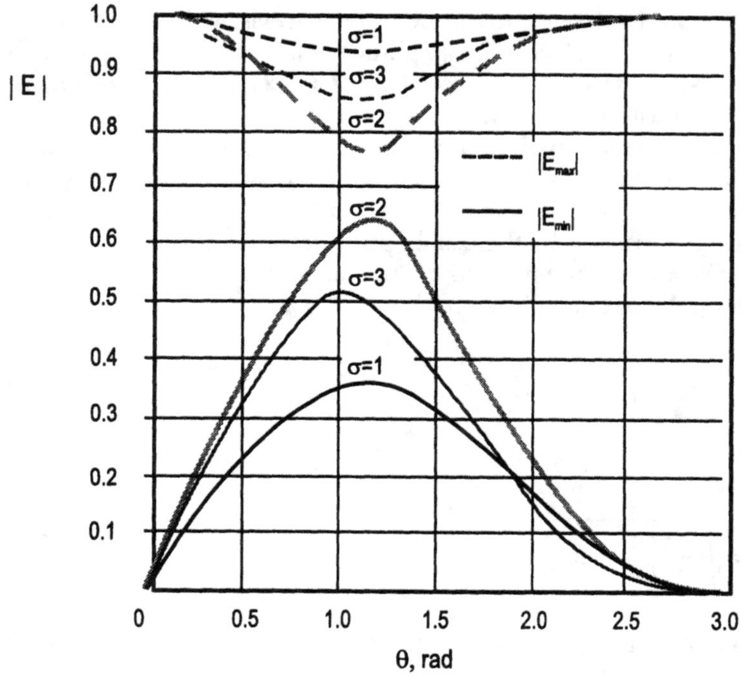

Figure 4.4 Axes of the polarisation ellipse against the total twist angle in radians

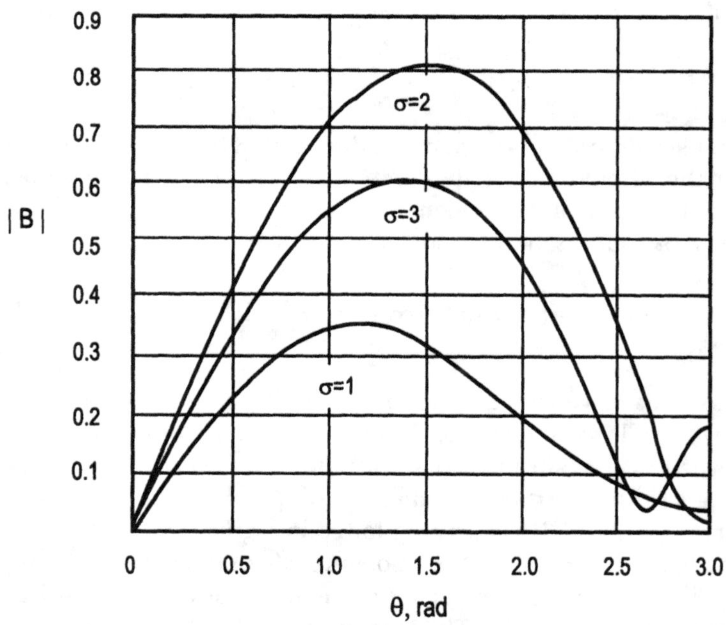

Figure 4.5 Cross polarisation at the twist output against the total twist angle in radians

$$B = \frac{i\sigma}{2} \cdot \frac{\sin^2(\theta)}{\theta} \tag{4.33}$$

The field perturbation is then very small and, since the waveguide is nearly circular, the cross-polarisation value of eqn. 4.33 can also be calculated by the simple methods of the perturbation theory. Doing so, the twisted waveguide must be described as a series of uniform slightly elliptical waveguides of the length Δz, with major axes which are rotated by the angle $\Phi(z)$ towards the incident wave polarisation, considering the direction of the incident wave polarisation as well as its amplitude being constant along the twist. The cross polarisation after each section of the slightly elliptical waveguide can easily be determined by the usual optical formulae, i.e. decomposing the incident wave into two orthogonal waves polarised with regard to the elliptical axes and taking into account the phase difference of these waves (for example, see Sandsmark (1955)). The resulting output value of the cross-polarised component B is the superposition of the corresponding small components and since $|\sigma| \ll 1$, the additional phase difference due to the propagation from a z-element until the output can be neglected. In such a way, for an arbitrary function $\Phi(z)$, one gets

$$B = \frac{i\left(h_j - h_m\right)}{2} \int_0^L \sin(2\Phi)\, dz \tag{4.34}$$

For a uniform helical twist eqn. 4.34 transforms itself into eqn. 4.33.

Eqn. 4.34 can be utilised for a statistical approach to the problem, when the direction of the elliptical axis of each cross section is a random function of z. According to eqn. 4.34, the total depolarisation effect of any nonuniform section is proportional to the mean value of $\sin[2\Phi(z)]$ along this section.

According to eqn. 4.33, when the value of σ is fixed and $|\sigma| \ll 1$, the magnitude of $|B|$ for a uniform helical twist increases with increasing θ until the value of $\theta = 67°$, when it is equal to its maximal value $0.36\ |\sigma|$. According to eqn. 4.34 this magnitude is only 28% less than the maximal value of $|B|$ for a given σ, the latter is equal to $0.5\ |\sigma|$. It is achieved in a uniform elliptical waveguide, with an axis which is twisted by the angle $\varphi = \pi/4$ with respect to the incident wave polarisation.

If the output phase difference σ has a finite (not small) value, the field at the output of the nonuniform section depends on the parameter q. Generally, if $q = 0$ (for σ = const it means that $\theta \to \infty$) the decomposition in two orthogonal waves becomes conventional. The polarisation does not become elliptical and does not rotate; according to eqns. 4.30, 4.31 and 4.32 $E_{min} = 0$, $B = 0$ and $\vartheta_0 = 0$. The total field coincides with the incident mode at each cross section. If $|q| \ll 1$ and σ has a finite value, there is a high coupling between these two modes and hence a significant power exchange between them. However, this exchange almost totally compensates the polarisation rotation of these modes in such a manner that the polarisation of the total field is almost the same as that of the incident field.

Only if both parameters of the problem σ and q have finite (not small) values, is the transmitted field significantly disturbed and eqns. 4.34 and 4.33 derived from the theory of small perturbations cannot be applied. An elliptical polarisation of the output field can be especially large for nonsmall $|\sigma|$ and finite $|q|$. The coupling between the two modes caused by the twist decreases with increasing $|q|$. For $|q|>>1$ the output mode is almost linearly polarised and its cross polarisation is small

$$|E_{min}/E_{max}| = \frac{2\sin^2(\kappa)}{|q|} << 1 \tag{4.35}$$

The output polarisation rotates by an angle which is only slightly less than the angle θ of the waveguide cross-section twist. Finally, for $|q| \to \infty$, the mode rotates as does the helix. The same behaviour occurs in a rectangular waveguide twist far from a mode degeneracy.

In conclusion for this subsection, let us remark that the coupling coefficients between two arbitrary modes in a slightly elliptical waveguide can easily be derived from eqns. 2.98 and 4.8. For example, the coupling coefficient of two waves similar to the TM_{11} waves in a circular waveguide and polarised along the elliptical axes of the cross section equals Φ', as it does for two TE_{11} waves. Hence, all the relationships found in the problem considered above for the TE_{11} wave incidence on a twist of a slightly elliptical waveguide are also valid for the incidence of a TM_{11} wave. Only the relationships between the difference $(h_j - h_m)$ of the wavenumbers and the deformation value l/a are different in these two problems. According to eqn. 4.12, this relationship for two TE_{11} waves is

$$h_j - h_m = 1.8\frac{\alpha_m^2}{h_m}\cdot\frac{1}{a} \tag{4.36}$$

although for the two TM_{11} waves the factor 1.8 must be exchanged by 1.

4.1.5 *Twisted dielectric waveguide—general considerations*

The problem of a twisted dielectric waveguide will be addressed in the following three sections. Formally, this problem is beyond the topics of this book, since a dielectric waveguide is an open system. However, this problem is very similar to the subject of this Section both by its method of solution and by its resulting effects. So we will present in subsections 4.1.5, 4.1.6 and 4.1.7 some general results of the theory of nonuniform open waveguides.

A complete set of surface eigenmodes does not exist in open waveguides, hence not only mode conversion but also radiation occurs in those waveguides. The set of surface modes (discrete spectrum modes) is completed in the theory of open waveguides (Shevchenko, 1971) by the set of modes depending on a continuous parameter (continuous spectrum modes) in such a way that the radiation is described as a mode conversion phenomenon. For this purpose, the fields of the continuous spectrum modes must decrease faster than $\rho^{-1/2}$ at large radial distance. This condition is the necessary and sufficient condition for describing all phenomena in nonuniform waveguides by the derived mode set. The cross-section method can be applied to this set of modes. The sums in eqn. 2.1 are then finite, but the integrals over the parameter numbering continuous spectrum modes are added to the sums of eqn. 2.1. The coupled-wave differential eqns. of 2.46 are replaced by coupled-wave integral-differential equations. These equations contain both the coupling coefficients depending on two discrete subscripts to characterise the coupling between surface modes and the coupling coefficients depending on continuous parameters to characterise the coupling of continuous spectrum modes to continuous spectrum and surface modes.

If the properties of the nonuniform waveguide are changed slowly, more exactly if the ratio of the wavelength to the distance, along which the properties vary essentially, is small, the influence of the continuous spectrum modes on the discrete spectrum mode conversion appears only in the second order terms in expansions by this small parameter (Dikarev, 1974). Let us limit ourselves to first-order terms in the expansions for this small parameter in the present subsection. Hence, we can neglect the integral terms in the coupled-wave integral-differential equations and solve only a finite number of the coupled-wave differential equations. In such a way, the mathematical technique of the cross-section method for open waveguides will be the same as for metallic waveguides. The formulae for the coupling coefficients of the surface modes are valid, in particular the general formula of eqn. 2.47. Since the coupling is caused by a dielectric permittivity dependence on the longitudinal co-ordinate, eqns. 2.98 are not valid anymore for these coupling coefficients. However, eqns. 2.57 can be applied.

4.1.6 Rotation of polarisation in twisted dielectric waveguides

A waveguide twist causes the rotation of the polarisation plane. This phenomenon becomes more significant as the difference between the uniform and the nonuniform waveguides increases. The asymmetry in a metallic waveguide is defined by the function characterising the cross-section shape $\rho(\beta)$ (for example, eqn. 4.7). In a symmetrical dielectric waveguide $\varepsilon = \varepsilon(\rho)$, hence the difference between the waveguide with $\varepsilon = \varepsilon(\rho,\beta)$ and the symmetrical waveguide is characterised by a function of two variables. Therefore, the coupling coefficients of various modes in a dielectric waveguide as well as the phase difference Δh of the wavenumbers of two fundamental modes depend on each other in a lower degree as in a metallic waveguide. As a consequence, the theory of twisted dielectric waveguides (Katsenelenbaum *et al.*, 1996) is slightly more complicated than the theory of twisted metallic waveguides. We will limit ourselves to the problems where $\varepsilon = \varepsilon(\rho,\beta)$ only slightly deviates from a symmetric function $\varepsilon = \varepsilon(\rho)$. In particular, this condition is satisfied in optical fibres, so the following results can first be applied to them.

The dielectric permittivity in the nonuniform waveguide obtained by a uniform twist (i.e. a twist corresponding to eqn. 4.18) of a uniform waveguide with $\varepsilon = \varepsilon(\rho,\beta)$ depends on all the three co-ordinates ρ, β and z. However, only a special combination of β and z can happen

$$\varepsilon = \varepsilon(\rho, \beta + 2\pi z/t) \tag{4.37}$$

where t is the twist step. The rotation angle is $\varphi = 2\pi z/t$ and the relation between the twist step and the angle of rotation θ after the length L is $t = 2\pi L/\theta$. The dimensionless parameter q from eqn. 4.20 is connected with t and with the phase difference Δh of the wavenumbers of two uniform waveguide fundamental modes with orthogonal polarisation by the formula

$$q = \Delta h\, t \,/\, (2\pi) \tag{4.38}$$

The coupling coefficient of these two waves for small Δh must be derived from eqn. 2.47, since eqn. 2.57 is not convenient for practical calculations. One yields

$$S_{12} = (2\pi/t)\, A \tag{4.39}$$

This formula differs from eqn. 4.13 by the factor A. In a symmetrical waveguide $A = 1$ and A is close to one in a waveguide which is nearly symmetrical. The deviation of A from one is small but essential. This deviation depends on the type of the function $\varepsilon = \varepsilon(\rho,\beta)$. In an optical waveguide $|1-A|$ is of the same (second) order of the small parameter causing the asymmetry as $(\Delta h)^2$.

The polarisation plane is rotated by the angle θ_0 for the waveguide twist angle θ where the angles θ and θ_0 are connected with each other by eqn. 4.31. According to eqn. 4.31, the polarisation plane in practice follows the waveguide twist for a very slow twist, when the step t is large compared to the beat wavelength $\Lambda_0 = 2\pi / \Delta h$ in the uniform waveguide, i. e. when parameter $q >> 1$. The same effect takes place in a metallic waveguide, see the text following eqn. 4.35. The polarisation plane rotates slower as t decreases. As follows from eqns. 4.31 and 4.22, the angle of rotation per unit length for the most interesting case $q << 1$, i.e. $L\Delta h >> 1$ is equal to

$$\frac{\theta_0}{L} = \left| \frac{2\pi(A-1)}{t} + \left(\frac{\Delta h}{2}\right)^2 \cdot \frac{t}{2\pi} \right|, \quad t\Delta h << 2\pi \tag{4.40}$$

If $A < 1$ the magnitude of θ_0/L decreases monotonously for decreasing t and it becomes equal to zero at

$$t_0 = \frac{4\pi\sqrt{2(1-A)}}{\Delta h} \tag{4.41}$$

According to the remark following eqn. 4.39, this t_0 has a finite magnitude (in the scale of the wavelength $2\pi/h$). For $t << t_0$ the magnitude of θ_0/L is linearly proportional to $(1/t)$ and increases as $\theta_0/L = 2\pi(A-1)/t$.

The dependence of θ_0/L on t is slightly more complicated for $A>1$, but in both cases the polarisation plane rotation per unit length is proportional to t (hence, it increases with increasing t) in the interval $2\pi/h << t << 2\pi/\Delta h$. A faster rotating waveguide corresponds to a slower rotation of the polarisation plane. At the same time, the beat wavelength Λ in the twisted waveguide increases. According to eqn. 4.40, $\Lambda = 4\Lambda_0^2 / t$, where Λ_0 is the beat wavelength in the uniform (untwisted) waveguide. The increment of Λ for decreasing t is widely used to obtain optical fibres with $\Lambda >> \Lambda_0$ (Barlow *et al.*, 1981).

4.1.7 Nontransparency band

When ε against z is characterised by eqn. 4.37, the fact that periodicity of waveguide properties is not essential for the polarisation plane rotation described above is the main factor for another phenomenon: the appearance of a narrow nontransparency band for one of the rotation directions of a circularly polarised mode. This phenomenon is caused by superposition of backward waves excited by all nonuniformities. If these small amplitude waves are in phase, they form a backward wave with a finite intensity and the forward wave will be completely reflected from the periodic section. The eigenmodes of a twisted waveguide with $\Delta h\, t << 1$ are the circularly polarised modes, and the waveguide becomes nontransparent for one of these eigenmodes. The centre of the nontransparency band corresponds to the wavelength being approximately equal to the twist step. Let us remember that the usual Bragg nontransparency band for periodic structures corresponds to a wavelength equal to two periods hence, in this regard, the twisted waveguide is identical to a periodic medium with the period equal to half of the twist step. As is known, an analogous nontransparency band exists in holler liquid crystals.

To analyse the phenomenon occurring in the nontransparency band, one should apply the four coupled-wave differential equations of the cross-section method instead of two coupled-wave eqns. 4.16. Two of these equations correspond to the forward waves and two to the backward waves. The algebraic equation containing the wavenumbers of the eigenmodes is then biquadratic. Far from the nontransparency band this equation splits into two quadratic equations corresponding to the forward and backward waves of both circular polarisations. The solution of one of those equations is given in eqn. 4.27. Two of the four roots of the biquadratic equation become imaginary in the nontransparency band, therefore two (the forward and the backward) waves of an eigenmode become evanescent waves and cannot propagate. The width of the nontransparency band as well as the whole reflection process is determined by the coupling coefficients of two waves propagating in opposite directions. These coupling coefficients, unlike those in a metallic waveguide, are of the first order of a small parameter characterising the waveguide asymmetry (see text following eqn. 4.13). However, the coupling coefficients of the opposite propagating modes are small when compared with the coupling coefficients of eqn. 4.39 of two forward waves, which do not contain this small parameter. The coupling coefficients of the opposite waves can be calculated using eqn. 2.57, where one must assume $\partial\mu/\partial z \equiv 0$.

The coupling between two opposite propagating modes of orthogonal polarisations is the main factor for some specified functions of ε against the angle β used in practical calculations. The relative width of the nontransparency band $\Delta k/k$ is in this case approximately twice that of the coupling coefficient of both waves. It was assumed in the optical fibre with a step-wise distribution of dielectric permittivity that $\varepsilon(\rho,\beta) = \varepsilon_0(\rho) + \delta \sin^2(\beta)$, where $\delta << \Delta$ and Δ is the difference between ε in the core and in the shell of the fibre, whereas $\Delta << 1$. For this example, the value of $\Delta k/k$ is approximately equal to $D\Delta\delta$, where the factor D depends on frequency, but it remains practically constant inside the two-mode band and it is approximately equal to 0.3. If we assume that $\Delta = 10^{-3}$ and $\delta = 10^{-5}$, then the width of the nontransparency band would be of the order of 3×10^{-9} of the light frequency carrier.

4.1.8 TE_{11} and TM_{11} circular polarisers in oversized waveguides

Mode coupling in elliptically deformed oversized circular waveguides was studied theoretically and experimentally by Thumm and Kumrić (1989) in order to optimise TE_{11} and TM_{11} polarisers for the 1 MW/70 GHz long-pulse (3s) ECRH system of the Wendelstein 7-AS stellarator at the Max-Planck Institute for Plasmaphysics at Garching in Germany. The polarisation converters essentially consist of smooth-wall circular waveguides (inner diameter $2a$ = 27.79 mm) which are gradually squeezed. A sine-squared function of the length co-ordinate is used to get an almost elliptical cross section in the middle and circular cross sections at both ends ($l(z) = a\,\delta(z)$ with $\delta(z) = \delta_{max} \sin^2(\pi z/L)$, see eqn. 4.7). The normalised amplitude of the elliptical deformation δ can be determined from measurements of the maximum and minimum waveguide diameters as: $\delta(z) = [\,D_{max}(z) - D_{min}(z)]\,/\,4a$. The maximum diameter squeeze $\Delta D_{max} = 2a\cdot\delta_{max}$ is at the centre of the deformed waveguide section ($z = L/2$). Starting from linear polarisation oriented 45° with respect to the major axis of the ellipse (Figure 4.6), the magnitude of the two perpendicularly polarised components E_r^0 and E_r^x will be equal when the coupling integral of the deformation satisfies (Doane, 1986)

$$\left| \int_O^L S_{x,o}(z)dz \right| = \pi/4 \tag{4.42}$$

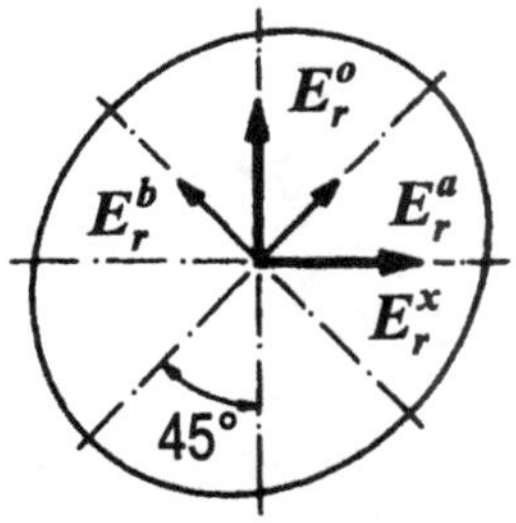

Figure 4.6 *Elliptical deformation $l = a\delta$ oriented to couple ordinary E_r^0 (vertical) and cross-polarised E_r^x (horizontal) radial electric field components and the components E_r^a and E_r^b with the greatest difference in propagation constants (phase velocities)*

where $S_{x,o}(z)$ is the relevant coupling coefficient and the integral covers the length L of the distorted waveguide section. Thus, an incident TE_{11} or TM_{11} mode with fixed (linear) polarisation can be converted to a circularly polarised mode and *vice versa*. Since the components E_r^0 and E_r^x are both oriented midway between the maxima and minima of the ellipse, they must travel with identical phase velocities (degenerate modes with $\pi/2$ phase difference). The condition of eqn. 4.42 is equivalent to the statement that there must be a $\pi/2$ relative phase shift for the independent polarisations E_r^a and E_r^b (Figure 4.6), corresponding to the fact that $S_{x,o}(z)$ is purely imaginary. The requirement of $\pi/2$ relative phaseshift leads to the condition

$$\left|\int_0^L |\Delta h|\, dz\right| = \pi / 2 \tag{4.43}$$

with Δh given in eqn. 4.36. For the TE_{11} mode we have $\Delta h = 6.3\, l / a^3 h$ and for the TM_{11} mode $\Delta h = 14.8\, l / a^3 h$ which means that the ellipticity coupling for the TE_{11} mode is 2.4 times smaller than that for the TM_{11} mode.

Figure 4.7 shows the dependence of the measured mode power in vertical and horizontal TM_{11} on the maximal diameter squeeze ΔD_{max} at the centre of the deformed waveguide section ($z = L/2$). Circular polarisation was achieved with $\Delta D_1 = 0.83$ mm. To rotate the polarisation by 90° requires a doubling of the distortion integral of eqn. 4.42 to $\pi/2$. This corresponds to a phase shift of π between the components parallel and perpendicular to the major axis of the ellipse. Figure 4.7 shows that the polarisation rotation was found at $\Delta D_2 =$ 1.65 mm. The dependence of the measured mode power in vertical and horizontal TE_{11} on the maximal diameter squeeze is plotted in Figure 4.8. As expected, because the coupling constant for the TE_{11} mode is 2.4 times smaller than that for the TM_{11} mode, circular polarisation was measured at $\Delta D_1 =$ 1.98 mm. By rotating the linear input polarisation, arbitrary elliptical polarisation states can be generated.

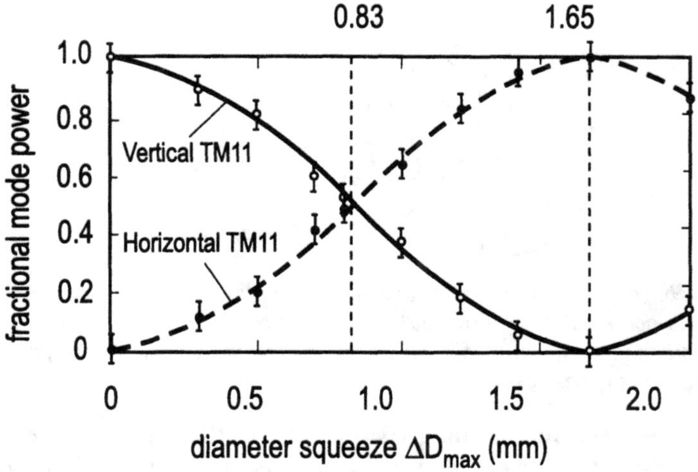

Figure 4.7 *Measured fractional mode power in vertical and horizontal* TM_{11} *as function of the maximum diameter squeeze* ΔD_{max} *of an improved elliptical polariser (70 GHz, 2a = 27.79 mm, L = 2 m, vertical input)*

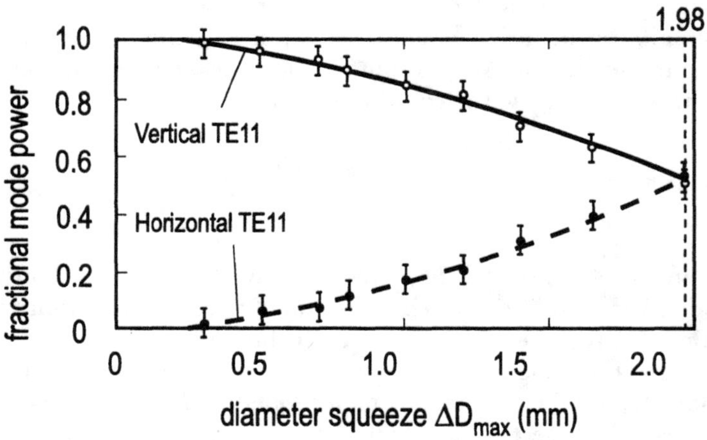

Figure 4.8 *Measured fractional mode power in vertical and horizontal* TE_{11} *as function of the maximum diameter squeeze* ΔD_{max} *of an improved elliptical polariser (70 GHz, 2a = 27.79 mm, L = 2 m, vertical input)*

4.2 Transitions between circular or rectangular waveguides of different cross sections

4.2.1 Coupling coefficients for circular and rectangular waveguides

Let us consider the excitation of parasitic waves along waveguide tapers, which are nonuniform waveguide sections with varying circular or rectangular cross section.

If a considered parasitic wave can propagate in a narrow waveguide, its amplitude is given by eqns. 2.75. If the parasitic wave cannot propagate, its amplitude is given by eqns. 3.56. If the waveguide contour has tilts at both end junctions of the nonuniform waveguide section, one yields a closed form analytical expression for the amplitudes, i.e. the two-term eqns. 2.78 or 3.58. A general expression for the coupling coefficients is given by eqn. 2.98.

Expressions for the coupling coefficients S_{jm} in the cases of circular and rectangular cross sections are derived in this subsection, using eqns. 2.98. As in eqn. 2.98, the first of the equations below *a* is for the coupling of two *TE* modes, the second equation *b* is for two *TM* modes and the third equation *c* is for a *TE* mode (m) and a *TM* mode (j). The upper sign corresponds to forward waves, $h_j > 0$, and the lower sign corresponds to backward waves, $h_j < 0$. These formulae assume no attenuation for both considered modes, i.e. h_m and h_j have real values.

For circular waveguides the coupling coefficients are given by

$$S_{jm} = \frac{\nu\left[\pm\mu_m^2\mu_j^2 + \left(|h_j|h_m \mp k^2\right)n^2a^2\right]}{|h_j|\left(|h_j| \mp h_m\right)a^3\sqrt{\mu_j^2 - n^2}\sqrt{\mu_m^2 - n^2}} \tag{4.44a}$$

$$S_{jm} = \frac{\nu\left(k^2 \mp |h_j|h_m\right)}{|h_j|\left(|h_j| \mp h_m\right)a} \tag{4.44b}$$

$$S_{jm} = \mp\frac{\nu k n}{|h_j|a\sqrt{\mu_m^2 - n^2}} \tag{4.44c}$$

According to eqn. 2.98, any incident wave excites only parasitic waves with its own first index, n, which determines the azimuthal field dependence. Obviously for $n = 0$ only modes of the same type (*TE* or *TM*) as the incident mode are excited. This case is an example of the general principle: the circularly symmetric *TE* and *TM* modes propagate independently from each other for any radius profile of a circular waveguide.

Coupling coefficients at an arbitrary location z are proportional to ν, i.e. to the tangent of the contour line angle at the given z. However, the expressions of the amplitudes P_j mentioned above are valid only for small values of ν and they include only terms of the first order of ν.

Hence, substituting the equations of this section in those expressions, ν can be assumed equal to the tangent of the angle between the contour line of the taper and the waveguide axis.

In the case of transitions between rectangular waveguides, let us consider, for the sake of simplicity, only those symmetric tapers where both angles between the axis and the two narrow walls are equal to $\arctan(\nu_1)$ and the angles between the axis and the two broad walls are equal to $\arctan(\nu_2)$. The narrow wall length is b and the broad wall length is a, so that $\nu_1 = (1/2)(da/dz)$ and $\nu_2 = (1/2)(db/dz)$.

At such a transition the coupling coefficient of two modes is not equal to zero only if one of the parasitic mode indices is the same as the corresponding index of the incident mode, and the other pair of indices is of the same parity. Then the coupling coefficient and the amplitude of a parasitic mode are proportional to ν_1, if the index q is the same for the coupled waves. The index q determines the field dependence on the co-ordinate parallel to the narrow side, i.e. it is the second index in eqn. I.21. If the first indices n are the same, the coupling coefficient S_{jm} is proportional to ν_2. Hence in eqns. 4.45 given below, one should calculate either only the first term proportional to ν_1 (if $q_j = q_m$ and $n_j + n_m$ is an even number) or only the second term proportional to ν_2 (if $n_j = n_m$ and $q_j + q_m$ is an even number). Only the coupling coefficients for backward waves contain both tangents of the angles. For example, if the TE_{10} wave is incident ($n_m=1$; $q_m=0$), the odd-order waves TE_{30}, TE_{50}, ... are excited in both directions with the amplitudes proportional to ν_1. The degenerate wave pairs TE_{12}/TM_{12}, TE_{14}/TM_{14}, ... etc. are also excited in both directions with amplitudes proportional to ν_2. The reflected TE_{10} wave is excited with the amplitude P_{-m} containing both ν_1 and ν_2.

The coupling coefficients are equal to

$$S_{jm} = \frac{N^j N^m}{2|h_j|(|h_j| \mp h_m)} \left\{ \nu_1 \left[\pm\alpha_j^2 \alpha_m^2 b \in_q + (|h_j| h_m \mp k^2) \frac{\pi^2 q^2}{a} \right] + \nu_2 \left[\pm\alpha_j^2 \alpha_m^2 a \in_n + (|h_j| h_m \mp k^2) \frac{\pi^2 n^2}{a} \right] \right\} \tag{4.45a}$$

$$S_{jm} = \frac{M^j M^m \pi^2 (k^2 \mp |h_j| h_m)}{2|h_j|(|h_j| \mp h_m)} \left(\nu_2 \frac{n_j n_m b}{a^2} + \nu_2 \frac{q_j q_m a}{b^2} \right) \tag{4.45b}$$

$$S_{jm} = \pm \frac{M^j N^m \pi^2 k}{2|h_j|} \left(\nu_1 \frac{q n_j}{a} + \nu_2 \frac{n q_j}{b} \right) \tag{4.45c}$$

where the normalisation factors M and N are given in eqn. I.23.

4.2.2 *Reflection coefficient for the TE_{10} mode in rectangular waveguide*

Let us consider in more detail the problem of the reflection of the fundamental wave TE_{10} from a rectangular waveguide section with varying cross section. According to eqn. 2.75 the reflection coefficient R is equal to

$$R = -\int_0^L S_{-mm} \exp\left(-2i\int_0^z h_m dz\right) dz \tag{4.46}$$

According to (4.45*a*) the coupling coefficient can be written in the form

$$S_{-mm} = \frac{1}{2b}\frac{db}{dz} - \frac{\pi^2}{2a^3 h_m^2}\frac{da}{dz} \tag{4.47}$$

It can easily be shown, that if the side direction is invariant, this expression for S_{-mm} is valid for the TE_{10} wave in an arbitrary varying cross section. In particular, the cross section may be not symmetric along any fixed axis.

According to eqn. 4.47 the TE_{10} mode reflection coefficient at a junction of two rectangular waveguides with nearly the same cross sections is equal to

$$R = \frac{1}{2}\left(\frac{\Delta b}{b} - \frac{\Delta a}{a}\frac{\pi^2}{a^2 h_m^2}\right) \tag{4.48}$$

Eqns. 4.46 and 4.48 may be written in the same form as the formulae for the reflection coefficient in a long inhomogeneous transmission line with wave impedance $W(z)$. As is known, using the same approximation as for eqn. 4.46, one yields for the reflection coefficient of a long transmission line section with a variable wave impedance

$$R = -\frac{1}{2}\int_0^L \frac{d\ln(W)}{dz} \exp\left(-2i\int_0^z h_m dz\right) dz \tag{4.49}$$

The reflection coefficient for a junction of two transmission lines with wave impedances which differ by ΔW is equal to:

$$R = \frac{\Delta W}{2W} \tag{4.50}$$

Eqn. 4.49 would be transformed into eqns. 4.46 and 4.47, whereas eqn. 4.50 would be simultaneously transformed into eqn. 4.48 if the wave impedance of the TE_{10} mode in a rectangular waveguide is taken as

$$W = C\frac{b}{h_m} \tag{4.51}$$

where it is assumed that C does not depend on a and b.

However, this expression can be used only for the calculation of the reflection coefficient and cannot be used to calculate the relation between the current, voltage and power flux in a waveguide, for example.[2]

Let us note that eqn. 4.46 for the reflection coefficient of an arbitrary mode in an arbitrary waveguide can be written in the form of eqn. 4.49, if the wave impedance is determined by

$$\frac{d\ln(W)}{dz} = -\frac{1}{2}S_{-mm} \tag{4.52}$$

Therefore, the reflection coefficient for the junction of two rectangular waveguides with nearly the same cross sections may in this way be written in the form of eqn. 4.50.

If the contour line of a rectangular waveguide taper has a tilt, the integral expression of eqn. 4.46 is transformed into a two-term formula similar to eqn. 2.78. Let us present the first term of such a formula

$$i\frac{S_{-mm}}{2h_m} \tag{4.53}$$

This term is caused by a tilt at the beginning of the transition. Sometimes this term is referred to in the literature as the reflection from a junction. According to eqn. 4.47, this value is equal to (Piefke, 1957)

$$i\frac{1}{2h_m b}\nu_2 - i\frac{\pi^2}{2h_m^2 a^3}\nu_1 \tag{4.54}$$

The first term in eqn. 4.54 describes the so-called reflection from an abrupt E-plane waveguide step and the second the reflection from an abrupt H-plane waveguide step. Eqn. 4.54 was derived many times by various methods. As it may be easily proven, the reflection coefficient from a waveguide step given by eqn. 4.53 for the fundamental TEM mode in a parallel plate waveguide is equal

[2] The expression $W = Cb/ah$ is routinely used to calculate powers. However, it cannot be used to calculate R using eqns. 4.49 and 4.50 as was proposed in Kienlin and Kürzl (1958), Johnson (1959), Matsumaru (1959) and Wray (1959).

to the first term in eqn. 4.54, where the free-space wavenumber k must be substituted instead of h_m.

4.2.3 Conical waveguide tapers

Let us consider several examples, where the coupling coefficient formulae derived above in subsection 4.2.1 are applied. By means of those formulae one can calculate the amplitudes of the waves scattered from an arbitrary system. The plots presented later illustrate only basic features of the general phenomena.

Let us first derive expressions for the phase factors $\gamma_j(z)$ in straight tapers, where the contour lines are straight lines, i.e. the taper tangent of the angle ν does not depend on z.

A simple calculation for a conical taper in a circular waveguide, where the cross-section radius is given by $a(z) = a(0) + \nu z$ and $\nu < 0$, yields

$$\begin{aligned} \gamma_m(L) &= C\{g_m(0) - \arctan(g_m(0)) - g_m(L) + \arctan(g_m(L))\} \\ g_m(z) &= \frac{h_m(z)}{\alpha_m(z)} \end{aligned} \tag{4.55}$$

Here, the coefficient $C = -\mu_m/\nu$ for *TE* modes and $C = -\nu_m/\nu$ for *TM* modes, where ν_m and μ_m are the roots of the corresponding Bessel function and of its derivative (see Section I.3). Eqn. 3.29 is a particular case of eqn. 4.55, it is derived from eqn. 4.55 for $h_m(L) = 0$.

In the general case of a pyramidal-tapered rectangular waveguide with straight contour lines (i.e. for $n_m \neq 0$; $q_m \neq 0$; $\nu_1 \neq 0$; $\nu_2 \neq 0$), the calculation of γ_m using eqns. I.13, I.22 leads to elliptical integrals and is not presented here. The wavenumber h_m of a TE_{n0} mode does not depend on b and eqn. 4.55 is valid with $C = -\pi n/(2\nu_1)$. A similar result is valid for TE_{0q} modes. If $\nu_1 = 0$ or $\nu_2 = 0$ (i.e. if the waveguide step is in one plane only), eqn. 4.55 is also valid for all modes with $C = -\pi q/(2\nu_2)$ if $\nu_1 = 0$ or $C = -\pi n/(2\nu_1)$ if $\nu_2 = 0$.

Finally, eqn. 4.55 is also valid for a transition where all cross sections are similar to each other, i.e. a/b does not depend on z. There $C = -\dfrac{\pi}{(2\nu_1)}\sqrt{n^2 + \dfrac{a^2 q^2}{b^2}}$.

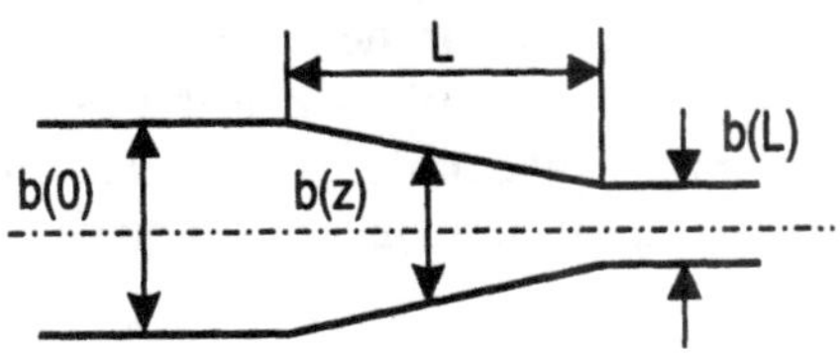

Figure 4.9 Schematic of linear E-plane taper (a = const) from b(0) to b(L) (b(0) > b(L))

As a first example, let us consider a transition from a quadratic waveguide ($b = a$) to a rectangular waveguide where $b/a = 1/3$. We assume that only the narrow side is tapered (E-plane taper), $b(z)=b(0)+2\nu_2 z$ and the broad side is constant, i.e. $\nu_1 = 0$ (Figure 4.9).

The TE_{10} wave is incident from the broad waveguide. The frequency is in the range $\sqrt{5}\pi < ka < 3\pi$, so that the mode pair TM_{12} and TE_{12} is excited there. At the highest frequency, i.e. when $ka=3\pi$, the cut-off cross section for these modes is located at a distance of approximately $0.6L$ from the broad waveguide end ($b(0)$). At the lowest frequency, i.e. when $ka = \sqrt{5}\pi$, the cut-off cross section is located near the broad waveguide end and those modes are not excited. The amplitude of the TE_{12} wave propagating backward in the broad waveguide ($b(0)$) is calculated by eqn. 3.58 without the integral term. The amplitude is equal to

$$P_{-j}(0) = i\sqrt{\frac{2}{5}}\nu_2 \left\{ \frac{1}{a\left(h_m - h_j\right)} \exp\left(-2i\left(\tilde{\gamma}_j + \pi/4\right)\right) + \frac{1}{a\left(h_m + h_j\right)} \right\}_{z=0} \tag{4.56}$$

Let us note that the coupling coefficient of the TE_{12} mode to the TE_{10} mode when $\nu_1 = 0$ does not contain the wavenumber h_j of the TE_{12} mode in its denominator and therefore this coefficient does not become infinite near the cut-off cross section where $h_j = 0$. Hence, eqn. 4.56 can be used, unlike eqn. 4.57 for the amplitude of the TM_{12} mode, for any small distance between the cut-off cross section and the beginning of the broad waveguide, i.e. for an arbitrary small $h_j(0)$. Obviously, other coupling coefficients of that mode become infinite if $h_j \to 0$, in particular $S_{-jj} \to \infty$, and all the considerations presented in Sections 3.1 and 3.2 are valid.

The amplitude of the TM_{12} wave is equal to

$$P_{-j}(0) = -2i\sqrt{\frac{2}{5}}\nu_2 \left\{ \frac{k}{h_j} \left[\frac{1}{a\left(h_m - h_j\right)} \exp\left(-2i\left(\tilde{\gamma}_j + \pi/4\right)\right) + \frac{1}{a\left(h_m + h_j\right)} \right] \right\}_{z=0} \tag{4.57}$$

This formula is valid if $h_j(0)$ is not very small. If this is not the case, the integral term in eqn. 3.59 cannot be neglected.

The absolute values of the amplitudes of eqn. 4.56 and 4.57 are shown in Figure 4.10 for $\nu_2 = -0.1$. The upper curve is for the TM_{12} wave, its amplitude is approximately two times higher than the amplitude of the TE_{12} wave.

Let us consider further a circular waveguide. The reflection coefficient $\rho = P_{-m}(0)$ of the TE_{11} wave from a straight circular cone is equal to (if there is no cut-off cross section for this mode)

$$\rho = \frac{i\nu}{4\left(\mu_m^2 - 1\right)} \left\{ \left[\frac{-\mu_m^4 + \left(k^2 + h_m^2\right)a^2}{h_m^3 a^3} \right]_{z=0} - \left[\frac{-\mu_m^4 + \left(k^2 + h_m^2\right)a^2}{h_m^3 a^3} \right]_{z=L} \exp\left(-2i\tilde{\gamma}_m\right) \right\} \tag{4.58}$$

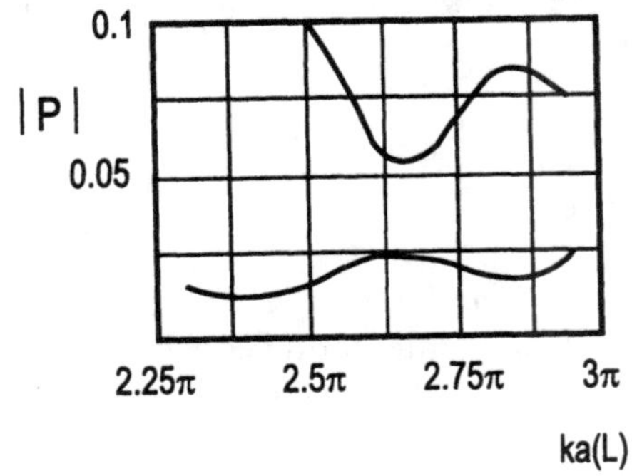

Figure 4.10 Normalised amplitudes of the parasitic backward waves TE_{12} (lower curve) and TM_{12} (upper curve) in rectangular E-plane down-taper

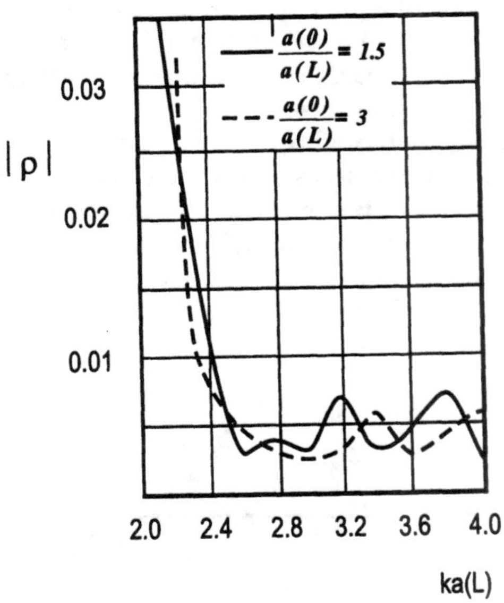

Figure 4.11 Reflection coefficient of the TE_{11} wave in conical up-tapers with $\nu = -0.1$

where $\mu_m = 1.84$. The absolute value $|\rho|$ against $ka(L)$ is presented in Figure 4.11 for $\nu = -0.1$ and for two values of the ratio between the waveguide radii. The curves oscillate, since, according to eqn. 4.58, ρ is equal to the sum of two interfering terms related with the beginning of the transition and with its end.

However, the term corresponding to the narrow part dominates, since the third power of the radius contained in the denominator of eqn. 4.58 is smaller there. The oscillations are fast enough because the phase of the second term is approximately equal to $2kL$ and it varies rapidly against k.

The reflection coefficient of the TE_{01} wave at an arbitrary circular symmetric transition can be calculated using the following integral formula, if there is no cut-off cross section in the transition

$$\rho = -\frac{1}{2}\int_0^L \frac{1}{h_m}\frac{dh_m}{dz}\exp\left(-2i\int_0^z h_m\,dz\right)dz \tag{4.59}$$

This equation is derived from eqns. 2.75 and 2.97. There is no second term in eqn. 2.97 for this wave. For a conical taper, integrating eqn. 4.59 by parts and substituting the analytical expression for the wavenumber, we yield after simple manipulations

$$\rho = \frac{-i\nu\mu_m^2}{4}\left\{-\left(\frac{1}{h_m^3 a^3}\right)_{z=0} + \left(\frac{1}{h_m^3 a^3}\right)_{z=L}\exp(-2i\bar{\gamma}_m)\right\} \tag{4.60}$$

where μ_m = 3.83. The absolute value of eqn. 4.60 is presented in Figure 4.12 against $ka(L)$ in the range from $ka(L)$=3.83 to $ka(L)$=5. In this plot the ratio of the input and output waveguide radii is equal to two, i.e. $a(0)/a(L)$ = 2, and the tangent of the contour tangent of the angle ν = – 0.1.

Obviously, if the product $ka(L)$ is only slightly greater than μ_m, eqn. 4.60 cannot be applied. In such a case one has to use eqns. 3.32 and 3.17. These formulae allow us to calculate ρ as a function of frequency from small values given either by eqn. 4.59 or by its equivalent, eqn. 3.31 until $|\rho|$ = 1. The parameter A in eqn. 3.18 is equal to $(-2\nu/\mu_m)^{1/2}$ for the TE_{01} mode. The substitution of the given value of ν yields A=0.23 and

$$-t_L = 3.74\left[k\cdot a(L) - \mu_m\right] \tag{4.61}$$

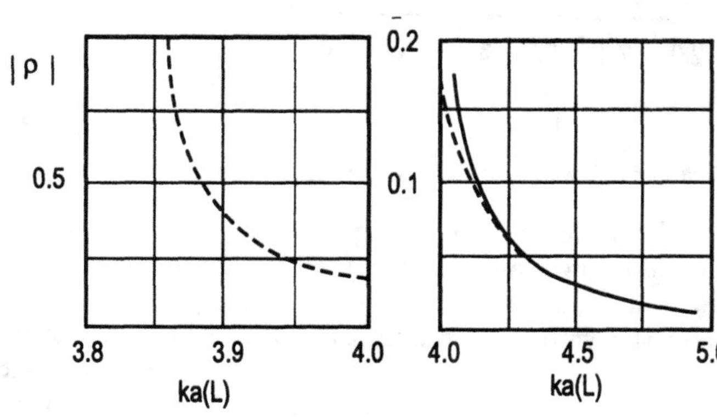

Figure 4.12 Reflection coefficient of the TE_{01} wave in conical up taper with ν= -0.1 *and $a(0)/a(L)$ = 2*

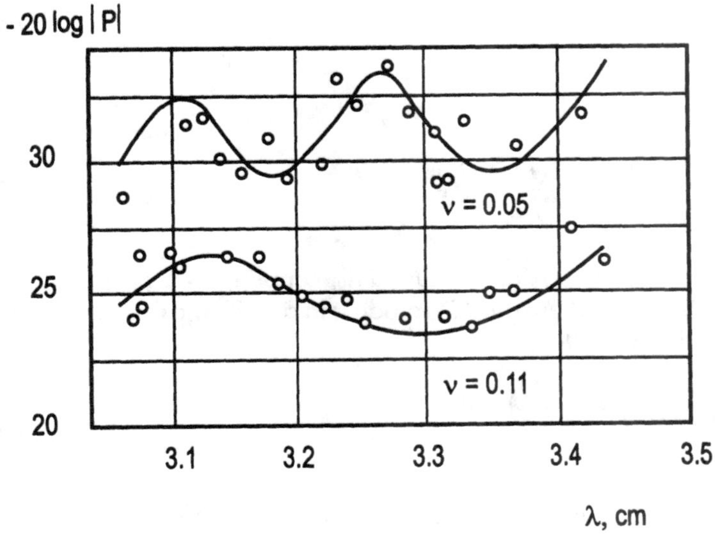

Figure 4.13 Relative power of the parasitic TM_{11} mode in different conical TE_{11} mode tapers

The curves calculated using eqn. 3.32 (the dashed line in Figure 4.12) and according to eqn. 4.60 (the solid line) smoothly overlap. If the more complicated eqn. 3.34 was used, the agreement would be better.

Figure 4.13 is presented in Vaganov (1960). If the TE_{11} wave is incident on a straight circular cone, the curves show the relative power of the TM_{11} wave in dB, i.e. in relation to the incident TE_{11} wave.

The radii of the two waveguides connected by the conical taper are equal to 1.5 cm and 2.5 cm, respectively. The upper curve corresponds to $\nu = -0.05$, i.e. the taper's length L is 20 cm and the lower curve corresponds to $\nu = -0.11$, i.e. when the taper's length is 9 cm. The curves are calculated employing eqn. 3.58 without the integral term. The points indicate experimental results. The results are in good agreement with the theoretical curves within the measurement tolerances.

4.2.4 Mode conversion in straight tapers

Here we summarise useful formulae for straight up tapers with the length L for rectangular waveguides carrying the TE_{10} mode and for circular waveguides supporting the TE_{01} mode. Solymar (1958, 1959b) has shown that if the input waveguide is below cut off for the corresponding parasitic modes, the reduced amplitude p_j of the TE_{10} mode considering coupling to the TE_{30} mode or the composite TE_{12}/TM_{12} modes in a symmetrical H-plane or E-plane taper, respectively, is given by

$$|p_j| = 0.12\frac{a(L)(a(L)-a(0))}{\lambda L} \tag{4.62a}$$

and

$$|p_j| = 0.23\frac{b(L)(b(L)-b(0))}{\lambda L} \tag{4.62b}$$

The corresponding expression for TE_{02}-TE_{01} coupling in a TE_{01} mode taper with input radius $a(0)$, where the TE_{02} mode is in cut off, and output radius $a(L)$ is (see also later eqn. 4.90)

$$|p_j| = 0.57\frac{a(L)(a(L)-a(0))}{\lambda L} \tag{4.63}$$

4.2.5 TE_{01} mode reflection from symmetrical dielectric transition

Eqn. 4.59 can also be applied to calculating the TE_{01} wave reflection from a symmetrical dielectric transition connecting a circular waveguide filled with a dielectric with an empty circular waveguide. The validity of eqn. 4.59 can be proven by comparison between eqn. 2.58 for h_m' and eqn. 2.57 written down for S_{-mm}. If the magnetic permeability is constant and $\mu = 1$, these expressions differ by a common factor $1/(2h_m)$ and by the sign of the term $(E_z^m)^2$. For circular symmetric TE modes and for a symmetrical filling $E_z^m \equiv 0$, and hence

$$S_{-mm} = -h_m'/2h_m \tag{4.64}$$

Eqn. 4.59 for the reflection coefficient from a dielectric insertion follows from eqn. 4.64 and from the general expression of eqn. 2.75.

A reference waveguide for the nonuniform waveguide in Figure 4.14 is the uniform waveguide with a dielectric rod shown in Figure 4.15.

The values of h_m for the TE_{01} mode in such a waveguide were determined in Kazantsev (1957) and Van Huan-Chzho (1960) for various values of the rod radius b. Using the function $h_m(b)$ determined in such a way, the reflection coefficients ρ were calculated for various transition shapes. The dependence of ρ on the shape of the transition and on its medium was studied, whereas Figure 4.16 is taken from Kazantsev (1957).

The plot shows the frequency dependence of the TE_{01} wave reflection coefficient from a polystyrene cone with the length of 100 mm situated in a circular waveguide of 25 mm radius. The experimental results are indicated by crossed points.

It should be noted, that the angle of the cone's contour line was constant along the whole transition. Despite this, the coupling coefficients of the circular symmetric TE modes decrease to zero near the end of the cone. The reason

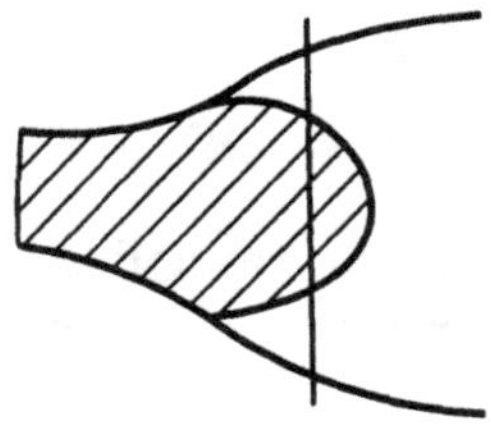

Figure 4.14 Symmetrical dielectric transition

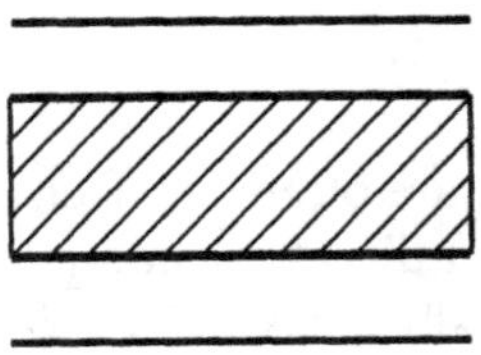

Figure 4.15 Reference waveguide for the transition of Figure 4.14

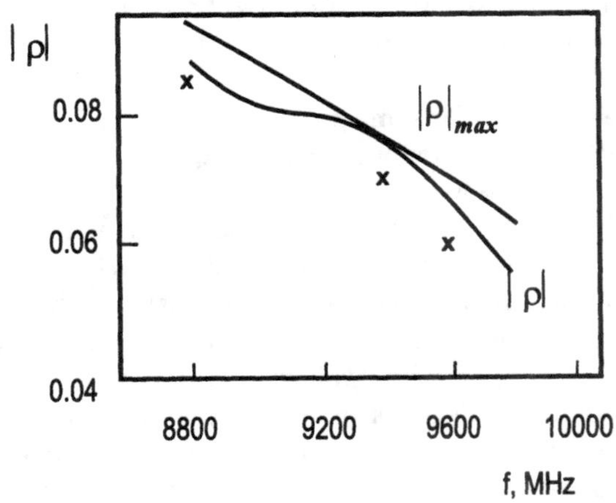

Figure 4.16 Reflection coefficient for a polystyrene cone

for this is that eqn. 2.63 for those modes contains only one component E_s of the electrical field approaching to zero near the metal. Hence, a dielectric cone is a smooth transition for TE_{0q} modes, i.e. S_{jm} is equal to zero at the ends of such a transition. Therefore, eqn. 2.78 cannot be used for the problem considered in this subsection. For other modes containing an electrical field component E_n normal to the metal, eqn. 2.78 gives a correct expression for the amplitudes of waves scattered from a dielectric cone or from another dielectric transition, because the values of the coupling coefficients at the transition end are not equal to zero. A simple expression for these coupling coefficients follows from eqn. 2.63

$$S_{jm} = \frac{1}{2h_j\left(h_j - h_m\right)} \frac{\varepsilon - 1}{\varepsilon} \oint \nu E_n^m E_n^j \, ds \tag{4.65}$$

This formula contains the fields on the metal in a waveguide completely filled by a dielectric. Those fields are easily determined using formulae similar to eqn. I.14.

4.3 Smoothed nonlinear tapers

4.3.1 General considerations

The conversion losses to parasitic modes are often relatively high in linearly tapered waveguide transitions with contour tilts as considered in the first three subsections of the previous section. Generally speaking, one could increase the transition length to decrease these conversion losses. If there are no stationary phase points along the transition then, according to eqn. 2.78, conversion losses are proportional to ν_0^2 and they decrease with increasing L. The conversion losses are approximately inversely proportional to L^2. If there is a stationary phase point along the transition, then conversion losses decrease more slowly, approximately as $1/L$.

However, this way is unacceptable for some problems, where very small conversion losses are required. It will be obvious from examples considered later that the required transition lengths would be too long. Another method is to improve the matching of a waveguide taper at the junctions with uniform waveguides. Without any calculations it is obvious from eqns. 2.78 and 3.58, as well as from the formulae of the previous section, that the losses decrease if the ν value at the transition ends is decreased. In such a way we design so-called smoothed nonlinear tapers, i.e. tapers with ν values at the transition ends that are either equal to zero or significantly lower than the ν values along the taper. More exactly, the coupling coefficients and not the ν values must possess this feature. However, in most cases, the dependence of S_{jm} is mainly determined by the ν value along a transition, apart from for several exceptions mentioned at the end of the previous section.

Eqns. 2.78 and 3.58 cannot be used for these transitions, since it was supposed in the derivation of these formulae that the terms of the order ν^2 and ν' are small compared with $\nu(0)$ and $\nu(L)$. This is not the case for smoothed transitions. The integral eqns. 2.75 or 3.56 must be used for calculation of smoothed tapers.

As a rule, low conversion losses are required from waveguide tapers in a broad frequency band. If broadband behaviour is not required, the interference structure of a parasitic mode amplitude can be used to decrease its amplitude at the taper output (Doane, 1984). The parasitic mode is excited in a transition with tilts of the waveguide contour. The taper length L can be designed in such a way that both terms in eqn. 2.78 have the opposite signs at a given frequency. At the same time, $\nu(0)$ and $\nu(L)$ must be designed so that these two terms in eqn. 2.78 have their absolute values equal to each other. For example, according to eqn. 4.60, the reflection coefficient of the TE_{01} mode is equal to zero for a circular symmetric taper between two circular waveguides, if the following two conditions are satisfied

$$\gamma_m(L) = n\pi, \quad n = 1,2,\ldots, \quad \left[\frac{\nu}{h_m^3 a^3}\right]_{z=0} = \left[\frac{\nu}{h_m^3 a^3}\right]_{z=L} \tag{4.66}$$

According to the second condition, the contour tilt angle must be significantly greater at the broad end of the transition than at its narrow end.

If low conversion losses to forward parasitic modes are required, the opposite is true: the tilt angle at the broad transition end must be designed to be smaller than that at its narrow end. Indeed, according to eqn. 2.78, the values of $S_{jm}/(h_j - h_m)$ must be equal at both ends. Since the ratio of the coupling coefficients is of the order of ν/a, the ratio of $\nu(0)$ and $\nu(L)$ must be approximately equal to the value of $a(h_j - h_m) = \dfrac{(a\alpha_m)^2 - (a\alpha_j)^2}{(h_m + h_j)a}$, i.e. roughly it is equal to $1/a$. The phase difference of the two modes at the output of the taper, defined by eqn. 2.82, must be equal to π, or to 2π, 3π etc. If σ is small $(|\sigma| < \pi)$, compensation is impossible. As we will see later, smoothed tapers designed to provide small conversion to forward parasitic modes in a broad frequency band, also must be smoother at a broad waveguide side and the $|\sigma|$ value of these transitions must be large enough.

As was shown by Solymar (1959c, 1959d), the frequency band can be broadened by successive connection of several cones with different contour angles. These systems are in fact similar to the smoothed tapers, considered later. The contour angle is tapered there continuously.

Obviously, a mutual compensation of the two terms in eqn. 2.78 or in eqn. 3.58 means that a parasitic mode amplitude P_j is equal to zero only in the first order of ν_0 and $P_j \neq 0$ in the next order. So, the remark in the footnote of subsection 2.2.8 must be taken into account.

4.3.2 Reflection coefficient for different contours

The main problem of waveguide tapers is to design the taper length and its contour shape. The same problem arises in the design of broadband matching sections between two transmission lines, e.g. between two coaxial lines of different wave impedances. We use this analogy later to design the shape of a waveguide taper.

According to eqn. 4.49, the reflection coefficient ρ of an inhomogeneous transmission line with a tapered wave impedance W and with a constant phase velocity is equal to

$$\rho(\sigma) = -\frac{1}{2}\int_0^1 \frac{d\ln(W)}{d\zeta}\exp(-i\sigma\zeta)d\zeta \tag{4.67}$$

Here, $W(\zeta)$ is the wave impedance, the integration variable is $\zeta = z/L$ and L is the taper length. The value of $2kL$ is denoted by σ to underline an analogy with the formulae for spurious mode amplitudes, which are written later in the same form. As we have already mentioned in Section 4.2, eqn. 4.67 is valid in the same approximation as eqns. 2.75 and 3.56.

The inverse problem was studied extensively for eqn. 4.67, i.e. the problem of an optimal tapering $W(\zeta)$. The function $W(\zeta)$ is determined in this problem by the requirement that $|\rho(\sigma)|$ is less than ρ_{max} for all values of the parameter σ greater than σ_{min}, which means

$$|\rho(\sigma)| < \rho_{max} \text{ for } \sigma > \sigma_{min} \tag{4.68}$$

The function $W(\zeta)$ is the optimal function if its ρ_{max} has the minimal value achievable for a given σ_{min} or if its σ_{min} has the minimal value for a fixed ρ_{max}. In fact, determining the optimal $W(\zeta)$ function, we achieve the minimal reflection coefficient for a minimal taper length, since σ is proportional to L.

The function $W(\zeta)$ is optimal in this sense and was determined in Klopfenstein (1956). There is a simple relationship between ρ_{max} and σ_{min} for $W_{opt}(\zeta)$

$$\left|\frac{\rho_{max}}{\frac{1}{2}\ln\left(\frac{W(1)}{W(0)}\right)}\right| = \frac{1}{\cosh\left(\frac{\sigma_{min}}{2}\right)} \tag{4.69}$$

We will employ the function $W_{opt}(\zeta)$ in the next section. Let us use in the present section two other functions $W(\zeta)$ proposed in Willis and Sinha (1955) to design the contour of a waveguide taper with the required low conversion losses. These functions give a solution of the optimisation problem within a class of functions with continuous derivatives in the range of $0 \le \zeta \le 1$, in contrast to the function $W_{opt}(\zeta)$ having discontinuities at the interval ends. A waveguide taper designed with the $W_{opt}(\zeta)$ function would have small steps at its ends.

The first of these functions corresponds to $\sigma_{min} = 2.7\pi$. It is determined from the following equation

$$\frac{d\ln(W(\zeta))}{d\zeta} = \sin(\pi\zeta) \tag{4.70}$$

The left-hand side of eqn. 4.69 for this function is equal to 0.07. According to eqn. 4.69, this value for W_{opt} is achieved for $\sigma_{min} = 2.14\pi$, i.e. the value is 20% less. The function

$$\left| \frac{\rho(\sigma)}{\frac{1}{2}\ln\left(\frac{W(1)}{W(0)}\right)} \right| = \left| \frac{\pi^2}{\pi^2 - \sigma^2} \cos\left(\frac{\sigma}{2}\right) \right| \qquad (4.71)$$

is presented in Figure 4.17*a*. Eqn. 4.71 is the integral of eqn. 4.67 normalised to one and calculated with this $W(\zeta)$ function.

The second function $W(\zeta)$ corresponds only to a slightly greater σ_{min} value, $\sigma_{min} = 3\pi$, but its left-hand side value of eqn. 4.69 is significantly less and equal to 0.03. This value for W_{opt} is achieved for σ_{min} which is only 11% less. The function is determined from the equation

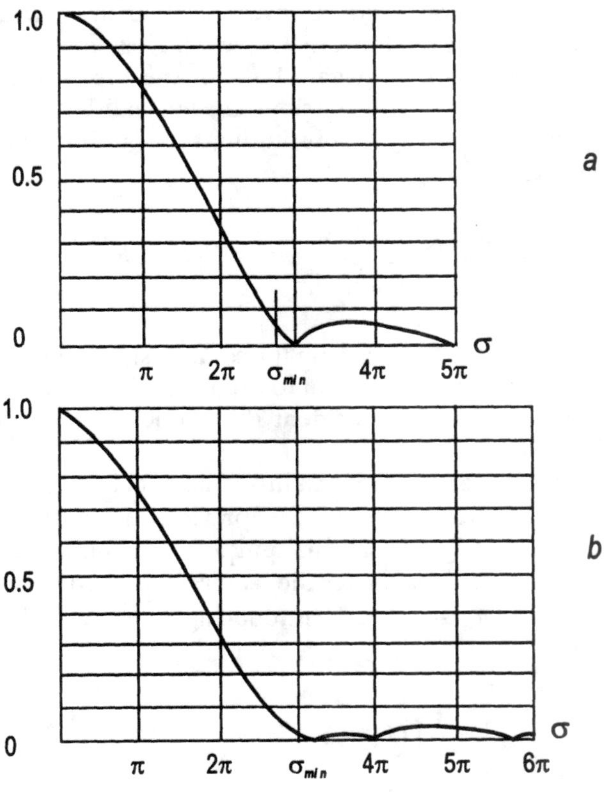

Figure 4.17 Functions of (a) eqn. 4.71 and (b) eqn. 4.73 against the phase parameter σ

$$\frac{d\ln(W(\zeta))}{d\zeta}=1-0.636\cos(2\pi\zeta) \tag{4.72}$$

and for this function

$$\left|\frac{\rho(\sigma)}{\frac{1}{2}\ln\left(\frac{W(1)}{W(0)}\right)}\right|=\left|\frac{\sin\left(\frac{\sigma}{2}\right)}{\frac{\sigma}{2}}\frac{4\pi^2-0.364\sigma^2}{4\pi^2-\sigma^2}\right| \tag{4.73}$$

Eqn. 4.73 is plotted in Figure 4.17*b*.

4.3.3 Nonlinear TE_{01} mode taper

To illustrate the design method of a smoothed waveguide taper with low losses in a broad frequency band, let us design a circular symmetric waveguide taper for the TE_{01} mode in a circular waveguide. Conversion losses in such a taper are mainly caused by conversion to the forward TE_{02} wave. First, let the frequency be so high that this mode can propagate both in the broad and narrow waveguides, i.e. there is no cut-off cross section. The normalised amplitude of the TE_{02} mode is given by eqn. 2.75. Substituting the coupling coefficients from eqn. 4.44*a*, we yield

$$\sqrt{\frac{h_j(0)}{h_m(0)}}p_j(L)=-\frac{2\mu_m\mu_j}{\mu_j^2-\mu_m^2}\int_0^L\frac{h_m+h_j}{2\sqrt{h_mh_j}}\frac{a'}{a}\exp\left(-i\int_0^{z_1}\left(h_m-h_j\right)dz\right)dz_1 \tag{4.74}$$

Here, $a(z)$ is the desired function, describing the contour shape. The square of the right-hand side absolute value is equal to the ratio of the power carried away by the TE_{02} mode to that of the incident TE_{01} mode, i.e. the conversion losses into the TE_{02} mode.

The integral of eqn. 4.74 can be rewritten in an identical form to eqn. 4.67. Hence, the mathematical technique of the previous subsection can be employed for a taper design. For this purpose we omit the factor $(h_m+h_j)/(2\sqrt{h_mh_j})$. Let us note that hence the requirement to the frequency is weaker than if we required $\hbar_m\cong 1$, $\hbar_j\cong 1$. Introducing a new variable

$$\zeta=\frac{\int_0^z\left(h_m-h_j\right)dz}{\int_0^L\left(h_m-h_j\right)dz}=\frac{1}{\sigma}\int_0^z\left(h_m-h_j\right)dz \tag{4.75}$$

and substituting $\mu_m = 3.83$ and $\mu_j = 7.02$ we yield a formula similar to eqn. 4.67

$$p_j(\sigma) = -1.56\int_0^1 \frac{d\ln(a)}{d\zeta}\exp(-i\sigma\zeta)\,d\zeta \tag{4.76}$$

Comparing eqn. 4.76 with eqn. 4.67, one notes that if the derivative of $\ln a(\zeta)$ is equal to the derivative of $\ln W(\zeta)$ (both are considered as functions of ζ) and $W(\zeta)$ provides a broadband matching of two long transmission lines with different wave impedances, then the waveguide taper has low conversion losses to the TE_{02} mode.

For further consideration, it is convenient to introduce the parameter a_{av}

$$a_{av}^2 = \int_0^1 \frac{\hbar_m + \hbar_j}{2} a^2(\zeta)\,d\zeta \cong \int_0^1 a^2(\zeta)\,d\zeta \tag{4.77}$$

This parameter is practically independent of frequency, it is measured in length units and it is in the range between $a(0)$ and $a(L)$. As follows from eqns. 4.75 and 4.77, the parameter σ, which plays the same role in the waveguide taper problem as the value $2kL$ in eqn. 4.67, is related with a_{av} by

$$\sigma = \frac{\mu_j^2 - \mu_m^2}{2}\,\frac{L}{ka_{av}^2} \tag{4.78}$$

and the differentials of the variables ζ and z are related by

$$\frac{dz}{L} = \frac{\hbar_m + \hbar_j}{2}\cdot\frac{a^2}{a_{av}^2}\,d\zeta \cong \frac{a^2}{a_{av}^2}\,d\zeta \tag{4.79}$$

The following procedure is employed to determine the desired function $a(z)$, describing a taper contour: one gives a ρ_{max} value and a σ_{min} value connected with it. Further, a $W(\zeta)$ function corresponding to those values is determined. Then $a(\zeta)$ is derived from the condition that $d(\ln a)/d\zeta$ has to be proportional to $d(\ln W)/d\zeta$. Two arbitrary constants, i.e. the proportionality factor between the functions $d(\ln a)/d\zeta$ and $d(\ln W)/d\zeta$ and the integration constant, are determined by the condition that $a(\zeta)$ has the required values at the taper ends, i.e. at $\zeta = 0$ and at $\zeta = 1$. After this, the parameter $(a_{av})^2$ is calculated by eqn. 4.77 and z/L is derived as a function of ζ using eqn. 4.79. So, one yields a parametric form of the dependence of a versus z/L. The taper length L is then determined by the requirement that the parameter σ is greater than σ_{min} in the whole frequency band, i.e. the equation that follows from eqn. 4.78 has to be satisfied

$$L = 0.73\sigma_{min}(a_{av})^2/\lambda_{min} \tag{4.80}$$

Here, λ_{min} corresponds to the smallest wavelength in the required frequency band.

The normalised amplitude of a parasitic mode $|p_j(0)/[1.56\ln(q)]|$, where q is the ratio $a(L)/a(0)$, is equal to the normalised reflection coefficient

$$\left|\frac{p_j(\sigma)}{1.56\ln(q)}\right| = \left|\frac{\rho(\sigma)}{\frac{1}{2}\ln\left(\frac{W(l)}{W(0)}\right)}\right| \tag{4.81}$$

corresponding to the same $W(\zeta)$ function. According to eqns. 4.78 and 4.80, the parameter σ is determined by the formula

$$\sigma = \sigma_{min}\frac{\lambda}{\lambda_{min}} \tag{4.82}$$

Let us note a significant difference between the problems connected with the integrals of eqns. 4.67 and 4.76. In the first integral (eqn. 4.67) the parameter σ is inversely proportional to λ. So, the operation of a matching section of a transmission line with a variable impedance is limited from the low frequency end. The length of the section is defined by

$$L = \frac{\sigma_{min}}{4\pi}\lambda_{max} \tag{4.83}$$

and for all frequencies where λ is lower than a given value of λ_{max}, σ will be larger than σ_{min} and $|\rho|$ less than $|\rho_{max}|$. In contrast to this, for a waveguide taper σ is directly proportional to λ, hence the operation of a waveguide taper is limited from the high frequency end. Eqn. 4.80 contains the minimal given wavelength, where for all lower frequencies the conditions $\sigma > \sigma_{min}$ and $|\rho| < |\rho_{max}|$ are satisfied.

From a physical point of view, this difference is caused by the following. The backward wave amplitude is determined in eqn. 4.67 and σ is a sum of electrical lengths for backward and forward waves. The electrical length for each of these waves as well as for their sum increases with increasing frequency. In contrast to this, for the second integral a forward wave amplitude is determined in eqn. 4.76 and σ is a difference of electrical lengths for two forward waves and despite the increment of h_m and h_j with the frequency, their difference decreases with the increasing frequency.

The phase velocities of all modes become closer to each other with increasing frequency. The amplitudes of forward parasitic modes excited along a nonuniform waveguide increase with the frequency, since the mutual compensation of modes excited along different taper sections decreases.

In other words, the advantages of smoothed nonlinear tapers against abrupt linear tapers disappear at very high frequencies. From a mathematical point of view, the exponential factors in the integrals like eqn. 2.75 or like the first term in eqn. 3.55 cannot provide a sufficiently fast sign change of the integrand. In this case, those integrals are of the order of the product of ν_0 and L/a, i.e. they have finite values. Obviously, the weak coupling approximation cannot be used in such a case.

4.3.4 Examples of nonlinear TE_{01} mode tapers

We illustrate the calculation method described above. First, let us take the $W(\zeta)$ function according to the simple eqn. 4.70. Deriving $a(\zeta)$ following the procedure described above we obtain

$$a(\zeta) = \sqrt{a(0)a(L)}\exp\left(-\frac{1}{2}\cdot\ln(q)\cdot\cos(\pi\zeta)\right) \tag{4.84}$$

Here, the parameter $(a_{av})^2$ is equal to

$$a_{av}^2 = a(0)a(L)I_0(\ln(q)) \tag{4.85}$$

where I_0 is the modified Bessel function. The dependence between z/L and ζ is given in an integral form

$$\frac{z}{L} = \frac{1}{I_0(\ln(q))}\int_0^{\zeta}\exp(-\ln(q)\cos(\pi\zeta))\,dz \tag{4.86}$$

According to eqn. 4.80, the taper length must be not less than

$$L = 3.1I_0(\ln(q))\frac{a(0)a(L)}{\lambda_{min}} \tag{4.87}$$

As is easy to prove, calculating the derivatives da/dz and $d^2a/(dz)^2$ by the parametric relations of eqns. 4.84 and 4.86, the designed taper contour has no tilts at its ends and

$$\frac{a''(0)}{a''(L)} = \frac{a^3(L)}{a^3(0)} \tag{4.88}$$

i.e. the contour steepness is significantly greater at the narrow end than at the broad one.

Let us consider a numerical example. The radii of the circular waveguides to be connected are $a(0) = 24.4$ mm and $a(L) = 9.0$ mm and the minimal

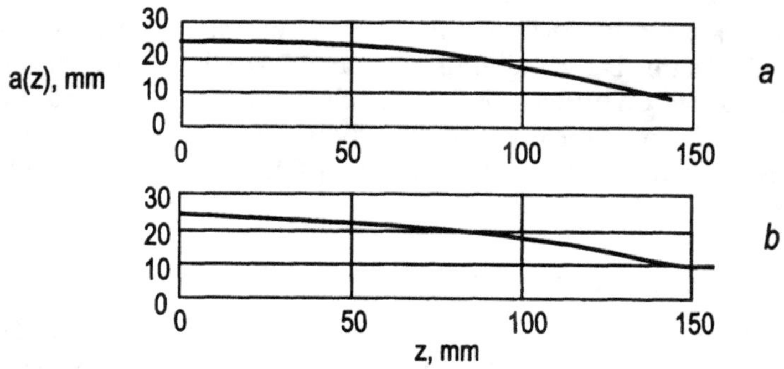

Figure 4.18 TE_{01} mode taper contours
a eqn. 4.70
b eqn. 4.72

operating wavelength required for the taper is λ_{min} = 6.0 mm (50 GHz). The taper length is equal to L = 144.5 mm. The taper contour is represented in an appropriate scale in Figure 4.18*a*. The amplitude of the TE_{02} mode versus λ is calculated using either the curve in Figure 4.17*a* or eqns. 4.81 and 4.71. The amplitude p_j of the parasitic mode is less than 0.11 for $\lambda > \lambda_{min}$, this corresponds to the conversion losses lower than −19 *dB*.

Let us compare this smoothed nonlinear taper with a straight conical taper (ν = const). The integral of eqn. 4.76 for a cone, when $\sigma >> 1$ is equal to

$$|p_j(\sigma)| = 1.56\frac{1}{\sigma}\left|\frac{(1-q)^4}{q^2} + 4\frac{(1-q)^2}{q}\sin^2\left(\frac{\sigma}{2}\right)\right|^{1/2} \tag{4.89}$$

Let us substitute in eqn. 4.89 subsequently 1 and 0 instead of $\sin^2(\sigma/2)$ and use the relation of eqn. 4.78 between σ and L. Taking into account the fact that for a cone $(a_{av})^2 = a(0)a(L)$, we have an evaluation of upper and lower limits of $p_j(\sigma)$. It is more convenient to use the limits instead of the exact formula of eqn. 4.89

$$0.57\frac{[a(0)-a(L)]^2}{\lambda L} \le |p_j(\sigma)| \le 0.57\frac{|a(0)^2 - a(L)^2|}{\lambda L} \tag{4.90}$$

For the values of $a(0)$, $a(L)$ and λ_{min} from the example given above the requirement $|p_j(\sigma)| < 0.11$ will be satisfied only for a length L greater than 0.5 m, this means that the conical taper must be three times longer than the simplest smoothed nonlinear taper.

Using for the taper design the other function $W(\zeta)$ corresponding to the more complicated eqn. 4.72, we obtain the taper length equal to 158 mm for the same values of $a(0)$, $a(L)$ and λ_{min}. There are small tilts at the ends of this

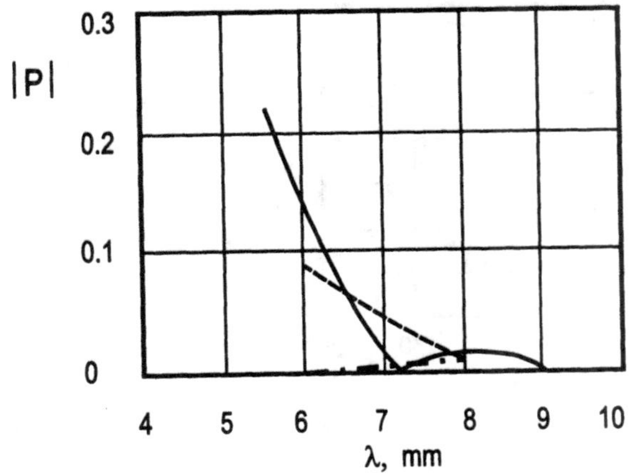

Figure 4.19 Conversion losses (normalised amplitude of parasitic TE_{02} mode) of the TE_{01} mode taper shown in Figure 4.17

taper. Its contour is presented in Figure 4.18*b*. According to eqn. 4.73, the losses of this taper are –26 dB. A conical taper with such losses is nearly 1 m long.

The described design method for the optimal shape of a nonlinear waveguide taper assumes that the derivation of eqn. 4.76 from eqn. 4.74 does not cause significant errors. The plot shown in Figure 4.19 illustrates to which degree this assumption is valid.

The normalised amplitude of the parasitic mode excited along the taper is presented in the plot. The taper contour is designed according to eqn. 4.72 with parameter values: $a(0) = 9$ mm, $a(L) = 30$ mm and $\lambda_{min} = 6.6$ mm (45.4 GHz). The length of this taper is equal to 193.7 mm, the contour angles with the z-axis at the ends are equal to 5.4° and 1.6°. The taper contour is presented in Figure 4.20.

The solid line in Figure 4.19 is calculated according to eqn. 4.73, it repeats a part of the curve shown in Figure 4.17*b*. The dashed line is calculated by numerical integration of the two-mode coupled-wave equations. These coupled equations are derived taking two equations corresponding to the forward and backward propagating TE_{02} mode from the coupled-wave eqns. 2.46. Only the terms with the amplitudes of this mode are required as well as the terms with the amplitude of the TE_{01} mode. The normalised amplitude of the TE_{01} mode is assumed to be constant. In such a way, the mutual interaction of the forward and backward TE_{02} waves is also taken into account in this calculation, hence the calculation is a little more exact than it would be by integration as in eqn. 4.74. Comparing these curves, the efficiency of the design method described above for a function $a(z)$ can really be appreciated. Conversion losses are only slightly higher than the losses calculated by eqn. 4.73 in a

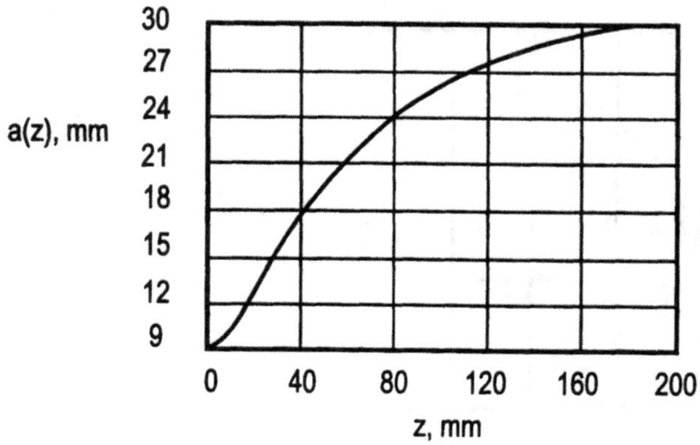

Figure 4.20 Contour of the TE_{01} mode taper with $a(0) = 9$ mm, $a(L) = 30$ mm and $\lambda_{min} = 6.6$ mm (45.4 GHz)

broad frequency band up to a given wavelength when a cut-off cross section is already inside the taper. However, the details of the p_j versus λ dependence cannot be calculated by this formula.

The dot-dashed curve shows the normalised amplitude of the backward TE_{02} wave. As can be concluded from general considerations, the backward propagating wave is very small.

Let us mention an auxiliary computation technique used for numerical integration. This technique originates in the boundary conditions for the amplitudes of the forward and backward propagating waves of eqn. 2.48 to be stated at the opposite interval ends. It is a usual feature of the coupled-wave eqns. 2.46–2.48. In terms of the ζ variable, the boundary conditions have the form

$$P_j(0) = 0, \quad P_{-j}(1) = 0 \tag{4.91}$$

The well known auxiliary technique solves two sets of coupled equations instead of the given inhomogeneous coupled equations with the boundary conditions of eqn. 4.91. The boundary conditions for these other coupled equations are stated at the same interval end, namely

$$P_j(1) = 1, \quad P_{-j}(1) = 0 \tag{4.92}$$

The first set of coupled equations is the same as in eqn. 2.46. The second set is the same as the first one but the terms containing the TE_{01} mode amplitude are omitted, so that one has homogeneous equations. These two problems are therefore Cauchy problems, convenient for a numerical solution, i.e. for computation of the functions $P_j^{(1)}(\zeta)$, $P_{-j}^{(1)}(\zeta)$, $P_j^{(2)}(\zeta)$ and $P_{-j}^{(2)}(\zeta)$. Linear combinations of the solutions of two Cauchy problems represent solutions of the given coupled-wave equations, i.e.

$$P_j = P_j^{(1)} - \frac{P_j^{(1)}(0)}{P_j^{(2)}(0)} P_j^{(2)}, \quad P_{-j} = P_{-j}^{(1)} - \frac{P_j^{(1)}(0)}{P_j^{(2)}(0)} P_{-j}^{(2)} \tag{4.93}$$

This auxiliary technique has a general application for the numerical solution of the coupled-wave eqns. 2.46 and 2.48 as well as for similar coupled-wave equations in a bent waveguide, eqns. 2.32, 2.10.

4.3.5 Other examples of nonlinear circular waveguide tapers

The properties of several smooth-wall nonlinear tapers developed for highly oversized high-power (200 kW, CW) transmission lines for electron cyclotron resonance heating (ECRH) of nuclear fusion plasmas are summarised in Table 4.1. The contours were synthesised according to eqn. 2.75 using the coupling coefficients from eqns. 4.44 and then analysed by numerical integration of the corresponding multimode coupled-wave equations (Thumm *et al.*, 1986; Möbius and Thumm, 1993, and references given there). The optimised tapers have been fabricated by electroforming methods. Figure 4.21 shows a photograph of the 70 GHz TE_{02} taper.

Due to its extremely low losses the hybrid HE_{11} mode (see for more details subsection 5.3.9) is also very often used for high power millimetre-wave transmission (Doane, 1982; Thumm *et al.*, 1985; Thumm *et al.*, 1986; Thumm, 1986). Due to this fact, Table 4.1 also contains the parameters of several circumferentially corrugated nonlinear tapers for the low-loss balanced HE_{11} hybrid mode. The coupling coefficients for the design of these transitions can be found in Doane (1985, 1984a) and Li and Thumm (1991b).

Table 4.1 Data of smoothed nonlinear tapers from a = 31.7 mm to 20.0 mm at 28 GHz and a = 31.7 mm to 13.9 mm at 70 and 140 GHz (Thumm et al., 1986)

Propagating mode	Main coupled modes	Frequency (GHz)	Length (mm)	Suppression of spurious modes (dB)	
				calculated	measured
TE_{01}	TE_{02}	28	310	40	(37±4)
TE_{01}	TE_{02}	70	705	40	(36±5)
TE_{01}	TE_{02}	140	1410	40	
TE_{02}	TE_{01}, TE_{03}	28	310	37	≥25
TE_{02}	TE_{01}, TE_{03}	70	705	37	≥25
TE_{03}	TE_{02}, TE_{04}	140	920	37	
TE_{11}	TM_{11}, TE_{12}	28	510	25	24±2
TE_{11}	TM_{11}, TE_{12}	70	1095	25	23±2
HE_{11}	HE_{12}, EH_{12}	28	310	30	≥25
HE_{11}	HE_{12}, EH_{12}	70	700	30	≥25
HE_{11}	HE_{12}, EH_{12}	140	1400	30	

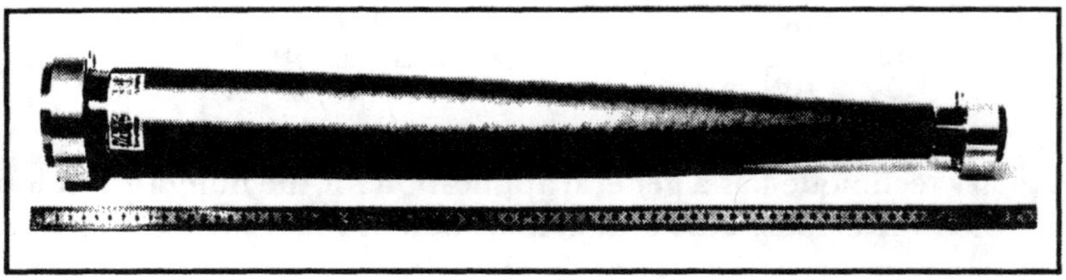

Figure 4.21 Electroformed nonlinear 70 GHz TE_{02} waveguide radius taper with a length of almost one meter (Thumm et al., 1986)

4.3.6 Nonlinear taper between arbitrary cross sections

The design method for a smoothed nonlinear taper was applied above to an example of a circular symmetric *TE* mode in a circular waveguide. This design method can be employed in an arbitrary cross-section waveguide. According to the results of Section 2.3, all coupling coefficients S_{jm} for forward modes are not frequency dependent at high frequencies, i.e. at $\hbar_m \cong 1$ and $\hbar_j \cong 1$. The coupling coefficients have the form $C\,\nu/a$, where a is a characteristic size of the cross section and C is a constant. Eqns. 2.75 for forward wave amplitudes have the following form at high frequencies

$$p_j = C\int_0^L \frac{\nu}{a}\exp\left(-i\left(\gamma_m - \gamma_j\right)\right)dz \tag{4.94}$$

The integral of eqn. 4.94 can also be transformed to a form identical to eqn. 4.67. For this purpose, let us introduce an f function by

$$\frac{df}{dz} = \frac{\nu}{a} \tag{4.95}$$

Then the integral p_j has the form:

$$p_j = C\int_0^1 \frac{df}{d\zeta}\exp(-i\sigma\zeta)\,d\zeta \tag{4.96}$$

where σ and ζ have the same meaning as in eqn. 4.75. Using the required function $W(\zeta)$, the function $f(\zeta)$ is calculated by the same method as the $\ln(a)$ function in subsection 4.3.3. When the functions $f(\zeta)$ and $f(a)$ are known, $a(\zeta)$ can be found. To calculate first the z/L ratio as a function of ζ and later the taper length L, one should employ the equations

$$\frac{dz}{L} = \frac{1}{\int_0^1 \frac{d\zeta}{\alpha_j^2 - \alpha_m^2}}\frac{d\zeta}{\alpha_j^2 - \alpha_m^2}, \qquad \sigma = \frac{1}{\int_0^1 \frac{d\zeta}{\alpha_j^2 - \alpha_m^2}}\frac{L}{2k} \tag{4.97}$$

These equations are an obvious generalisation of eqns. 4.79, 4.78 and 4.77.

In contrast to the example of the TE_{02} mode described above, an additional problem arises: the function f has to be derived. However, this function can be found in a simple way for some problems. For example, for a twist considered in Section 4.1, the function $f(z)$ is proportional to the angle $\Phi(z)$ of the cross-section twist. So, the function $\Phi(z)$ for a twist with low conversion losses far from mode degeneracy can easily be found.

4.3.7 Nonlinear taper with cut-off cross section of a parasitic mode

If there is a cut-off cross section for a parasitic mode in the taper, then the shape of the contour designed by eqn. 4.76 or in a general case by eqn. 4.94, can be absolutely unsatisfactory. Let us return to the problem of the TE_{02} mode in a circular symmetric waveguide taper. As an illustration of the statement above, one could point out to the curve *III* in Figure 4.22.

This contour shape $a(z)$ was computed using the method described above, whereas a function satisfying eqn. 4.72 was used. A control analysis was performed at a single frequency inside the band by eqn. 3.55. The parasitic mode amplitude is unacceptably high, i.e. equal to 0.26. The conversion losses are small for other contours, represented in Figure 4.22 by the curves *I, II, IV*, i.e. $|P| < 0.02$.

However, the method for the calculation of the function $a(z)$ described above can be generalised in such a way that the design would also be satisfactory for a cut-off cross section. The first term of eqn. 3.55 is usually the largest of its three terms. If we omit the nonimportant factors for further discussion, then this term of eqn. 3.55 corresponds to the integral of eqn. 4.74 — the only difference is the upper limit of this term (further denoted as z_L), which is not equal to the whole taper length L but is a little bit less than the distance from the broad taper end to the cut-off cross section. Now, the integral of eqn. 4.76 cannot be substituted instead of the integral of eqn. 4.74, since the factor $(h_m + h_j)/(2\sqrt{h_m h_j})$ changes rapidly near the cut-off cross section.

To generalise the method, the integral of eqn. 4.74 — and not the integral of eqn. 4.76 — has to be immediately transformed to the form of eqn. 4.67, i.e. the largest term of eqn. 3.55 has to be transformed. For this purpose a new function $F(a)$ is introduced by

$$dF = \frac{da}{a} \frac{h_m + h_j}{2\sqrt{h_m h_j}} \tag{4.98}$$

Substituting this function and the variable ζ of eqn. 4.75 instead of z, one yields the integral of eqn. 4.74 in a form identical to the integral of eqn. 4.67. After this one assumes that F is proportional to $\ln W(\zeta)$, where $W(\zeta)$ satisfies e.g. eqn.

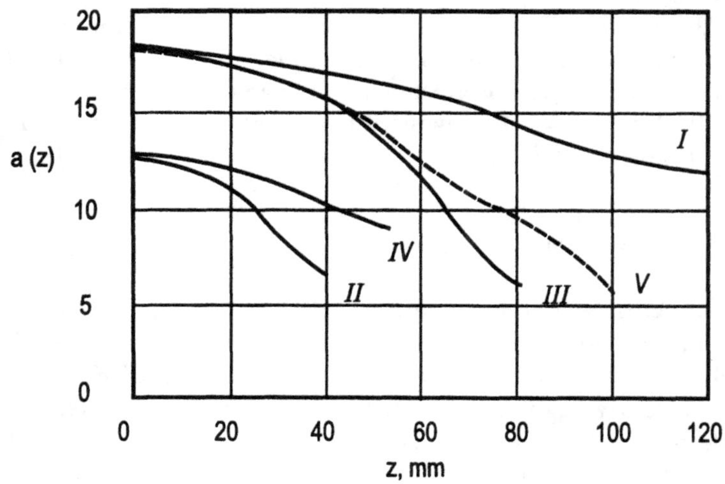

Figure 4.22 Shape of taper contours. TE_{02} mode in symmetric taper

4.72. The proportionality coefficient is determined in the same way as in subsection 4.3.3. When $F(a)$ and $F(\zeta)$ are found, one could determine $a(\zeta)$ in an implicit form. Later, z versus ζ is determined by an equation generalising eqn. 4.79

$$\frac{dz}{z_L} = \frac{h_j + h_m}{2k} \frac{a^2(\zeta)d\zeta}{\int_0^1 \frac{h_j + h_m}{2k} a^2(\zeta)d\zeta} \tag{4.99}$$

The taper length until the cut-off cross section as well as a taper contour $a(z)$ are calculated in such a way. The dashed line V in Figure 4.22 represents the result of such a design. The parasitic mode amplitude computed for this contour by the rigorous formula of eqn. 3.55, where all three terms were taken into account, is eight times less than for the contour *III*.

Obviously, this design method starting immediately from the integral of eqn. 4.74 can be employed in the absence of a cut-off cross section as well. However, the significantly simpler design employing the integral of eqn. 4.76 produces probably satisfactory results in this case.

4.3.8 Gyrotron output tapers

An example of nonlinear tapers having a cut-off cross section of the main parasitic mode are the up tapers in gyrotrons. In this subsection we describe several gyrotron taper design approaches and hot test measurements conducted by different gyrotron developers and users (Möbius and Thumm, 1993). The specific difficulties in the design of gyrotron output tapers in comparison with highly oversized waveguide tapers for ECRH transmission systems (see subsection 4.3.5) are as follows:

(*i*) At the taper input the working mode is very close to cut off. Input reflection as well as forward and backward-scattered waves have to be considered.
(*ii*) In the case of conventional gyrotrons, where the electron beam collector serves as an output waveguide, the ratio a_2/a_1 of the collector to the cavity output radius is very large so that there is strong excitation of many parasitic modes.
(*iii*) Gyrotron working modes are higher-order modes which couple to higher and lower undesired mode neighbours with $\Delta n = 0$ (circular symmetry). TE_{np} modes with $n > 0$ couple not only to TE_{nq} modes, but also excite TM_{nq} modes.

Employing the near optimum mode conversion distribution including reconversion and reflection, we have synthesised several nonlinear gyrotron up taper contours, which were subsequently analysed by numerically solving the corresponding coupled-wave equations. The taper input contours are matched to the flare angle of the resonator output port. This section is included in the analysis to describe the generation of unwanted modes in the whole nonuniform waveguide. The results are summarised in Table 4.2 together with experimental values obtained by measurements of the gyrotron output mode content using wavenumber spectrometers (Behm and Jensen, 1986; Pochelon *et al.*, 1991; Borie *et al.*, 1992).

The parasitic mode levels predicted by the synthesis and analysis codes, respectively, are always in good agreement. The high-power experimental results are slightly higher, but this is not surprising since insulation gaps and

Table 4.2 Geometrical parameters and calculated and measured levels of parasitic modes at the output of optimised nonlinear gyrotron tapers

Mode	Frequency (GHz)	Input diameter (mm)	Output diameter (mm)	Length (mm)	Parasitic mode level (dB) theoretical: synthesis	theoretical: analysis	experimental
TE_{02}	39	20.0	62.6	300	–27	–28	–(24±2)
TE_{02}	70	12.2	63.5	570	–20	–19	–(15±2)
TE_{03}	140	8.5	70.0	750	–20	–20	–(16±2)
$TE_{10.4}$	140	17.4	40.0	130	–19	–19	–(19±2)

Table 4.3 Calculated unwanted parasitic mode levels (analysis) of optimised nonlinear 140 GHz, $TE_{10.4}$ mode taper (inside diameter, 17.4 – 40.0 mm). Output mode purity is 99 per cent

Parasitic mode	Amount of parasitic mode (per cent)	Parasitic mode level (dB)
$TM_{10.3}$	0.69	–21.6
$TM_{10.4}$	0.13	–28.9
$TE_{10.3}$	0.04	–34.0
$TE_{10.5}$	0.02	–37.0

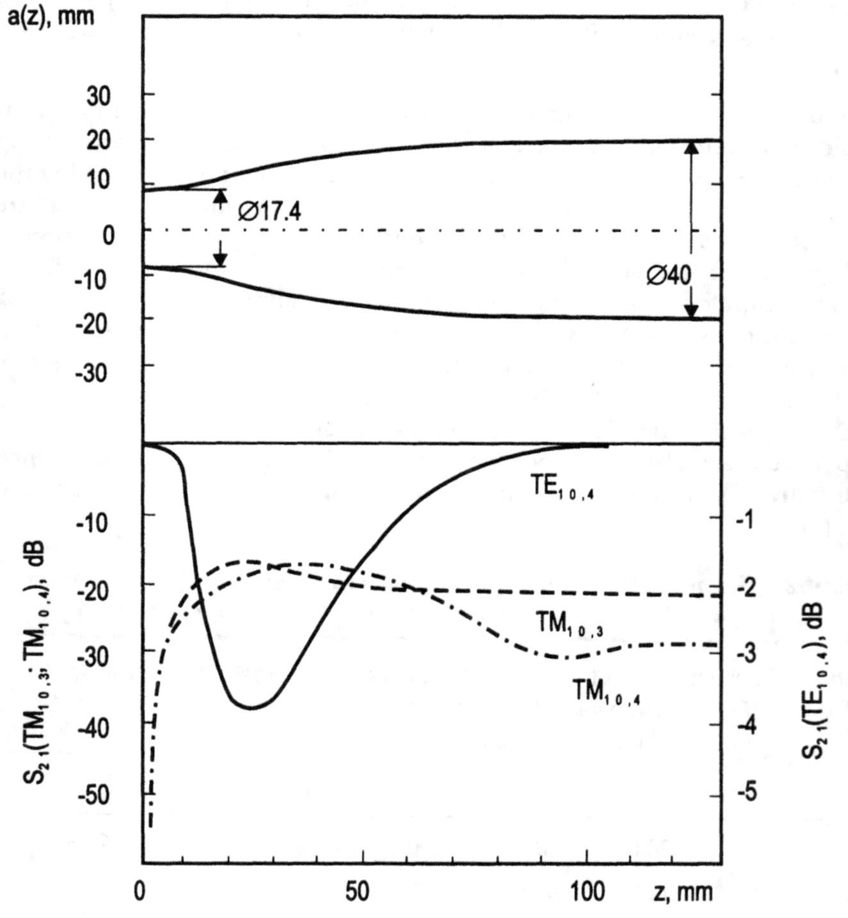

Figure 4.23 Optimum contour of a $TE_{10.4}$ gyrotron up taper (input diameter = 17.4 mm, output diameter = 40.0 mm) together with the excitation of the $TM_{10.3}$ and $TM_{10.4}$ modes and the $TE_{10.4}$ transmission along the taper

the RF output window in the gyrotron are additional sources of unwanted modes.

The calculated levels of undesired modes at the output of the 140 GHz $TE_{10.4}$ mode up taper are listed in Table 4.3. In spite of the fact that the coupling coefficients for coupling to *TM* modes are a factor of about ten weaker than those for coupling to *TE* modes, the levels of undesired *TM* modes are higher than those of *TE* modes. This effect is due to the small differences of the wavenumbers of the *TM* modes from those of $TE_{10.4}$, which means that the required taper length is determined by the corresponding long beat wavelength. The excitation of $TM_{10.3}$ and $TM_{10.4}$ together with the $TE_{10.4}$ transmission loss along the taper and the optimal taper contour are plotted in Figure 4.23. It is evident that the effect of reconversion into the operating $TE_{10.4}$ mode is quite important.

Conversion of the operating gyrotron mode to unwanted parasitic modes at discontinuities of an ordinary gyrotron cavity can be reduced substantially by introducing smooth transition regions between the individual parts of the open resonators (Zarudneva *et al.*, 1988). Cavity and adjacent tapered output waveguide should be optimised as one single unit (Dumbrajs *et al.*, 1992).

4.4 Diffraction by a periodic surface—absorbing and nonreflecting surfaces

4.4.1 Reduction of 3D diffraction by a periodic surface to a rectangular waveguide problem

In this section we employ the cross-section method for the problem of plane wave normal incidence on a surface which has a periodic structure along two directions. This problem can be reduced to a waveguide problem. One should only add two sets of planes perpendicular to the given surface and put such boundary conditions on those planes where the plane wave with $E_z \equiv H_z \equiv 0$ would be the fundamental mode of the rectangular waveguide formed by those planes.

The given surface is a set of cones with tapered contour (thorn shape) situated within the layer $0 < z < L$ and the periods along x and y-axes being equal to a and b, respectively (Figure 4.24). For the sake of simplicity (it would significantly influence our further considerations) let us assume that each of the thorns has two planes of symmetry, namely x =const and y = const. The plane wave with the components $E_x = H_y = \exp(-ikz)$ is incident normally on this surface. One should determine the reflection coefficient of this wave as well as the intensity distribution of the lateral spectrum. If $ka > \pi$ or $kb > \pi$ there would be several nonevanescent diffraction orders, analogous to the spectrum of a diffraction grating.

Let us add two sets of parallel planes to the given surface, i.e. $x = ta$ and $y = sb$ (where $t, s = \ldots, -1, 0, 1, \ldots$). These planes must be situated in such a way that each

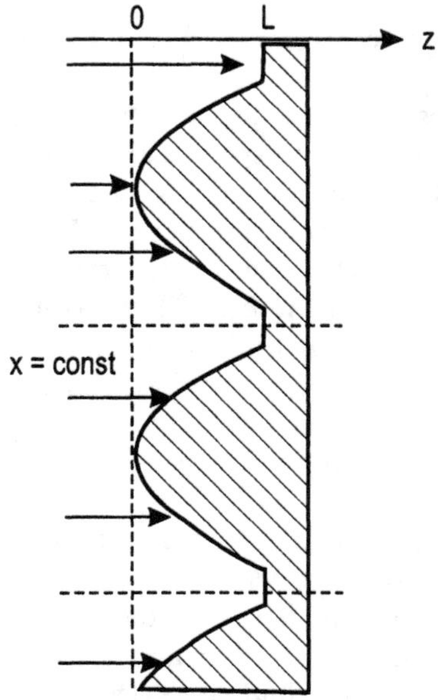

Figure 4.24 Schematic of nonreflecting surface with periodic (thorn shaped) structure

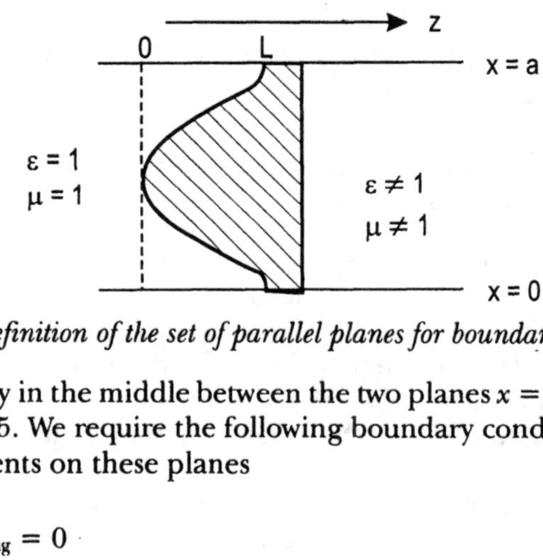

Figure 4.25 Definition of the set of parallel planes for boundary conditions

thorn is exactly in the middle between the two planes x = const and y = const, see Figure 4.25. We require the following boundary conditions for tangential field components on these planes

$$E_{\text{tang}} = 0 \tag{4.100a}$$

on the planes x = const, i.e. $E_z = E_y = 0$

$$H_{\text{tang}} = 0 \tag{4.100b}$$

on the planes y = const, i.e. $H_z = H_x = 0$. Such planes do not disturb the incident plane wave.

A rectangular waveguide is formed by four neighbouring planes. As Figure 4.25, shows, this waveguide has a uniform section for $z < 0$ and a nonuniform section for $0 < z < L$. The fundamental mode of the uniform section has the components $E_x = H_y = \exp(-ikz)$. This mode is incident on the nonuniform section in the waveguide problem, so in the uniform section, i.e. for $z < 0$, the reflected fundamental mode is excited together with higher order modes propagating backward. Several of these higher order modes are nonevanescent if either $ka > \pi$ or $kb > \pi$.

In this problem, the scattered fields have two symmetry planes, i.e. $x = (t + 1/2)a$ and $y = (s + 1/2)b$, since both the incident wave and the nonuniformities have the same planes of symmetry. This means that the fields on two opposite waveguide walls are identical. Hence, the field in all waveguides is continuous on these artificially defined walls. Therefore, this waveguide field solves the problem of scattering from the periodic surface. The amplitude of the fundamental mode reflected in the waveguide is equal to the reflection coefficient from the periodic surface, and the magnitudes of the lateral spectrum components are equal to the amplitudes of the corresponding higher order modes.

The eigenmode set of the rectangular waveguide with the walls described by eqn. 4.100 contains *TE* modes ($E_z \equiv 0$) and *TM* modes ($H_z \equiv 0$). According to eqn. I.14, these modes can be expressed through the eigenfunctions ψ^j and φ^j satisfying eqns. I.11. The boundary conditions on the corresponding rectangle sides (Figure 4.25) have the form

$$\varphi^j = 0, \quad \frac{\partial \psi^j}{\partial x} = 0 \tag{4.101a}$$

on the sides $x = 0$ and $x = a$

$$\psi^j = 0, \quad \frac{\partial \varphi^j}{\partial y} = 0 \tag{4.101b}$$

on the sides $y = 0$ and $y = b$. These boundary conditions are different from eqn. I.12. These functions are equal to

$$\psi^j = N^j \cos\left(\frac{\pi n x}{a}\right)\sin\left(\frac{\pi q y}{b}\right) \text{ for } n = 0,2,\ldots \text{ and } q = 2,4,\ldots \tag{4.102a}$$

$$\varphi^j = M^j \sin\left(\frac{\pi n x}{a}\right)\cos\left(\frac{\pi q y}{b}\right) \text{ for } n = 2,4,\ldots \text{ and } q = 0,2,\ldots \tag{4.102b}$$

The normalisation factors are given by eqn. I.23. The wavenumbers h_j are given by eqns. I.13 and I.23. The plane wave (i.e. the *TEM* mode) must be added to complete this mode set. Its field components can be derived from eqn. I.14*a* assuming for example

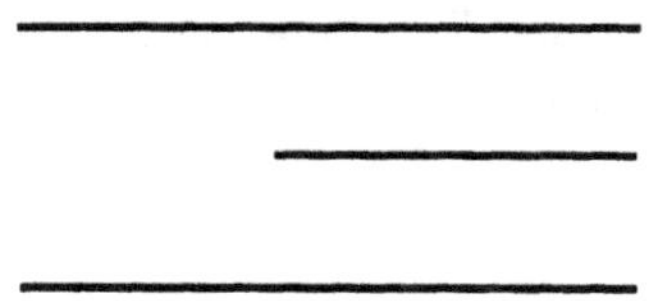

Figure 4.26 Reference waveguide for Figure 4.25 at $z = 0$

$$\varphi = \frac{x}{\sqrt{ab}} \text{ and } \psi \equiv 0$$

where the normalisation factor is determined from eqn. I.8.

If the thorns on the surface are metallic, the field in the nonuniform waveguide, i.e. for $0 < z < L$, is described by the coupled-wave eqns. 2.46, where the coupling coefficients S_{jm} can be derived from eqns. 2.98. One should take into account only the fact that the transversal eigenfunctions satisfy the boundary conditions of eqn. 4.101 and not the boundary conditions of eqn. I.12 on the rectangle sides. According to eqn. I.14, the eigenfunctions of the reference waveguide satisfy the usual boundary conditions $\varphi^j = 0$ and $\partial\psi^j/\partial N = 0$ on the metallic surface of the inner rod in the reference waveguide.

The boundary conditions for the coupled-wave eqns. 2.46 at $z = L$ are in our problem (Figure 4.25) different from the boundary conditions of eqn. 2.48. The latter conditions were taken to provide no incidence of waves from the area $z > L$. This means that the waveguide is closed at $z = L$ in the problem considered here, i.e. the transversal electric field components are equal to zero in this plane. According to eqn. 2.42, $Q_j = 0$ must be valid at $z = L$ for all $j \geq 1$ and according to eqn. 2.2 this means that

$$P_j(L) = P_{-j}(L), \text{ for } j = 1,2,\ldots \tag{4.103}$$

The geometrical topology of the cross-section changes at $z = 0$, hence the boundary conditions will also be different from eqn. 2.48. Within the framework of the cross-section method, one could use the technique described in subsection 3.5.2, i.e. one can avoid infinite S_{jm} values caused by the field concentration near the edge defining an auxiliary rapidly varying variable ξ instead of z.

However, one can solve an independent problem: the fundamental mode of the empty waveguide with the walls described by eqn. 4.100 incident on its junction with the same waveguide containing an inner rod (Figure 4.26), i.e. with a coaxial waveguide. The amplitudes of the coaxial waveguide eigenmodes determined in this problem are the boundary conditions for $P_j(0)$ for $j \geq 1$. Such problems are usually solved by employing an integral equation for the transversal electrical field components at the junction plane. The fields for $z > 0$ and for $z < 0$ are then expressed through these transversal components and the transversal magnetic field continuity condition is applied afterwards at $z = \pm 0$ to compose this integral equation.

4.4.2 Absorbing and nonreflecting surfaces

The method described above reduces the problem of 3D diffraction by a periodic surface to a waveguide problem. It can be applied to the analysis of a nonreflecting surface. More exactly, we consider surfaces containing a double periodic set of cones from a magnetodielectric medium with losses (Figure 4.24). The cross sections of the cones may be nonlinearly tapered versus z (thorns), however they are usually either circular cones or pyramids. The cones are situated in the matching layer $0 < z < L$ between the free space ($z < 0$) and the protected body ($z \geq L$). This layer absorbs the incident power and matches this body to the free space. Usually, a metallic surface is situated directly after the absorbing layer, i.e. at $z = L$ or near it. Hence, the main function of the layer is to absorb the power without reflection. The cone thickness is tapered along z, it varies from zero at $z = 0$ (peak of cone) to a maximal value at $z = L$ (foundation of the cone). Besides this, the complex values of ε and μ may also be tapered along z.

One would face essential difficulties in trying to apply to this problem the cross-section method developed in the subsections 2.2.1 to 2.2.8 and applied in the rest of this book. This technique employs eigenmodes of reference waveguides. To determine their fields and their wavenumbers h_j, one should solve an auxiliary problem of eigenmodes of a rectangular waveguide with a circular or rectangular dielectric inner rod.

The variables are not separated in this 2D vector problem, hence this problem cannot be reduced to transcendental equations. The problem is further complicated if ε and μ of the cones depend on x and y.

Hence, the cross-section method mentioned in subsection 2.2.9 may be very useful for this problem. Fields and wavenumbers in reference waveguides are not required by this approach. Only the fields and h_j of an empty waveguide are required, i.e. the fields obtained from eqns. I.14 and 4.102, and h_j from eqns. I.13 and I.22. The coupled-wave equations for $P_j(z)$ have the form of eqns. 2.86 with the coupling coefficients of eqn. 2.87. The fields and wavenumbers of empty waveguide modes are in fact those with the superscript 0 and h_j^0 participating in eqn. 2.87. These fields can be expressed through eqns. 4.102 and further, eqn. 2.87 can be transformed to a form similar to eqn. 2.98.

On the other hand, the coupled-wave eqns. 2.86 converge worse than eqns. 2.46, hence more of these equations have to be taken into account. One cannot study metallic surfaces using eqns. 2.86, because all the coupling coefficients M_{jm} become infinite for $|\varepsilon| \to \infty$. The reason is clear, since the field in a reference waveguide for large $|\varepsilon|$ and $|\mu|$ is qualitatively different from the empty waveguide field. The coupling coefficients are not equal to zero for $\nu \equiv 0$ and they are not small for smooth nonuniformities, so the coupled-wave eqns. 2.86 have no adiabatic solutions. However, all these difficulties are probably less important than those we would face when applying to this problem the coupled-wave eqns. 2.46, unless the problem contains a small parameter. The boundary conditions at $z = 0$ are identical to those at $z = 0$ in eqn. 2.48. The boundary conditions at $z = L$ depend on the distance from the protected metallic surface to the foundation of the cones. If the cones are situated directly on this surface, i.e. if the plane $z = L$ is metallic, the boundary conditions are identical to those of eqn. 4.103. They are slightly more complicated if the metallic plane is situated at $z = L + d$, i.e. if there is a layer with thickness d between the cones' foundation and the metal. The layer has a dielectric permittivity ε_L and a magnetic permeability μ_L. Obviously, $E_x = E_y = 0$ at $z = L + d$, i.e. $Q_j = 0$ for all j when $j \geq 1$. There is a uniform waveguide between $z = L$ and $z = L + d$, so there $h_j^2 = k^2\varepsilon_L\mu_L - \alpha_j^2$.

As a consequence, the function $Q_j(z)$ is proportional to $\sin\left[h_j(z - L - d)\right]$ in this interval. Therefore

$$Q_j(L) = -\frac{\tan\left(k_j d\right)}{h_j} Q_j'(L) \tag{4.104}$$

The desired boundary condition for the variables P_j has the following form

$$P_j(L) - P_{-j}(L) = -\frac{\tan\left(h_j d\right)}{h_j}\left[P_j'(L) - P_{-j}'(L)\right] \tag{4.105}$$

This condition can easily be generalised, if ε and μ are not constants within the layer $L < z < L + d$ but depend on z, i.e. if the layer is inhomogeneous. Then the value Q_j/Q_j' at $z = L$ has to be derived from the equation

$$Q_j'' + h_j^2(z)Q_j = 0 \tag{4.106}$$

and from the condition $Q_j(L+d) = 0$.

The boundary condition of eqn. 4.105 and its particular kind eqn. 4.103 take into account the reflection of the mode not totally absorbed during its forward and backward propagation through the protecting layer. According to eqn. 4.105, the reflection coefficient depends on the distance between the metal and the protecting layer.

Coupled-wave equations for $Q_j(z)$ and $R_j(z)$ similar to eqns. 3.1 can be easily derived from eqns. 2.86. When N modes are taken into account, one should solve N Cauchy problems with the initial (boundary) conditions $Q_j(L) = 0$ and $R_j(L) = 1$ for $j \geq 1$. Further, $2N$ numbers q_{mj} and r_{mj} have to be determined for each of these Cauchy problems, whereas $q_{mj} = Q_{jm}(0)$ and $r_{mj} = R_m(0)$ for $m \geq 1$. The solution satisfying the boundary condition $P_m(0) = \delta_{1m}$ is a linear combination of these Cauchy problem solutions. Its coefficients A_j are derived from this boundary condition, i.e. from the coupled-wave equations

$$\sum_j \left(q_{mj} - r_{mj}\right)\frac{A_j}{2} = \delta_{1m}, \text{ for } m = 1,2,\ldots \tag{4.107}$$

The reflection coefficient is expressed through A_j by the following formula

$$P_{-1}(0) = \sum_j \left(q_{1j} + r_{1j}\right)\frac{A_j}{2} \tag{4.108}$$

As is known, one faces numerical calculation difficulties in a Cauchy problem for $h_j^2 < 0$. The methods for treating these numerical problems are also well known, hence we do not consider them here.

Chapter 5
Bent waveguides

In this chapter we apply the cross-section method to calculations on bent rectangular and circular waveguides.

5.1 Bent rectangular waveguides

5.1.1 Curvature coupling coefficients for rectangular waveguides

Let us restrict ourselves to calculations of two-mode coupling coefficients F_{jm} for tilts or bends with the axis parallel to one of the cross-section sides. The determination of F_{jm} for other axial topologies can be performed using eqns. 2.18 and 2.20 – 2.22 which do not contain any complicated integrals.

A tilt couples only modes with special index relationships to each other. These index relationships are similar to those, for which the coupling coefficients S_{jm} are not equal to zero. The coupling coefficients S_{jm} were derived in the first subsection of Section 4.2 for special kinds of transitions; as was mentioned in Section 2.3, a tilt is a special case of such a transition. But a tilt is different from the transitions considered in Section 4.2, since opposite waveguide walls have angles with different signs to the z-axis. Hence, the relationships between the indices of two modes, whose F_{jm} is not equal to zero, are different by index parity conditions from the corresponding relationships of Section 4.2.

According to the condition assumed in Sections 1.1 and 2.1, the x-axis is directed towards the centre of curvature. The mutual location of the cross section and the tilt axis is presented in Figures 5.1 and 5.2. The broad side of the rectangular cross-section in Figure 5.1 is denoted by a and the narrow side by b. In contrast, let us denote the length of the narrow side in Figure 5.2 by a and the length of the broad side by b, so that the notations of Section I.3 are the same for both cases. The indicated orientation of the electric field corresponds to the fundamental mode. The x dependence of the field is given either by

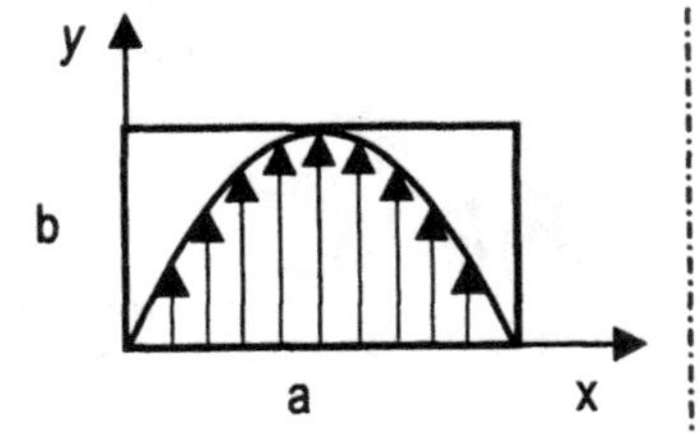

Figure 5.1 Geometry of H-plane bend

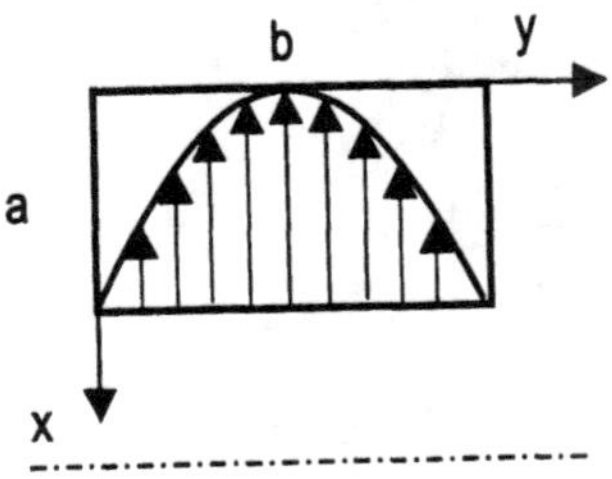

Figure 5.2 Geometry of E-plane bend

$\genfrac{}{}{0pt}{}{\sin}{\cos}\left(\frac{\pi x}{a}n_m\right)$ or by $\genfrac{}{}{0pt}{}{\sin}{\cos}\left(\frac{\pi x}{a}n_j\right)$ for modes with indices m or j, respectively. The y dependence of the field is given either by $\genfrac{}{}{0pt}{}{\sin}{\cos}\left(\frac{\pi y}{b}q_m\right)$ or by $\genfrac{}{}{0pt}{}{\sin}{\cos}\left(\frac{\pi y}{b}q_j\right)$.

Having calculated the integrals of eqn. 2.18, it is easy to prove that the necessary conditions for F_{jm} to be different from zero are

$$q_m = q_j \text{ and } n_m + n_j = \text{odd number} \tag{5.1}$$

In other words, a tilt excites modes with the same dependence on the co-ordinate parallel to the tilt axis as the incident mode. However, the excited modes have the other parity (even + odd = odd and odd +even = odd) of the index giving the dependence on the co-ordinate perpendicular to the tilt axis.

There are absolutely no difficulties in the derivation of the coupling coefficients F_{jm} from eqns. 2.18, 2.22 and I.21–I.23. Therefore, we present only the final results in eqns. 5.2. Eqn. 5.2*a* is for the case when both modes are *TE* modes, eqn. 5.2*b* for when both modes are *TM* modes and eqn. 5.2*c* is for the case when m is a *TE* mode and j is a *TM* mode. The conditions of eqn. 5.1 are to be satisfied for all cases. We write down the formulae in such a form to underline the fact that at high frequencies, when all normalised wavenumbers $\hbar$ are close to unity, the coupling coefficients are proportional to the frequency. For *TE* to *TE* and *TM* to *TM* coupling, the coupling coefficient is also proportional to a, whereas for TE to TM coupling it is inversely proportional to b (more accurately, one should take into account the dependence of the wavenumbers α_m and α_j on a and b).

$$F_{jm} = -\frac{4i}{\pi^2}\frac{1}{\sqrt{\epsilon_{n_j}\epsilon_{n_m}}}\frac{\left(\hbar_m\alpha_j^2 n_m^2 + \hbar_j\alpha_m^2 n_j^2\right)}{\hbar_j\alpha_j\alpha_m}\frac{\left(\hbar_j + \hbar_m\right)}{2}\frac{ka}{\left(n_j^2 - n_m^2\right)^2} \tag{5.2a}$$

$$F_{jm} = -\frac{4i}{\pi^2}\frac{\left(\hbar_m\alpha_j^2 + \hbar_j\alpha_m^2\right)}{\hbar_j\alpha_j\alpha_m}\frac{\left(\hbar_j + \hbar_m\right)}{2}\frac{n_j n_m ka}{\left(n_j^2 - n_m^2\right)^2} \tag{5.2b}$$

$$F_{jm} = \frac{4i}{\sqrt{\epsilon_{n_m}}}\frac{\left(\hbar_j + \hbar_m\right)}{2\hbar_j}\frac{qn_j}{\left(n_j^2 - n_m^2\right)}\frac{k}{\alpha_m\alpha_j b} \tag{5.2c}$$

The transversal wavenumbers α_m and α_j are determined by eqn. I.22.

5.1.2 H- and E-plane tilts and bends for the TE_{10} mode

Let us apply these general formulae to the problem of fundamental mode incidence on a tilt or on a bend.

For the H-plane tilt (Figure 5.1) the fundamental mode is denoted by TE_{10}, hence $n_m = 1$, $q_m = 0$. According to eqn. 5.2c, no TM modes are excited, only $TE_{n,0}$ modes can appear with an even $n \equiv n_j$. The coupling coefficient derived from eqn. 5.2a is equal to

$$F_{jm} = -\frac{8i}{\pi^2}\frac{n_j}{\left(n_j^2 - 1\right)^2}\frac{\left(\hbar_j + \hbar_m\right)^2}{4\hbar_j}ka \tag{5.3}$$

The power losses in an H-plane tilt are mainly caused by the TE_{20} mode ($n_j = 2$). If $\hbar \cong 1$ this mode takes away a power approximately equal to $0.032(ka)^2\,\theta_0^2$, where θ_0 is the angle of the tilt in radians and the incident mode power is normalised to one as everywhere later. The power taken away by the TE_{20} mode is more than 99% of the power loss caused by all $TE_{n,0}$ modes. So, the modes TE_{40}, TE_{60} etc. have to be taken into account only near their cut-off frequencies.

The power of the propagating TE_{10} mode in an H-plane bend with a large, constant radius of curvature is coupled periodically to the spurious modes. The amplitude of the transmitted spurious TE_{20} mode is equal to

$$\left|P_j\right| = 1.2\times10^{-2}\frac{\left(\hbar_j + \hbar_m\right)^3}{8\hbar_j}(ka)^2\frac{a}{r}\left|\sin\frac{\sigma}{2}\right| \tag{5.4}$$

Here, σ is the phase difference between the TE_{10} and TE_{20} modes along the entire bend. The phase is to be counted along the waveguide axis since, as it may be easily proven, the value of r_j introduced by the condition of eqn. 2.25 is for all modes equal to the distance from the bend axis to the cross-section centre. The mode conversion in such bends with constant curvature fluctuates widely with frequency and can be tolerated only in very narrow bands.

If $|\sigma|$ is not much less than one and σ is not small compared to one and it is required to achieve low losses for a broad frequency band, then the curvature radius should be chosen such that the value of $1.4\times10^{-4}\,(ka)^4\,(a^2/r^2)$ has to be as small as possible, since $|\sin(\sigma/2)| \le 1$. For example, if for the highest frequency of the band $ka = 12$, then the relative power losses are equal to $2.9a^2/r^2$. To achieve losses below -30 dB one should design a bend with a curvature radius 55 times longer than the broad side, a.

Figure 5.2 corresponds to an E-plane tilt and the fundamental mode is denoted as TE_{01}, hence $n_m = 0$, $q_m = 1$. According to eqn. 5.2, degenerate mode pairs $TM_{n,1}$ and $TE_{n,1}$ with odd numbers $n \equiv n_j$ are excited. The coupling coefficients for $TM_{n,1}$ modes are

$$F_{jm} = \frac{2\sqrt{2}i}{\pi^2}\frac{\left(\hbar_j + \hbar_m\right)}{2\hbar_j}\frac{1}{n_j\sqrt{1+\left(n_j^2\dfrac{b^2}{a^2}\right)}}kb \tag{5.5}$$

The coupling coefficients for $TE_{n,1}$ modes are

$$F_{jm} = -\frac{2\sqrt{2}i}{\pi^2}\frac{\left(\hbar_j + \hbar_m\right)}{2\hbar_j}\frac{1}{n_j^2\sqrt{1+\left(n_j^2\dfrac{b^2}{a^2}\right)}}ka \tag{5.6}$$

Let us remember that, in the last two equations, b is the broad side of the cross section and a the narrow side.

Far from the cut-off frequencies of the modes $TE_{3,1}/TM_{3,1}$, $TE_{5,1}/TM_{5,1}$, ... it is sufficient to take into account only the power taken away by the linear combination of the nearby TM_{11} and TE_{11} modes. The ratio of the power in these modes is approximately equal to the square of the ratio of the broad side to the narrow side. If the ratio of the sides is equal to 2, i.e. $b/a = 2$ in the notation of Figure 5.2, then the TM_{11} mode carries away a power approximately equal to $0.017(kb)^2\theta_0^2$ whereas the TE_{11} mode takes away approximately four times less power. The whole losses are approximately 1.5 times less than for a tilt of the same angle in the H-plane.

The amplitude of the TM_{11} mode excited by an E-plane bend (for $b/a = 2$) is

$$|P_j| = 1.3\cdot10^{-2}\frac{\left(\hbar_j + \hbar_m\right)^2}{4\hbar_j}(ka)^2\frac{a}{r}\left|\sin\frac{\sigma}{2}\right| \tag{5.7}$$

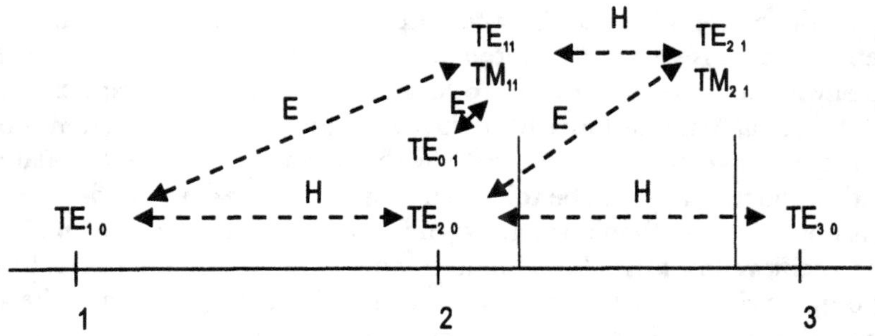

Figure 5.3 *Normalised cut-off frequencies in oversized rectangular waveguides with a/b = 2. Arrows show modes coupled by H and E-plane bends (Doane, 1988)*

According to eqns. 5.7 and 5.4, the conversion losses and the requirements on the curvature radius are very close for both kinds of bends at a fixed value of *ka*.

Let us now restrict ourselves to the notation of Figure 5.1. Bends in the plane of the broad waveguide wall are unfavourable because of the closeness of the most strongly coupled unwanted modes to the propagating main mode (see Figure 5.3).

Shorter waveguide bends can be realised with a radius of curvature $r(z)$ which varies along the bend. Since the mode conversion is a monotonous function of the differential phase constant when the differential attenuation is negligible, the design of gradual bends with hyperbolic secant spatial variation of the coupling is relatively simple (Doane, 1984b). When a broadband component is required, the hyperbolic secant coupling variation also frequently provides lower mode conversion for a given length than other common coupling distributions (linearly tapered or triangular coupling, cosine coupling, etc.). We take the curvature $c(z)=1/r(z)$ to have the special form

$$c(z) = \frac{1}{r_0}\operatorname{sech}(kz/2) \tag{5.8}$$

with $z = 0$ at the centre of the coupling region, i.e. of the bend, where $c(0) = 1/r_0$. A plot of the hyperbolic secant function sech $(kz/2)$ is presented in Figure 5.4.

The hyperbolic secant coupling may be truncated and still yields virtually the same transmission as for infinite coupling length. In two-mode approximation far from cut-off frequencies, the relative power lost from the TE_{10} mode (mode m) to an unwanted mode j in such a bend is given by the analytic expression (Doane and Anderson, 1989).

$$\left|P_j^2(\theta_0)\right| = \sin^2\left(\left|F_{jm}\right|\theta_0\right) / \cosh^2\left[(h_j - h_m)r_0\theta_0 / 2\right] \tag{5.9}$$

where θ_0 is the total bend angle in radians.

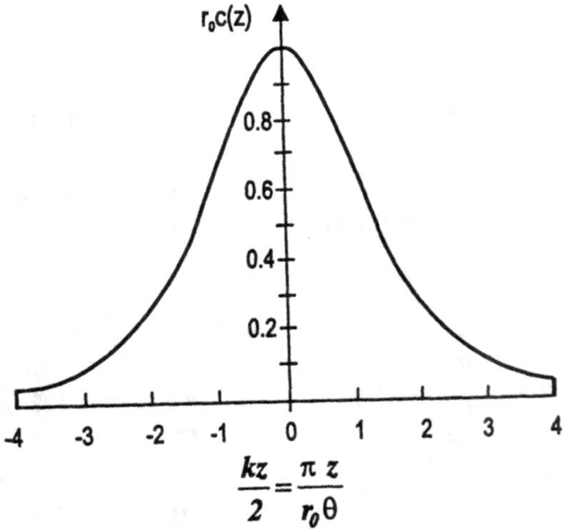

Figure 5.4 Hyperbolic secant coupling (Doane, 1984b)

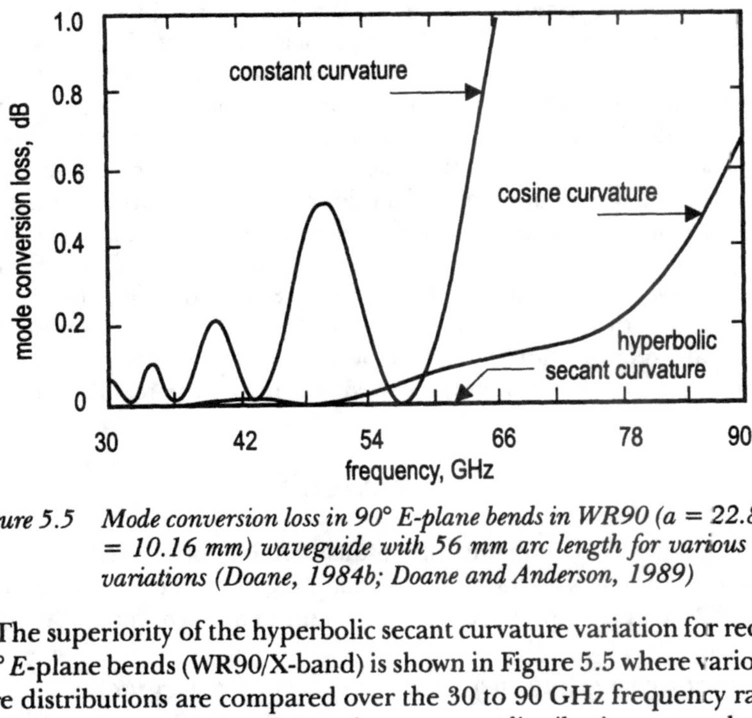

Figure 5.5 Mode conversion loss in 90° E-plane bends in WR90 (a = 22.86 mm × b = 10.16 mm) waveguide with 56 mm arc length for various curvature variations (Doane, 1984b; Doane and Anderson, 1989)

The superiority of the hyperbolic secant curvature variation for rectangular 90° *E*-plane bends (WR90/X-band) is shown in Figure 5.5 where various curvature distributions are compared over the 30 to 90 GHz frequency range.

The mode conversion for each curvature distribution was calculated by numerical integration of the coupled-wave equations describing the problem. To estimate the bend arc length L_B required in a hyperbolic secant bend to ensure less than 1% mode conversion, one can use the design rules

$$L_B \geq 5\,a^2 / \lambda_{min} \text{ for } H\text{-plane bends} \qquad (5.10a)$$

and

$$L_B \geq 16\, b^2 \,/\, \lambda_{\min} \text{ for } E\text{-plane bends} \tag{5.10b}$$

where λ_{min} is the free-space wavelength at the upper end of the frequency band (Doane and Anderson, 1989). We have confirmed that generalised Fourier-series curvature distributions (Sporleder and Unger, 1979) result in approximately the same bend arc lengths.

In ordinary waveguides with $a = 2b$, the required bend lengths L_B of eqn. 5.10 are comparable. In a WR 430 (a = 109.22 mm × b = 54.61 mm) waveguide, the necessary H and E-plane lengths at 8 GHz are 1.6 and 1.3 m, respectively. In the case of a tall waveguide with $b>a$, however, the difference in the required bend lengths becomes large. Notice that an H-plane bend in a tall waveguide appears mechanically like an E-plane bend in an ordinary waveguide.

Let us note in conclusion of this section that if a waveguide is excited within the fundamental mode frequency band, i.e. the frequency is less than the cut-off frequencies of parasitic modes, then the losses are caused only by wave reflection. The amplitude of the reflected wave is inversely proportional to the square of the curvature radius for a bend or proportional to the square of the tilt angle for a tilt. The amplitude can be found by the following procedure. The amplitudes of the forward and backward evanescent parasitic waves excited by the transmitted fundamental mode are determined using eqn. 2.27. The amplitude of the reflected fundamental wave excited by those parasitic modes is determined from eqn. 2.24 thereafter. The amplitude of the reflected fundamental wave is then expressed as a series. Since coupling coefficients decrease rapidly with the mode number of the parasitic modes, only a few series terms are required for the calculations. Of course, this procedure is restricted neither to only rectangular waveguides nor to the main mode.

5.2 TE_{11} mode in bent circular waveguides

This section is written using mainly the results of Kerzhentseva (1958). Other publications which deal with the curvature coupling of the TE_{11} mode are e.g. Morgan (1957), Unger (1957), Doane (1985) and Li and Thumm (1991a).

5.2.1 *Curvature coupling coefficients for circular waveguides*

The coupling coefficients of two modes are different from zero in a circular waveguide only if some special relationship between the mode indices and mode polarisation is satisfied. Let us consider only such asymmetrical modes, with transversal eigenfunctions ψ and φ which describe their polarisation dependence on the azimuthal angle β with the factor either $\sin(n\beta)$ or $\cos(n\beta)$. Since the modes of two basic polarisations differ by their behaviour on a tilt or bend, a mode with mixed polarisations described by the factor cos(β-β₀) is to be considered as a superposition of two modes with basic polarisations and each term is to be treated separately. Let us remember that the angle β is counted from the *x*-axis, which is oriented to the centre of curvature, i.e. the *x*-axis is in the plane of the bend.

The difference between the azimuthal indices of two modes coupled by a tilt has to be equal to one. This means that the necessary condition for F_{jm} not being equal to zero is the equation

$$n_j = n_m \pm 1 \tag{5.11}$$

If both coupled modes are of the same type *TE* or *TM*, then F_{jm} is not equal to zero only if the transversal eigenfunctions contain the same trigonometric function, i.e. both eigenfunctions are proportional either to $\cos(n\beta)$ or to $\sin(n\beta)$. If one of the coupled modes is a *TE* mode and the other a *TM* mode, then F_{jm} is not equal to zero if the transversal eigenfunctions are proportional to different trigonometric functions. For circular symmetric modes is $n = 0$, so that one should consider ψ and φ being proportional to $\cos(n\beta)$.

If the two conditions mentioned above, i.e. the polarisation condition and the azimuthal index condition, are satisfied, the coefficients F_{jm} are not equal to zero for any two modes. An exception to this rule is a combination of the modes TE_{0,q_m} and TM_{1,q_j} where F_{jm} is not equal to zero only if $q_m = q_j$ and the two waves have the same direction of propagation.

The special aspect of this case is connected both with the structure of eqn. 5.12*c* represented later, and with the degeneracy of modes the $TE_{0,q}$ and $TM_{1,q}$.

The coupling coefficient of two *TE* modes is

$$F_{jm} = -i\frac{\hbar_m\mu_j^2\left(\mu_m^2 - n_j n_m\right) + \hbar_j\mu_m^2\left(\mu_j^2 - n_j n_m\right)}{\left(\mu_j^2 - \mu_m^2\right)^2\sqrt{\mu_j^2 - n_j^2}\sqrt{\mu_m^2 - n_m^2}}\sqrt{\epsilon_{n_j}\epsilon_{n_m}} \times \frac{\hbar_j + \hbar_m}{2\hbar_j}ka \tag{5.12a}$$

The coupling coefficient of two *TM* modes is

$$F_{jm} = -i\frac{\hbar_m\nu_j^2 + \hbar_j\nu_m^2}{\left(\nu_j^2 - \nu_m^2\right)^2}\sqrt{\epsilon_{n_j}\epsilon_{n_m}}\frac{\hbar_j + \hbar_m}{2\hbar_j}ka \tag{5.12b}$$

If *m* is a *TE* mode and *j* is a *TM* mode, then their coupling coefficient is equal to

$$F_{jm} = -i\frac{n_m}{\sqrt{\mu_m^2 - n_m^2}\left(\mu_m^2 - \nu_j^2\right)^2}\sqrt{\epsilon_{n_j}}\frac{\hbar_j + \hbar_m}{2\hbar_j}ka \tag{5.12c}$$

The last equation is valid when

$$\psi^m \sim \cos(n_m\beta),\ \varphi^j \sim \sin(n_j\beta) \tag{5.13}$$

but if

$$\psi^m \sim \sin(n_m\beta),\ \varphi^j \sim \cos(n_j\beta) \tag{5.14}$$

then the sign in eqn. 5.12*c* changes to the opposite sign.

In particular, as follows from eqn. 5.12*c*, the TE_{0q} ($n_m = 0$) mode is not coupled by the tilt to the *TM* modes. Special cases are TM_{1q} modes with those values of q for which $\nu_j=\mu_m$ (TM_{1q} and TE_{0q} modes), in this case eqn. 5.12*c* cannot be applied. We analyse this case in the next section.

5.2.2 *Mode conversion at tilts in a circular waveguide carrying the TE_{11} mode*

Let us consider in detail the TE_{11} mode. Its transversal eigenfunction is

$$\psi^m = N^m \cdot J_1(\alpha_m\rho)\cos(\beta) \tag{5.15}$$

or

$$\psi^m = N^m \cdot J_1(\alpha_m\rho)\sin(\beta) \tag{5.16}$$

We denote a mode with the transversal eigenfunction of eqn. 5.15 as a mode with the first kind of polarisation. The transversal electrical field of this mode is represented in Figure 5.6. In contrast, a mode with the transversal eigenfunction of eqn. 5.16 is denoted as a mode with the second kind of polarisation. Its transversal electrical field is represented in Figure 5.7. If the TE_{11} mode with the first kind of polarisation is incident, then $TE_{0,q}$ modes, $TE_{2,q}$ modes (for which $\psi^j \sim \cos(2\beta)$) and $TM_{2,q}$ modes (for which $\varphi^j \sim \sin(2\beta)$) are excited at the tilt. On the other hand, if the incident mode is the TE_{11} mode with the second kind of polarisation, then $TM_{0,q}$ modes, $TE_{2,q}$ modes (for which $\psi^j \sim \sin(2\beta)$) and $TM_{2,q}$ modes (for which $\varphi^j \sim \cos(2\beta)$) are excited at the tilt. However, the coupling coefficients decrease rapidly with the number q, hence the greatest part of the power taken away by the parasitic modes from the first polarisation TE_{11} mode is carried away by the TE_{21} and TE_{01} modes. In contrast to this, the greatest part of the power taken away from the second polarisation TE_{11} mode is carried by the modes TM_{01} and TE_{21}. Table 5.1 shows the relative power losses (in % of the incident power) at a 1° tilt.

The total power losses at a tilt can be calculated simply for high frequencies, when $\hbar \cong 1$ can be substituted in eqns. 5.12. For the first kind of polarisation one yields

$$0.135(ka)^2\vartheta_0^2 \tag{5.17}$$

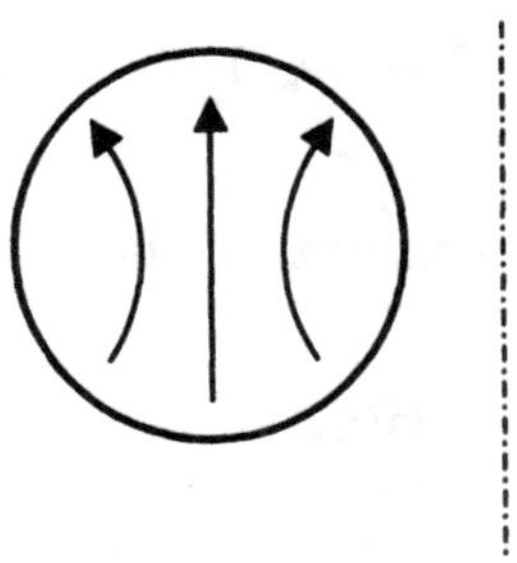

Figure 5.6 TE_{11} mode with first kind of polarisation

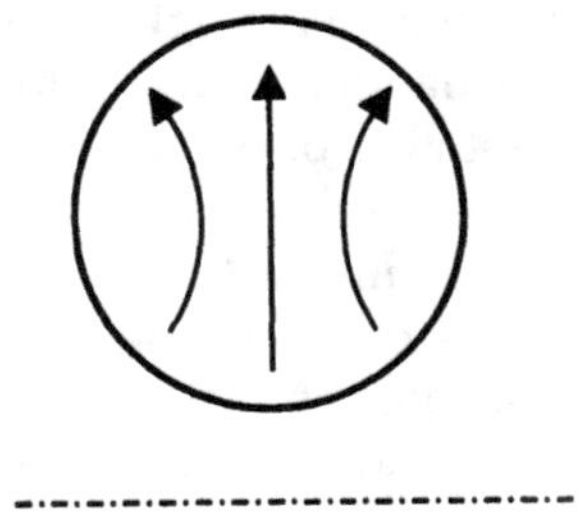

Figure 5.7 TE_{11} mode with second kind of polarisation

Table 5.1 TE_{11} power losses (in %) at a 1° tilt

mode j		ka									
		3.000	3.700	4.910	7.070	11.330	12.270	19.640	26.600	39.900	49.080
1st	TE_{21}	–	0.035	0.055	0.139	0.372	0.437	1.136	2.096	4.698	7.151
polari-	TE_{01}	–	–	0.015	0.043	0.118	0.147	0.392	0.727	1.646	2.490
sation	TM_{21}	–	–	–	0.001	0.003	0.004	0.009	0.017	0.038	0.057
	total	–	0.035	0.070	0.180	0.560	0.590	1.540	2.840	6.380	9.700
2nd	TM_{01}	0.040	0.084	0.102	0.211	0.542	0.636	1.630	3.014	6.732	10.179
polari-	TE_{21}	–	0.035	0.055	0.139	0.372	0.437	1.136	2.096	4.698	7.151
sation	TM_{21}	–	–	–	0.001	0.003	0.004	0.009	0.017	0.038	0.057
	total	0.04	0.120	0.170	0.350	0.930	1.080	2.180	5.130	11.470	17.390

where ϑ_0 is counted in radians. Almost three quarters of this power is taken away by the TE_{21} mode, approximately one quarter is carried away by the TE_{01} mode, whereas the TM_{21} mode takes away approximately 0.6%. Other modes can be neglected. The power losses of the second polarisation mode on a tilt are

$$0.247(ka)^2\vartheta_0^2 \tag{5.18}$$

Almost three-fifths of the losses are carried away by the TM_{01} mode and the other two-fifths by the TE_{21} mode.

5.2.3 Mode conversion in gradual curvature circular waveguide bends carrying the TE_{11} mode

The conversion losses of an incident first polarisation mode in a gradual bend with a large curvature radius are approximately one order smaller than the losses of the corresponding second polarisation mode. For example, calculations using eqn. 2.29 for a second polarisation mode show that the curvature radius of the bend with constant curvature must be at least 50 times greater than the waveguide radius, independent of the bend angle, to provide power losses less than 0.5 dB in a broad frequency band. In contrast to this, $r/a > 10$ would be sufficient for a mode with the first kind of polarisation.

If the curvature radius r is not very large, the amplitude of a parasitic mode can achieve a value close to one. At the same time, all other parasitic modes amplitudes can be still neglected. In this case, the normalised main mode amplitude $p_m(\vartheta)$ is not constant along the whole bend. Hence, eqns. 2.28 and 2.29 cannot be applied. However, the coupled-wave eqns. 2.24 can be simplified, retaining only two equations of the largest order, namely the equation for $p_m(\vartheta)$ and the equation for the normalised amplitude of the strongest parasitic mode $p_j(\vartheta)$. In such a way one yields two-wave coupled equations. Solving these, it is possible to determine the restrictions for the application of the more simple solutions of eqns. 2.28 and 2.29 and, if necessary, to calculate the amplitudes of the remaining parasitic modes. Doing this, one should take into account only two terms on the right-hand side of the coupled-wave eqns. 2.24, containing p_m and p_j. A similar consideration was carried out in Section 4.1. However, there we were mainly interested in the conditions close to mode degeneracy. We do not consider such conditions in the present section.

Let us give an example of this calculation representing a more accurate case of the general calculation carried out in subsection 2.1.4. We consider the example of a second polarisation TE_{11} mode. According to subsection 5.2.2, the main parasitic mode, i.e. the mode j, is the forward propagating TM_{01} mode. Taking to eqn. 2.24, the coupled-wave equations for $p_m(\vartheta)$ and $p_j(\vartheta)$ are

$$\begin{aligned} p_m' &= F_{mj} p_j \exp\left(-i\left(h_j - h_m\right) r\vartheta\right) \\ p_j' &= F_{jm} p_m \exp\left(-i\left(h_m - h_j\right) r\vartheta\right) \end{aligned} \tag{5.19}$$

Since only forward propagating waves are retained in these equations, the boundary conditions are

$$p_m(0) = 1, \quad p_j(0) = 0 \tag{5.20}$$

To present the solution in a form convenient for further analysis, let us introduce the critical angle ϑ_{cr}, determining it by the equation

$$\vartheta_{cr} = \frac{\pi}{r(h_m - h_j)} \frac{1}{W}, \quad W = \sqrt{1 - \frac{4F_{mj}F_{jm}}{r^2(h_m - h_j)^2}} \tag{5.21}$$

The solution of eqn. 5.20 of the coupled-wave eqns. 5.19 has the following form (compare with eqns. 4.21)

$$p_m = \left(\cos\left(\frac{\pi\vartheta}{2\vartheta_{cr}}\right) - i\frac{1}{W}\sin\left(\frac{\pi\vartheta}{2\vartheta_{cr}}\right)\right)\exp\left(i\frac{1}{W}\frac{\pi\vartheta}{2\vartheta_{cr}}\right) \tag{5.22a}$$

$$p_j = \frac{F_{jm}}{r(h_m - h_j)W/2}\sin\left(\frac{\pi\vartheta}{2\vartheta_{cr}}\right)\exp\left(-i\frac{1}{W}\frac{\pi\vartheta}{2\vartheta_{cr}}\right) \tag{5.22b}$$

Of course, eqns. 5.22 are nearly the same as eqns. 2.28 and 2.29*a*; eqns. 5.22 transform to eqns. 2.28 and 2.29*a* for increasing r.

A periodic power exchange from the main mode to the parasitic mode and *vice versa* occurs when ϑ is varied. The period of this exchange is equal to ϑ_{cr}. According to eqn. 2.29*a*, the amplitude $|p_j|$ also varies periodically between zero and its maximal value. This exchange period is different from ϑ_{cr} given by eqn. 5.21, because the factor $1/W$ is absent. The maximal value of $|p_j|$ calculated according to eqns. 2.29*a* and 5.22*b* also is different by this factor.

According to eqn. 5.22*a*, the main mode amplitude $|p_m|$ varies periodically from 1 when $\vartheta = 0, 2\,\vartheta_{cr}, \ldots$ etc. to the value of $1/W$ when $\vartheta = \vartheta_{cr}, 3\vartheta_{cr}, \ldots$ etc. Therefore, the quantitative estimation of the error which would occur if the reconversion of the parasitic mode to the main mode was not be taken into account is the deviation of the factor W from 1. More exactly, if $(W - 1) << 1$, this reconversion causes a relative variation of the maximal value of the parasitic mode amplitude equal to

$$C = \frac{2|F_{mj}F_{jm}|}{r^2(h_m - h_j)^2} \tag{5.23}$$

The relative variation of the power exchange period is equal to the same expression.

The parameter C in the bend problem has approximately the same physical meaning as the value of $1/q^2$ in the problem of a twisted waveguide (Section 4.1).

Let us return to our example of the modes TE_{11} and TM_{01}. For these modes, when $\hbar \cong 1$, the value of C is equal to

$$C = 0.2\frac{(ka)^4}{(r/a)^2} \tag{5.24}$$

For example, if we want to determine the maximal power of the TM_{01} mode with an accuracy of 10%, eqn. 2.29*a* can be used if $(r/a) \geq 0.8\,(ka)^2$, whereas for smaller r eqn. 5.22 has to be applied.

To assume $C << 1$ and to neglect the reconversion of a parasitic mode to the main mode, the curvature radius of the bend has to be larger for higher frequencies. This fact is explained by the increasing coupling coefficients and decreasing differences between the mode wavenumbers at higher frequencies.

If a first polarisation mode TE_{11} is incident on a bend, the main parasitic mode is the TE_{21} mode. Then, the parameter C of eqn. 5.23 can be given by eqn. 5.24, substituting the coefficient 0.2 by 0.1.

Of course, eqns. 5.22 and the condition $C << 1$, where C is determined by eqn. 5.23, can be applied to the analysis of a bend of an arbitrary cross-section waveguide. For example, if the mode TE_{10} is incident on an H-plane rectangular waveguide bend, the mode TE_{20} plays the main role, according to eqn. 5.3 and the parameter C for those modes is given by the same eqn. 5.24, where a is the broad side and the numerical coefficient is 3×10^{-4}.

5.2.4 Phase shift in TE_{11} mode bends with constant curvature

According to eqns. 2.28 $p_m(\vartheta) = 1$, i.e. the phase of the mode is determined only by the factor $\exp(-ih_m r\vartheta)$. According to eqn. 5.22*a*, the phase of $p_m(\vartheta)$ is not equal to zero, therefore the output mode phase is not equal to $h_m r\vartheta_0$. These phenomena may be considered as a variation of the wavenumber of the mode m related to the conversion into a parasitic mode j. As can easily be obtained from eqn. 5.22*a*, if $C << 1$, this equivalent variation of the value h_m is equal to

$$\Delta h_m = -\frac{1}{2}C(h_m - h_j) \tag{5.25}$$

This formula is similar to eqn. 4.26 for the twisted rectangular waveguide problem far from the mode degeneracy condition.

The wavenumber variation is of the order $1/r^2$, and the additional phase value is of the order $1/r$.

It is important from a practical point of view that the additional phase value is different for the two TE_{11} modes with orthogonal polarisations. The additional phase is substantially less for a first polarisation mode than for a second polarisation mode. At high frequencies ($\hbar \cong 1$) the phase difference in radians between both polarisation modes is equal to

$$0.24\frac{a}{r}\vartheta_0(ka)^3 \tag{5.26}$$

If an angle formed by the symmetry plane of an incident TE_{11} mode and the bend plane is not equal to zero or $\pi/2$, this phase difference causes elliptical polarisation of the transmitted TE_{11} mode. This effect also occurs in a fundamental mode circular waveguide, i.e. if $ka < 2.4$ and the TM_{01} mode does not propagate. In this case, h_j in eqn. 5.22*a* is a pure imaginary number and as the analysis of this equation yields the phase difference in a fundamental mode system is small and does not exceed several degrees. For example, if $ka = 2.2$; $r = 10a$ and $\vartheta_0 = \pi/2$, this phase difference is equal to 2.2°.

5.2.5 Gradual TE_{11} mode bends with tapered curvature distributions

To design short gradual TE_{11} bends, tapered curvature distributions need to be used as in Section 5.1. Coupled-wave equation calculations on a high power 70 GHz 90° bend (second polarisation TE_{11} mode) with an inner waveguide diameter of 27.78 mm (C76 standard waveguide) show that a gradual bend with triangular curvature variation must have a minimum arc length of 2 m in order to introduce less than 2% mode conversion. We have done these computations by taking into account the coupling of the incident main TE_{11} mode to the forward propagating unwanted modes TE_{01}, TE_{12} and TM_{11} (first-order coupling) and the coupling of the parasitic TE_{12} and TM_{11} modes to the modes TE_{21} and TM_{21} (second-order coupling for TE_{11}).

5.3 TE_{01} mode in bent circular waveguides

5.3.1 Curvature coupling coefficients for TE_{01} mode in circular waveguides

A tilt couples the TE_{01} mode with the modes $TE_{1,q}$ of both propagation directions. The polarisation of these $TE_{1,q}$ modes is such that

$$\psi^j = N^j J_1(\alpha_j\rho)\cos(\beta) \tag{5.27}$$

The coupling coefficients are equal to

$$F_{jm} = -i\frac{4\mu_j\mu_m}{\left(\mu_j^2-\mu_m^2\right)^2}\frac{\left(\hbar_j+\hbar_m\right)^2}{4\hbar_j}\frac{1}{\sqrt{2\left(1-1/\mu_j^2\right)}}ka \tag{5.28}$$

Apart from this, the coupling coefficient between the modes TE_{01} and TM_{11} propagating in the same direction is not equal to zero. The polarisation of the TM_{11} mode is such that

$$\varphi^j = M^j J_1(\alpha_j \rho)\sin(\beta) \tag{5.29}$$

The coupling coefficient is equal to

$$F_{jm} = -i\frac{ka}{\sqrt{2}\mu_j} \tag{5.30}$$

The coupling coefficient to the TM_{11} mode with $\varphi^j \sim \cos(\beta)$ is equal to zero.

Shushpanov (1961), derived similar formulae for a coaxial waveguide; the inner rod does not prevent degeneracy between the modes TE_{01} and TM_{11}.

The coupling between the modes TE_{01} and TM_{11} (more generally, between the modes $TE_{0,q}$ and $TM_{1,q}$) of the same propagation direction is not equal to zero. This phenomenon can be explained using the considerations given at the end of Section 2.3. Those modes are degenerate modes, i.e. their phase velocities are identical. According to Section 2.3, mode conversion at a small tilt is caused by the superposition of two effects: the conversion at a small step of a special kind (i.e. the conversion at an offset) and the addition of parasitic modes excited along a semiinfinite waveguide with these steps. Since the coupling coefficient, S_{jm}, between the $TE_{0,q}$ mode and all TM modes is equal to zero, the amplitudes of the TM modes excited on these steps are very small; their amplitudes are proportional to the square of the step height. The total amplitude of TM modes is also small. However, the TM_{11} mode has the same phase velocity as the TE_{01} mode so that the parasitic TM_{11} modes excited by all these steps are adding with the same phase. Their total amplitude is proportional to the tilt angle. This means that the coupling coefficient F_{jm} between the modes TE_{01} and TM_{11} is not equal to zero.

5.3.2 *Mode conversion at tilts in a circular waveguide carrying the TE_{01} mode*

The power scattered from a tilt to a $TE_{1,q}$ mode is mainly propagating in the forward direction. The ratio of the powers of backward and forward waves is

$$\left(\frac{h_m - h_j}{h_m + h_j}\right)^4 \tag{5.31}$$

This ratio is always less than one. It is close to one only near the cut-off frequency of a $TE_{1,q}$ mode; this special case is completely considered in Section 3.3. The relative power taken away by a $TE_{1,q}$ mode in forward and backward directions is equal to

$$\frac{8\mu_m^2\mu_j^4}{\left(\mu_j^2 - \mu_m^2\right)^4\left(\mu_j^2 - 1\right)}\frac{\left(\hbar_m + \hbar_j\right)^4 + \left(\hbar_m - \hbar_j\right)^4}{16\hbar_m\hbar_j}(ka)^2\,\vartheta_0^2 \tag{5.32}$$

The power taken away by the TM_{11} mode is

$$\frac{1}{2\mu_m^2}(ka)^2\,\vartheta_0^2 \tag{5.33}$$

To determine the total power losses one should add together all propagating modes $TE_{1,q}$ using eqn. 5.32. Further, eqn. 5.33 has to be added to the power losses as well. Eqn. 5.33 is proportional to the square of the parameter ka. The parameters h_m and h_j of eqn. 5.32 also include the frequency, so that the number of terms like those of eqn. 5.32 increases at higher frequencies. As the frequency approaches the cut-off frequency of a $TE_{1,q}$ mode, the power taken away by this mode increases. Near the cut-off frequency the amplitudes of the $TE_{1,q}$ modes are comparable with the incident mode amplitude (Section 3.3). Far from the cut off, at $h \cong 1$ the total power losses are equal to $0.16\,(ka)^2\,\vartheta_0^2$ (here ϑ_0 is in radians). Three-fifths of the total power losses are taken away by the TE_{12} mode, and each of the modes TM_{11} and TE_{11} takes away approximately one-fifth of the total power losses.

Table 5.2 presents the values of relative power losses at a 1° tilt calculated according to eqns. 5.32 and 5.33. Each row for $TE_{1,q}$ (q = 1,2,3) presents the power losses caused by the two $TE_{1,q}$ modes.

Table 5.2 TE_{01} power conversion losses (in %) at a 1° tilt

mode *j*	*ka*								
	4.910	7.070	11.330	12.270	12.770	19.610	26.600	39.900	49.080
TM_{11}	0.025	0.052	0.133	0.156	0.169	0.400	0.734	1.650	2.500
TE_{11}	0.016	0.043	0.126	0.150	0.163	0.398	0.739	1.670	2.550
TE_{12}	–	0.084	0.316	0.381	0.417	1.075	2.030	4.640	7.060
			(0.340)		(0.410)		(2.000)	(4.800)	
TE_{13}	–	–	0.002	0.002	0.003	0.008	0.015	0.036	0.055
total	0.040	0.190	0.580	0.690	0.750	1.880	3.510	8.000	12.200

The values in parentheses are given in the paper of Miller (1954)[1].

Table 5.3 is presented in Persikov *et al.* (1960). The measured results are compared with the calculated values using the described formulae. The relative power of the forward propagating parasitic modes TE_{11} and TM_{11} coupled to the incident TE_{01} mode are shown for a = 25mm; λ = 3.2cm and ϑ_0=2.3°.

[1]This paper contains no formulae for the tilt theory, only the numerical results mentioned above are described.

The results are in dB. The agreement between measurements and calculations is excellent.

Table 5.3 Mode conversion of TE_{01} mode at a 2.3° tilt

TE_{11}		TM_{11}	
measured	calculated	measured	calculated
-28.2	-28.6	-28.3	-28.9

So, the main part of the power losses is caused by the TE_{12} mode. However, if the electrical length of a tilt increases (for example, if r increases at a constant ϑ_0) then, according to eqn. 2.29 the amplitudes of all the modes decrease, but the amplitude of the TM_{11} mode remains unchanged. Therefore, the power losses of a bend with a large radius and a small ϑ_0 are mainly caused by excitation of the TM_{11} mode. This effect is especially important for bends of a finite angle ϑ_0. Let us consider those bends.

5.3.3 *Curvature coupling of the modes TE_{01} and TM_{11}*

Let us assume that $r >> a$ and that ϑ_0 has a finite value. 90° bends are most interesting from a practical point of view. Elementary TM_{11} modes excited along the various parts of the bend interfere with identical phases, therefore the TM_{11} mode amplitude achieves finite values. As was done in Section 4.1, two-wave coupled equations from general eqns. 2.24 have to be solved to study a degeneracy phenomenon. These equations contain the amplitudes of the forward propagating modes TE_{01} (this is $p_m(\vartheta)$) and TM_{11} (this is $p_j(\vartheta)$). These coupled-wave equations are

$$p'_m = Fp_j, \quad p'_j = Fp_m \tag{5.34}$$

Here, F is the coupling coefficient of eqn. 5.30 between the TE_{01} and TM_{11} modes; and $F_{jm} = F_{mj}$ because $h_m = h_j$.

The solution of the coupled-wave eqns. 5.34 with the obvious boundary conditions $p_m(0) = 1$, $p_j(0) = 0$ has the form

$$p_m = \cos\left(\frac{\pi\vartheta}{2\vartheta_c}\right) \tag{5.35a}$$

$$p_j = -i\sin\left(\frac{\pi\vartheta}{2\vartheta_c}\right) \tag{5.35b}$$

Let us call the value ϑ_c introduced above the Jouguet critical angle

$$\vartheta_c = -i\frac{\pi}{2F} \tag{5.36}$$

The angle ϑ_{cr} described in the previous section by eqn. 5.21 can be considered as a generaliastion of eqn. 5.36. If $h_m = h_j$ then eqn. 5.21 is transformed to eqn. 5.36.

According to eqn. 5.30

$$\vartheta_c = \frac{\pi\mu_m}{\sqrt{2}ka}[\text{rad}] = 77.6\frac{\lambda}{a}[\text{deg}] \tag{5.37}$$

The power is periodically transformed from the TE_{01} to the TM_{11} mode and *vice versa*. The full power conversion is achieved at $\vartheta = \vartheta_c$. This result was obtained by Jouguet (1947b).

So, for finite ϑ the amplitude of the TM_{11} mode can be of the order of one and the amplitude of the TE_{01} mode can deviate significantly from one. Therefore, the solution of eqn. 5.35 is to be substituted into the right-hand side of eqns. 2.24 to determine the amplitudes of parasitic modes excited together with the TM_{11} mode on a smooth bend with a finite angle ϑ_0. The amplitude of a mode with the index s is determined from the equation

$$p_s' = F_{sm}\cos\left(\frac{\pi\vartheta}{2\vartheta_c}\right)\exp(-i(h_m - h_s)r\vartheta) - iF_{sj}\sin\left(\frac{\pi\vartheta}{2\vartheta_c}\right)\exp(-i(h_j - h_s)r\vartheta) \tag{5.38}$$

p_s is determined by simple integrals. Parasitic modes with $n \neq 1$ appear already in the first order of $(1/r)$. Those modes are coupled to the TM_{11} mode and their coupling coefficient F_{sj} is different from zero.

5.3.4 *Circular waveguide TE_{01} mode bend in 3D space*

Jouguet's result of eqn. 5.35 was generalised by Viktorova and Sveshnikov (1958). They consider the incidence of the TE_{01} mode on a circular waveguide bend in 3D space. The waveguide axis forms a helical line with curvature radius r and torsion κ. As has been shown, the TE_{01} mode amplitude varies according to the following formula

$$p_m(\vartheta) = \frac{\kappa^2 r^2}{\kappa^2 r^2 + \dfrac{\pi^2}{4\vartheta_c^2}} + \frac{\dfrac{\pi^2}{4\vartheta_c^2}}{\kappa^2 r^2 + \dfrac{\pi^2}{4\vartheta_c^2}}\cos\left(\sqrt{\kappa^2 r^2 + \frac{\pi^2}{4\vartheta_c^2}}\,\vartheta\right) \tag{5.39}$$

Eqn. 5.39 is a generaliastion of eqn. 5.35*a* whereas eqn. 5.35*a* is valid only for a 2D (i.e. plane) bend.

According to eqn. 5.39 it is possible to avoid substantial conversion of the TE_{01} mode into a TM_{11} mode for a nonplanar (i.e. 3D) bend with an arbitrary angle ϑ_0 (valid only for one frequency). The axial torsion κ must satisfy the relation

$$\sqrt{\kappa^2 r^2 + \frac{\pi^2}{4\vartheta_c^2}}\vartheta_0 = 2\pi \cdot n, \quad n = 1,2\ldots \tag{5.40}$$

Eqn. 5.39 can be derived easily using the general procedure of Section 2.1. Let us consider for this purpose three modes coupled to each other: a TE_{01} mode and two orthogonally polarised modes TM_{11}. Let as assume that the polarisations of these modes are constant, e.g. the polarisations are horizontal and vertical. The x-axis of the co-ordinate system described in Section 2.1 is directed to the centre of curvature. In this co-ordinate system, the transversal eigenfunctions of these two modes are proportional to

$$\sin(\beta - \beta_0) \text{ and } \cos(\beta - \beta_0) \tag{5.41}$$

respectively. Here, β_0 is an angle in the cross-section plane between the x-axis and a fixed direction. For example, the fixed direction may be the direction of the x-axis at the beginning of the bend. The angle β_0 is proportional to the torsion

$$\beta_0 = \kappa r \vartheta \tag{5.42}$$

Let us denote the normalised amplitudes of the TE_{01} mode and of the two TM_{11} modes, with angle dependence indicated above, as p_m, p_q and p_s, respectively. The coupling coefficients between the TE_{01} mode and the modes with the numbers q and s are proportional to an integral of the product of a trigonometric factor of eqn. 5.41 and $\sin(\beta)$. The coupling coefficients are equal to

$$F\cos(\beta_0) \text{ and } F\sin(\beta_0) \tag{5.43}$$

respectively. Here, F is the coupling coefficient of eqn. 5.30 between the TE_{01} mode and the TM_{11} mode (eqn. 5.29). The transversal eigenfunction of this TM_{11} mode is proportional to $\sin(\beta)$. The two TM_{11} modes are not directly coupled to each other. The wavenumbers of all three modes are equal. The coupled-wave eqns. 2.24 for three variables $p_m(\vartheta)$, $p_q(\vartheta)$ and $p_s(\vartheta)$ have the following form

$$\begin{aligned}
&\frac{dp_m}{d\vartheta} = -\frac{i\pi}{2\vartheta_c}\left(p_q \cos(\kappa r\vartheta) + p_s \sin(\kappa r\vartheta)\right) \\
&\frac{dp_q}{d\vartheta} = -\frac{i\pi}{2\vartheta_c} p_q \cos(\kappa r\vartheta), \quad \frac{dp_s}{d\vartheta} = -\frac{i\pi}{2\vartheta_c} p_m \sin(\kappa r\vartheta)
\end{aligned} \tag{5.44}$$

Similar coupled-wave equations were derived in Viktorova and Sveshnikov, (1958) using another method. The function $p_m(\vartheta)$ in eqn. 5.39 is the solution of the coupled-wave eqns. 5.44 with the obvious boundary conditions $p_m(0) = 1$; $p_q(0) = p_s(0) = 0$.

5.3.5 TE_{01} to TM_{11} mode converters

As already mentioned, full power conversion from the TE_{01} mode to the TM_{11} mode is achieved at the Jouguet angle ϑ_c. The TM_{11} mode polarisation is perpendicular to the plane of the bend. This feature can be used to provide highly efficient TE_{01} to TM_{11} mode converters (Doane, 1982). Owing to the additional coupling of the TE_{01} mode to parasitic TE_{1q} modes and of the generated TM_{11} mode to TE_{2p} and TM_{2p} modes, the bend must be long enough and should have an optimised curvature distribution in order to convert virtually all of the TE_{01} power to TM_{11}. On the other hand, the mode converter ought to be as short as possible to minimise ohmic (wall) losses from the high loss TM_{11} mode and to be compact. Bent 70 GHz and 140 GHz TE_{01} to TM_{11} mode converters with a waveguide radius a = 13.9 mm (C76 standard waveguide) were optimised by means of numerical integration of the coupled-wave equations for the six forward waves: TE_{01}, TM_{11}, TE_{11}, TE_{12}, TE_{21} and TM_{21} (Kumrić and Thumm, 1986). Due to the multi-mode coupling, the conversion angle ϑ_{opt} depends on the curvature distribution and differs by a small amount from the Jouguet angle ϑ_c.

In both cases, the lowest level of undesired spurious modes together with the highest power of transmission efficiency (shortest conversion arc length L_c and thus lowest ohmic attenuation) is obtained with a sinusoidal curvature profile $c(z) = 1/r(z) = (\vartheta_{opt}\pi/2L_c)\sin(\pi z/L_c)$ instead of other tapered curvature distributions or constant curvature. The calculated conversion efficiencies for optimum 200 kW transducers of 98.0% at 70 GHz ($\vartheta_{opt} = 24.0°$, $L_c = 1.37$ m) and 95.2% at 140 GHz ($\vartheta_{opt} = 12.5°$, $L_c = 2.49$ m) are in excellent agreement with the measured values of $(97.6 \pm 0.4)\%$ and $(95 \pm 1)\%$, respectively (ohmic losses in copper waveguide are included). Figure 5.8 shows a photograph of the optimum mode transducers.

The far-field pattern of the pure 70 GHz TE_{01} input mode for low power test measurements is shown in Figure 5.9. In Figure 5.10 the E-plane and H-plane far-field radiation patterns of the generated TM_{11} mode, measured at the output of the optimal 70 GHz TE_{01} to TM_{11} mode converter are plotted.

5.3.6 $TE_{01} - TM_{11}$ mode coupling in helical waveguide with dielectric backing

If a bend with low conversion of the TE_{01} mode to TM_{11} in a broad frequency band is required, then their degeneracy has to be somehow removed, i.e. a waveguide system must be developed where their wavenumbers are not identical, in order to prevent mode conversion from TE_{01} to TM_{11}.

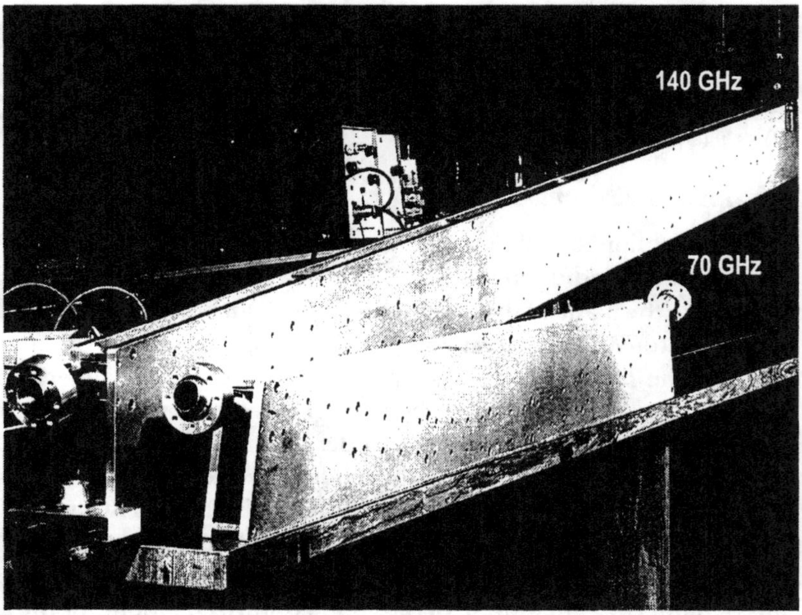

Figure 5.8 Photograph of 70 GHz and 140 GHz TE_{01} to TM_{11} mode converters with optimum sinusoidal curvature distribution (a = 13.9 mm) (Kumrić and Thumm, 1986)

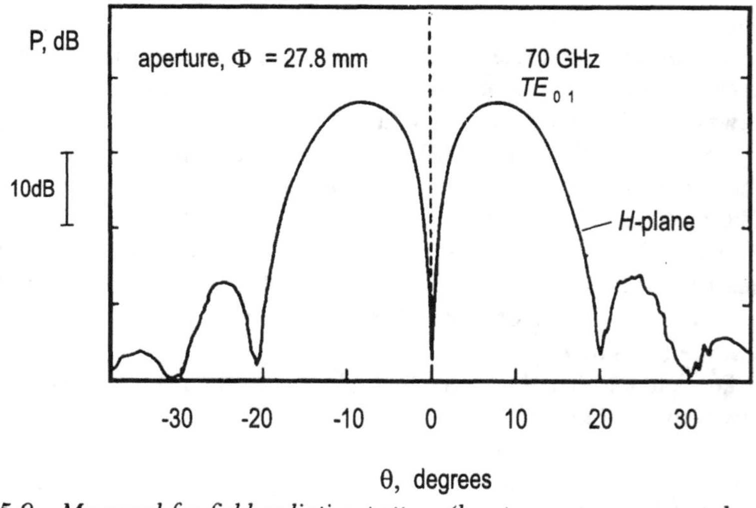

Figure 5.9 Measured far-field radiation pattern (low-power measurement, logarithmic scale) of the pure 70 GHz TE_{01} input mode. Separation of transmitting and receiving apertures is $4D^2/\lambda_0$ = 720 mm (Kumrić and Thumm, 1986)

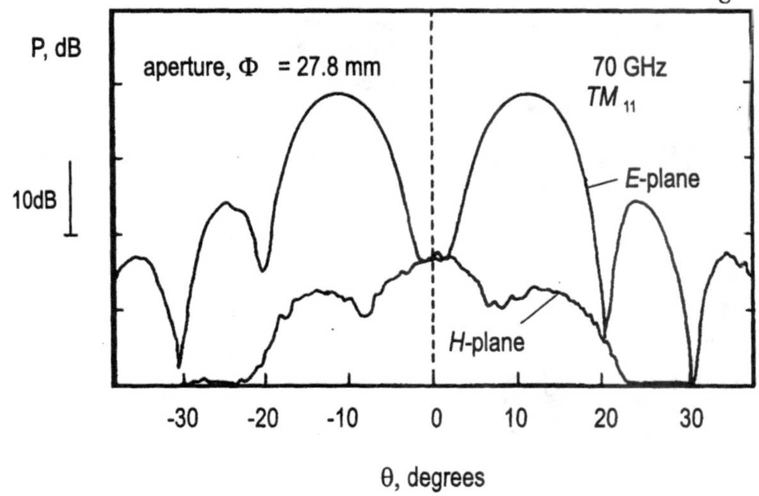

Figure 5.10 Measured E and H-plane far-field radiation patterns (low-power measurement, logarithmic scale) of the TM_{11} mode at the output of the optimum 70 GHz TE_{01} to TM_{11} mode transducer ($D = 2a = 27.8$ mm). Separation of transmitting and receiving apertures is $4D^2/\lambda_0 = 720$ mm (Kumrić and Thumm, 1986)

A simple method of degeneracy removal is to use a dielectrically coated metallic waveguide as Unger (1957). But much more efficient is to employ a helical waveguide with a composite jacket structure instead of the smooth-wall metallic circular waveguide (Figure 5.11). Its wall consists of closely wound insulated copper wire with a very small helical increment. A dielectric layer is situated outside the helical wire for $\rho > a$, where a is the radius of the helical cylinder. This dielectric layer, in its turn, is surrounded by a cylindrical metallic shield from the outside. The distance between helix and metallic shield should be about a quarter of the radial wavelength in the dielectric material. The fields of the TE_{0q} modes are practically totally shielded by the helix and these modes propagate in nearly the same manner as in a metallic circular waveguide with the same radius a. The fields of other modes penetrate through the helix; hence their wavenumbers differ from their values h_j in the metallic waveguide. In particular, the wavenumber of the TM_{11} mode is influenced by this design.

However, at the same time, the coupling coefficients of various modes in this design differ from F_{jm} in the metallic waveguide. Now one has to take into account in the formulae for F_{jm} in Section 2.1, that the field structure of all modes (except the TE_{0n} modes) in the helical waveguide differs from that of the modes described in Section I.3. This calculation is performed in Unger (1958b) and Kerzhentseva (1959). We will give here the numerical values of the coupling coefficient between the TE_{01} and TM_{11} modes from these references; for the metallic waveguide F is given by eqn. 5.30.

The plot in Figure 5.12 is taken from Unger (1958b).

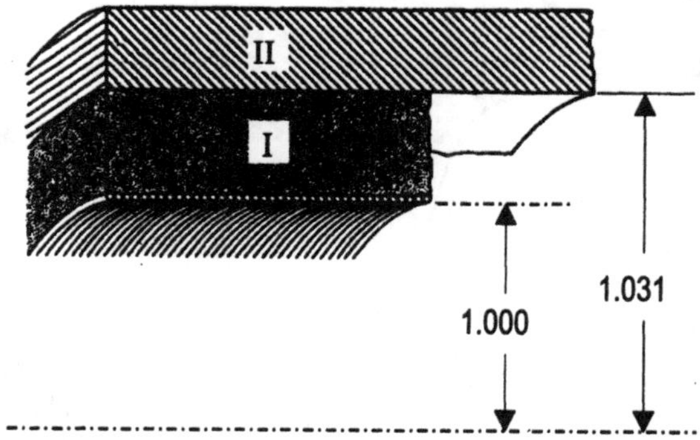

Figure 5.11 *Helix waveguide with dielectric jacket (I) and outer metallic shield (II) (Unger, 1958b); the radii are given in inches*

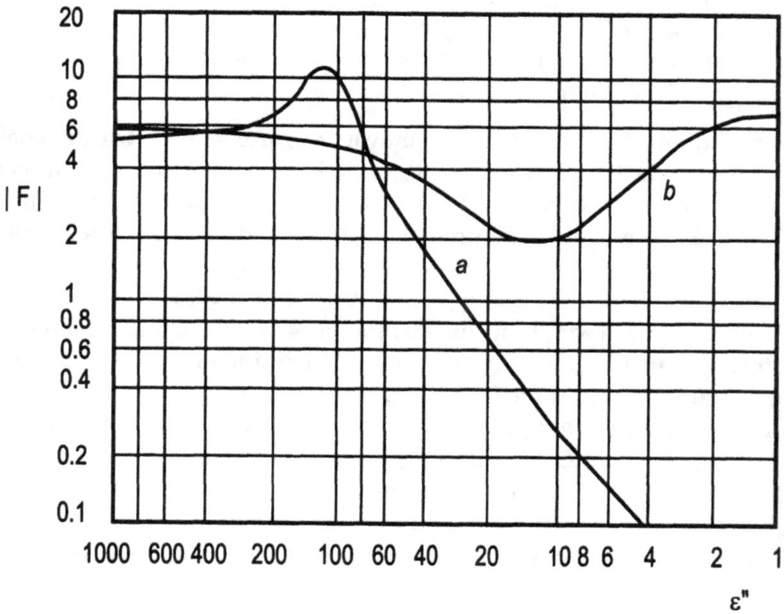

Figure 5.12 *TE_{01}-TM_{11} mode curvature coupling coefficient in helical waveguide according to Unger (1958b) for $ka = 29.5$ and $\varepsilon' = 4$ (a), $\varepsilon' = \varepsilon''$ (b)*

The absolute value of the coupling coefficient $|F|$ is presented there as a function of the imaginary part ε'' of the dielectric permittivity $\varepsilon = \varepsilon' - i\varepsilon''$ of the medium situated between the helix and the external metallic shield. The curve *a* corresponds to $\varepsilon' = 4$ and the curve *b* to $\varepsilon' = \varepsilon''$, both curves are computed for

$ka = 29.5$. Table 5.4 is presented in Kerzhentseva (1959). The $|F|$ values are given in this table for several values of ka, ε', and ε'' when there is no external metallic shield, i.e. the whole space $\rho > a$ is filled by the dielectric (unshielded helix waveguide). The $|F|$ value for a metallic waveguide is given in the last row, following eqn. 5.30.

Table 5.4 Absolute value of the TE_{01} - TM_{11} coupling coefficient

ka	6.5	6.5	6.5	6.5	12.9	12.9	12.9	29.5	29.5	29.5	29.5	29.5	29.5	29.5
ε'	4	4	4	4	40	2	1	1000	64	25	10	1	4	4
ε''	250	64	10	1	40	2	1	1000	64	25	10	1	90	8
$\|F\|$	1.17	1.13	1.07	1.24	2.2	1.10	1.7	5.2	4.2	6.5	1.6	2.8	4.1	0.01
$\|F\|$ from eqn. 5.30	1.20				2.43			5.45						

As follows from these data, the coupling coefficient in the helical waveguide may be significantly different from eqn. 5.30 and, hence, this difference has to be taken into account by calculating the conversion losses following from the formulae of Section 2.1. It seems as if the metallic shield has a little influence on the value of F, at least for the large values of ε''. For applications of the helix waveguide as a mode filter, the dielectric jacket should be lossy which causes a high transmission loss for all unwanted modes (Unger, 1958b).

5.3.7 Gradual helical waveguide TE_{01} mode bends with optimum curvature distribution

The wavenumbers of all modes are nearly equal to each other at high frequencies, where the parameter ka has large values. Then, according to eqn. 2.28*b*, the amplitudes of the forward parasitic waves excited along the constant curvature bend have a small value in their denominator $\Delta h = h_j - h_m$. Therefore, to provide a sufficiently low level of conversion losses, very long bends have to be employed. However, designing a varying curvature bend, the bend length can be decreased for a given Δh. This idea was proposed by Unger (1957) and realised in Andreasen (1958b). However, the mathematical technique employed in these references did not allow the authors to consider bends with small tilts at their ends. But exactly those bends can achieve minimal conversion losses at the minimal length. This was proven by Kerzhentseva (1960). We present the main results of this reference in this subsection.

The parasitic mode amplitude at the output of a varying curvature bend is given by eqn. 2.35. We rewrite it in the following form

$$p_j(\vartheta_0) = F_{jm} \int_{-L/2}^{L/2} \frac{1}{r(l)} \exp(-i\Delta h \cdot l)\, dl \tag{5.45}$$

Here, the origin of the co-ordinates was moved to the bend centre.

To select the optimal curvature radius distribution $r(l)$, let us first consider the value of

$$\frac{p_j(\vartheta_0)}{F_{jm}} = \int_{-L/2}^{L/2} \frac{1}{r(l)} \exp(-i\Delta h \cdot l)\, dl \tag{5.46}$$

This value has the same analytical form as the reflection coefficient eqn. 4.67 of the transmission line section with varying impedance $W(\zeta)$. One can solve the selection problem for the optimal function r(l), applying the optimal $W(\zeta)$ function. Unlike the problems considered in Section 4.3, it is worth employing the true optimal function $W_{opt}(\zeta)$, i.e. the limiting Chebyshev function determined by Klopfenstein (1956).

According to the calculation, the bend curvature must vary as following

$$\frac{1}{r(l)} = \frac{1}{\cosh\left(\frac{\sigma_{min}}{2}\right)} \left\{ \frac{\sigma_{min}}{2L} \cdot \frac{I_1\left(\sigma_{min}\frac{\sqrt{1-(2l/L)^2}}{2}\right)}{\sqrt{1-(2l/L)^2}} + \frac{1}{2}\delta\left(l-\frac{L}{2}\right) + \frac{1}{2}\delta\left(l+\frac{L}{2}\right) \right\}, \tag{5.47}$$

$$-\frac{L}{2} \le l \le \frac{L}{2}$$

where I_1 is the modified first kind Bessel function and δ is the delta function. The bend parameter is σ_{min}, it determines the relation between the product $\sigma = L \cdot \Delta h$ and the value given by eqn. 5.46. For all values of σ greater than the selected σ_{min}, the value of $\left|\frac{p_j(\vartheta_0)}{F_{jm}}\right|$ will be less than $\frac{1}{\cosh\left(\sigma_{min}/2\right)}$. If, for example, σ_{min} was selected to be equal to 10.6, then the value of eqn. 5.46 would be less than 0.01. The parasitic mode amplitude in the bend with the curvature varying according to eqn. 5.47 is equal to

$$|p_j| = |F_{jm}| \cdot \frac{1}{\cosh\left(\frac{\sigma_{min}}{2}\right)} \left|\cos\left(\frac{\sqrt{\sigma^2 - \sigma_{min}^2}}{2}\right)\right| \tag{5.48}$$

All the maxima of $|p_j|$ considered as a function versus σ are equal to each other and situated equidistantly; it is the feature of such a bend.

The value of $L/r(l)$ proportional to the curvature of eqn. 5.47 is presented in the Figure 5.13 as a function of l/L for two values of the parameter σ_{min} (the

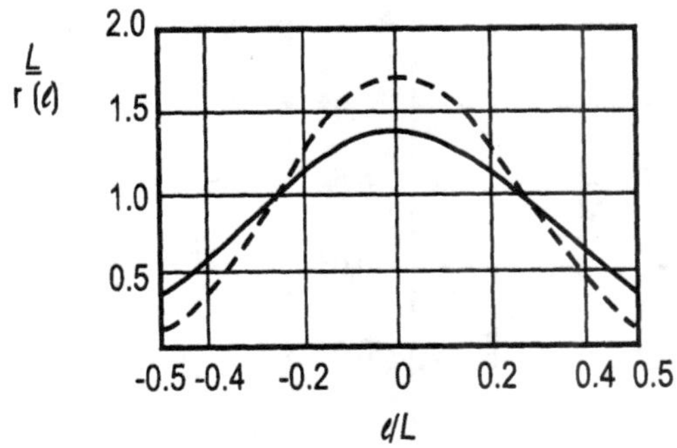

Figure 5.13 Curvature distribution in TE_{01} bend with optimum Chebyshev coupling distribution (solid curve: σ_{min} = 7.14, dashed curve: σ_{min} = 10.6)

solid curve corresponds to $\sigma_{min} = 7.14$ and the dashed one to $\sigma_{min} = 10.6$). There are small tilts at the bend ends. According to eqn. 5.47 the tilts are equal to $\dfrac{1}{2\cosh(\sigma_{min}/2)}$. The maximal curvature achieved at the bend centre is equal to

$$\frac{1}{r_{min}(l)} = \frac{1}{L} \cdot \frac{\sigma_{min}}{2\cosh\left(\dfrac{\sigma_{min}}{2}\right)} I_1\left(\frac{\sigma_{min}}{2}\right) \tag{5.49}$$

The lengths of three bends designed by three different methods with the same maximal value of the parameter of eqn. 5.46, i.e. with the same conversion losses, are compared in Figure 5.14. The solid line in this figure represents $\left|\dfrac{p_j(\vartheta_0)}{F_{jm}}\right|$ versus σ for the bend developed according to eqn. 5.47, the dashed line corresponds to the bend proposed by Unger, (1957) with $1/r(l)$ increasing linearly from zero to its maximal value at the centre, and the dotted-dash line to a constant curvature bend. The lower the required conversion losses are, the more advantageous is the bend with small tilts designed according to eqn. 5.47. At $L \cdot \Delta h = 20$, for example, $|p_j|$ is 400 times less than in the bend of the same length with linearly varying curvature.

The maximal value of the amplitudes $p_j(\vartheta_0)$ varies with frequency only because of the variation of the coupling coefficient F_{jm}. According to discussion above, the frequency variations of Δh do not influence the maximal conversion losses. The bend length should be selected by the following requirement

$$L \geq \frac{\sigma_{min}}{(\Delta h)_{min}} \tag{5.50}$$

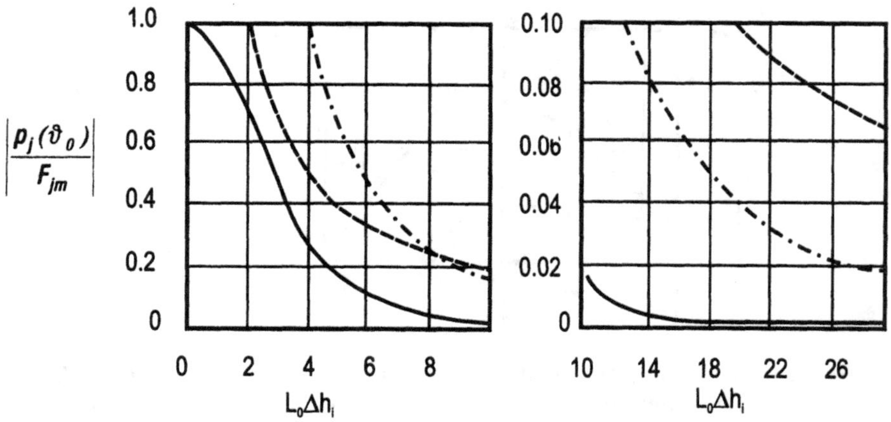

Figure 5.14 Conversion losses for TE_{01} bends with different curvature distributions (optimum —, linear - - -, constant – · – ·)

This requirement is similar to the condition of eqn. 4.80.

5.3.8 TE_{01} mode bends with partially dielectric filling

Another method of reducing conversion losses in bends applies partial filling of the cross section of the curved waveguide with a dielectric medium. This idea as well as the design of the corresponding compensator devices for the TE_{01} mode are described in Morgan (1957). The field in the curved waveguide with inhomogeneous waveguide filling is studied in this reference as the superposition of the modes of a straight empty waveguide. The inhomogeneous filling medium and the curvature are the two reasons for coupling of those modes. The total coupling coefficients of the TE_{01} mode to several other modes, first of all to the TM_{11} mode, can be forced to be equal to zero or significantly decreased by proper selection of the function $\varepsilon\,(x,y)$. Single and three-sector compensators with suitably shaped dielectrics (e.g. polystyrene) are designed in Morgan (1957). The single-sector device employs a dielectric sector situated on the side closest to the bend axis, i.e. at $\beta < \pi/2$. The three-sector compensator is also situated at $\beta < \pi/2$ (Figure 5.15).

Let us restore the notation of subsection 2.2.9. Then, the coupling coefficient between the modes (E^{0j}, H^{0j}) and (E^{0m}, H^{0m}) caused by the inhomogeneous filling medium is determined by the general formula of eqn. 2.87, whereas the coupling coefficient caused by the curvature was derived in eqns. 5.28 and 5.30. For example, the sum of those coupling coefficients for the TM_{11} mode is equal to zero in the single-sector compensator if its parameters ε and β_0, the sectorial angle, are related by the following expression

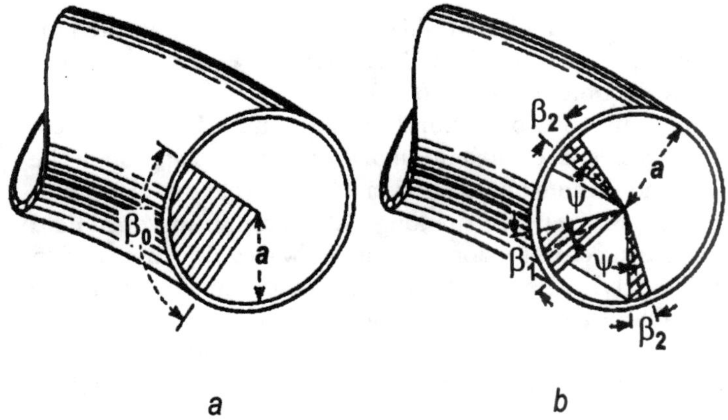

Figure 5.15 Single and three-sector compensators (Morgan, 1957)

$$(\varepsilon - 1)\sin\left(\frac{\beta_0}{2}\right) = 1.53\,\frac{a}{r} \tag{5.51}$$

The expansion in the modal fields of an empty waveguide employed in subsection 2.2.9 leads to a somewhat faster solution of this problem than the expansion in the modal fields of the uniform waveguide with the same dielectric filling. The reason for this is in the proximity of the modal field in the filled waveguide to the field of the TE_{01} mode of the straight plain waveguide, and the bend is connected to exactly those waveguides. The TE_{01} mode is an eigenmode of the straight section and, simultaneously, it is nearly identical with an eigenmode of the bend, therefore this mode is transmitted through the bend with little distortion.

Designing a more complicated dielectric shape one could decrease the coupling of the TE_{01} mode to several other modes and not only to TM_{11}. However, no specific selection of the scalar function $\varepsilon(x,y)$ (assuming $\mu = 1$) can provide zero coupling of the TE_{01} mode to all other modes and its propagation through the bend without any conversion losses. The bend compensators designed in Morgan (1957) have low losses due to the assumption $|\varepsilon - 1| << 1$. This limitation is caused both by the calculation method employed there and by the requirement to provide a small value of ε'', i.e. low dielectric losses.

If one designed the same device following the method of Section 2.1, first one would have to derive the eigenmodes of the dielectrically filled straight waveguide. For example, they could be derived following the method of subsection 2.2.9. For $|\varepsilon - 1| << 1$ and taking into account the coupling of the modes TE^0_{01} to TM^0_{11} only, the eigenmodes are in fact two linear combinations of those modes, nearly identical to the modes TE^0_{01} and to TM^0_{11}, respectively. The propagation of these modes through the bend is described by two differential coupled-wave equations similar to eqns. 2.9. Their boundary conditions are derived from the requirement that these two eigenmodes form the TE^0_{01} mode at the bend beginning. This calculation would be a little more complicated for $|\varepsilon - 1| << 1$ than the calculation in Morgan (1957), however it seems as if it would be significantly simpler for finite values of $(\varepsilon - 1)$.

Let us give, in conclusion, an explanation from the physical point of view, why it is impossible to synthesise a distribution of $\varepsilon(x,y)$ for propagation of the TE_{01} mode through a bend without any distortion. It seems that this fact can be explained by the relationship between the longitudinal and transversal field components at neighbouring cross sections of a curved waveguide discussed in the conclusion of Section 1.1. Selecting $\varepsilon(x,y)$, one could equalise the phase velocities of all points of the wave front at the cross section and provide the turn of the wave front of the mode without distortion. However, additional field components would appear due to the relationship mentioned above, therefore parasitic modes with other mode numbers would be inevitably excited. The longitudinal components become less essential with increasing frequency, hence the conversion losses in a bend containing an inhomogeneous dielectric can be decreased.

5.3.9 *Circumferentially corrugated TE_{0q} mode bends*

In bends for high-power microwave transmission lines, where pure metallic waveguide structures are preferable, the introduction of circular corrugations in the wall transverse to the axis of propagation also removes the degeneracy of TE_{0q} and TM_{1q} modes (Miller, 1952; Doane, 1984b; Thumm *et al.*, 1986). In cases where the waveguide radius a becomes fairly large compared to the free-space wavelength λ, and when the period p of the annular slots is assumed small compared with the guide wavelength, no space harmonics need to be considered and a circumferentially corrugated waveguide wall may be modelled well by an effective anisotropic surface reactance (wall impedance concept, i.e. Leontovich boundary condition) where the circumferential component is zero ($w_{ss} = 0$) and the longitudinal component is finite: $w_{zz} = iw$. For $ka >> 1$ the surface reactance may well be approximated as

$$w = \frac{s}{p} \cdot \tan(k \cdot d) \tag{5.52}$$

where d and s are the mechanical slot depth and width, respectively. The factor s/p reflects the fact that the finite width of the corrugation teeth reduces the reactance, and the $\tan(kd)$ dependence indicates that the annular slots act like short-circuited transmission lines. The longitudinal surface reactance may be reduced further if the slots and teeth are rounded in order to increase the power carrying capability.

The propagation behaviour of HE_{nq} and EH_{nq} hybrid modes (n > 0) in such circumferentially corrugated waveguides is described by the characteristic (dispersion) equation for the roots $X_{nq} = \alpha_{nq} \cdot a$

$$\frac{1}{wka} = \frac{1}{X_{nq}^2} \frac{J_n(X_{nq})}{X_{nq} J'_n(X_{nq})} \left[\frac{n^2 h_{nq}^2}{k^2} - \left(\frac{X_{nq} J'_n(X_{nq})}{J_n(X_{nq})} \right)^2 \right] \tag{5.53}$$

where α_{nq} is the radial separation constant (transverse wavenumber) and $h_{nq} = (k^2 - \alpha_{nq}^2)^{1/2}$ is the longitudinal propagation constant in the waveguide. The numerical solution of the characteristic eqn. 5.53 for a corrugated waveguide with the diameter $2a = 27.79$ mm at $f = 70$ GHz is plotted in Figure 5.16. The eigenvalues $\alpha_{nq} \cdot a$ of normal HE_{nq} and EH_{nq} hybrid modes are monotonically varying functions of the relative slot depth $d/(\lambda/4)$. For $d/(\lambda/4) = 0,2,4...(w = 0)$ the eigenvalues of hybrid modes are equal to the eigenvalues of the modes of a smooth waveguide (it is assumed that there are no ohmic losses in the waveguide). Hybrid modes consist of linear combinations of TE_{nq} and TM_{nq} field components which are determined by the hybrid factor (Clarricoats and Olver, 1984)

$$\overline{\Lambda}_{nq} = -\frac{J_n(X_{nq})}{X_{nq} J'_n(X_{nq})} \tag{5.54}$$

The balanced hybrid condition $(\overline{\Lambda})^2 = 1$ (at $d/(\lambda/4) = 1,3,5,...$ with $w = \infty$) has the two roots

$$\overline{\Lambda} = +1 \tag{5.55}$$

corresponding to *HE* modes and

$$\overline{\Lambda} = -1 \tag{5.56}$$

corresponding to *EH* modes. Here, $|\overline{\Lambda}| = \infty$ corresponds to *TE* modes and $\overline{\Lambda} = 0$ to *TM* modes.

In corrugated waveguide bends a pure propagating mode can be made the dominant mode by proper choice of the surface reactance (Figure 5.16). TE_{0q} modes are not affected by the corrugation. In the case of TE_{on} mode bends, an appropriate corrugation removes the degeneracy of the mode pairs TE_{op}/TM_{1q} and prevents a degeneracy with the hybrid HE_{2q} modes at corrugation depths of about $\lambda/4$ (Figure 5.16). Curvature coupling coefficients for corrugated waveguides where derived in Doane (1985) and Li and Thumm (1991a). The properties of actual bends have been optimised by numerical integration of the corresponding forward coupled-wave equations.

Optimal electrical slot depths appear at 0.2 $\lambda/2$, 0.3 $\lambda/4$ and 0.4 $\lambda/4$ for the modes TE_{01}, TE_{02} and TE_{03}, respectively. The arc length of the bends has been reduced by optimisation of the curvature distribution. The data of various corrugated waveguide bends with parasitic mode level at the output of approximately -20 dB are given in Table 5.5. Figure 5.17 shows a photograph

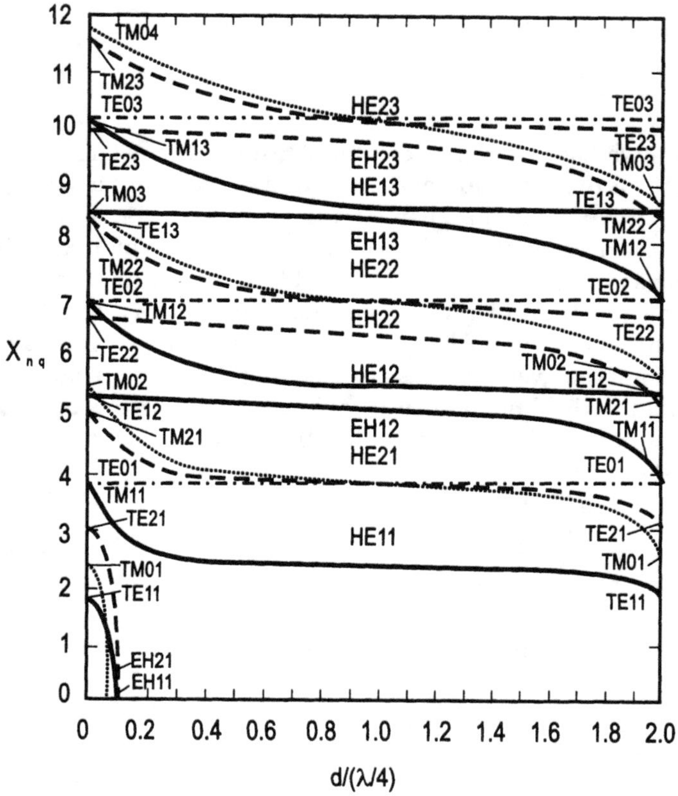

Figure 5.16 Eigenvalues X_{nq} of the characteristic equation against normalised electrical slot depth $d/(\lambda/4)$ for various modes in a circumferentially corrugated waveguide with inner radius a = 13.9 mm at 70 GHz (ka = 20.39)

of the 70 GHz TE_{02} mode bend (90°, electroformed corrugated Cu waveguide). Unintentional mode conversion losses in circumferentially corrugated TE_{oq} bends scale with $a^3/(\lambda^2\mu_{oq})$. Therefore, bends can be made shorter if they propagate higher TE_{oq} modes.

The optimal electrical slot depth for HE_{11} mode bends evidently is $\lambda/4$ (see Figure 5.16). The data of a 47° H-plane bend is also listed in Table 5.5. In the case of an E-plane bend propagating the HE_{11} mode, TM_{02} replaces coupling to TE_{01}. However, optimal H-plane bends also can be used as E-plane bends because the curvature coupling constants for HE_{11}-TE_{01} coupling and HE_{11}-TM_{02} coupling are identical.

Table 5.5 Data of various corrugated waveguide bends with inner diameter 2a = 63.4 mm at 28 GHz and 2a = 27.8 mm at 70 GHz and 140 GHz (Thumm et al., 1986)

Propag. mode	Bend angle	Frequency (GHz)	Arc length (m)	Curvature distrib.	Coupled modes 1st order	Coupled modes 2nd order
TE_{01}	90°	70	2.4	triangular	HE_{11}, EH_{12}	HE_{21}
TE_{02}	90°	28	3.3	sin	HE_{12}, EH_{12}, EH_{13}	HE_{22}, EH_{22}
TE_{02}	90°	70	1.74	sin	HE_{12}, EH_{12}, EH_{13}	HE_{22}, EH_{22}
TE_{03}	90°	140	3.5	sin	HE_{13}, EH_{13}, EH_{14}	HE_{23}, EH_{23}
HE_{11}	47°	70	1.5	$\sin^2$	HE_{21}, TE_{01}	HE_{12}, EH_{12}, HE_{31}

Figure 5.17 Corrugated 70 GHz TE_{02} mode bend (90°, a = 13.9 mm, arc length = 1.74 m)

Chapter 6

Mode converters with periodically nonuniform waveguide walls

6.1 Principles of rippled-wall mode converters

High-power millimetre microwave sources as gyro devices operate in higher-order modes of circular waveguides. For interfaces of these devices to high-power microwave transmission lines operating in waves with simple field structure, for feeding of gyro and Cherenkov amplifiers and for low power (cold test) measurements of electrodynamic circuits of such high power tubes, mode converters for higher-order modes are required (Thumm, 1993). In overmoded circular waveguides a selective transformation of one specific mode with azimuthal mode index m_1 into another mode with azimuthal mode index m_2 can be achieved by means of a periodic structure of the inner waveguide wall (rippled wall)

$$a(z,\beta) = a_0 + a_1 \cos(\Delta h \cdot z + \Delta m \cdot \beta), \quad a_1 = a_0 \delta \tag{6.1}$$

under the condition that the geometric period λ_w of the wall perturbations and the unperturbed wavenumbers h_1 and h_2 of the interacting modes have to satisfy the resonance relationship (Kovalev *et al.*, 1968)

$$\Delta h = h_1 - h_2 = 2\pi / \lambda_B = 2\pi / \lambda_w \tag{6.2}$$

and

$$\pm \Delta m = m_1 - m_2 \tag{6.3}$$

where λ_B is the beat wavelength of the two modes. The radial wavenumber of the mode is changed by scattering the incoming wave at the periodic structure of the waveguide wall which acts in a similar manner to a diffraction grating. Fourier integral transformation theory yields the general conclusion, that the length of the mode converter must be at least of the order of λ_B of the two considered modes. The coherence condition of eqn. 6.2 guarantees that conversion to other unwanted modes which are also coupled by the waveguide perturbations (same Δm) suffers destructive interference. The requirements for low parasitic mode amplitudes are $\Delta h \neq h_{1,2} - h_p$ or $\Delta m \neq m_{1,2} - m_p$, where h_p is the wavenumber and m_p the azimuthal mode index of the parasitic mode.

In the case of negative wavenumbers h_2, i.e. for coupling to backward travelling waves, eqn. 6.2 is the design equation for conventional ($h_2 = -h_1$) Bragg reflectors (Chong *et al.*, 1992) and mode converting ($|h_2| \neq |h_1|$) Bragg reflectors (Wagner, 1996).

Mode transformers consisting of a large number of geometrical periods create low levels of unwanted parasitic modes but inherently exhibit narrow bandwidth. The optimum number of geometrical perturbation periods is determined by the need to limit the conversion to parasitic modes and by limitations on the conversion rate to the desired mode as well as by bandwidth, power flux density and length requirements.

Due to the competition from unwanted parasitic modes, the efficiency of rippled wall mode converters with simple perturbation profiles (purely sinusoidal with constant perturbation amplitude) depends on the starting phase of the wall distortions at the input. The level of undesired modes can be further reduced by using various combinations of (Thumm, 1986)

1. Superposition of additional, small, phase-matched periodical perturbations to the main wall distortion. The geometric periods of these additional perturbations correspond to the beat wavelengths between the input and output mode and their related, strongest coupled parasitic modes.
2. Slightly changed perturbation period (main wall perturbation period $\lambda_w \neq \lambda_B$) resulting in a continuous phase rematching within the converter which causes a reduction of the remaining input mode content.
3. Proper placement of phase delay sections of a straight waveguide to adjust the phase so that coupling to parasitic modes is minimised.

Depending on specific values of the wavenumbers of desired and parasitic modes, undesired second-order resonances can spoil the conversion efficiency (Del Rio *et al.*, 1994).

In the case of small waveguide distortions (low coupling per wavelength) and far from cut off, reflections can be ignored and the optimisation of improved converter structures can be studied by numerical integration of the coupled-wave equations for the complex forward-travelling wave amplitudes (Moeller, 1982; Thumm, 1984).

6.2 TE_{0p} to TE_{0q} mode converters

For low-loss propagation and further conversion into the linearly polarised TE_{11} mode, the output mode of TE_{0p} mode gyrotrons should be converted to the fundamental circular symmetric TE_{01} mode. Mode transducers with axi-symmetric, periodic radius perturbations ($\Delta m = 0$) convert TE_{0p} into $TE_{0,p-1}$ power, provided the converter radius is periodic with the beat wavelength $\lambda_B(p, p-1)$ of the two modes. The waveguide radius is approximately given by

$$a(z) = a_0[1 - \delta_0 \cos(2\pi z / \lambda_w(p, p-1))] / (1 - \delta_0) \tag{6.4}$$

The optimum wavelength $\lambda_w(p,p-1)$ of the wall perturbations differs slightly from $\lambda_B(p,p-1)$ resulting in a reduction of the remaining TE_{0p} mode content. The conversion efficiencies η_0 can be further improved by superimposing small amounts of additional, phase matched, higher harmonics on the wall structure and/or by inserting cylindrical phase shifters in order to reduce the amplitudes of the undesired modes $TE_{0,p+1}$ and $TE_{0,p-2}$ at the converter output. Results of optimisation calculations for several single-step 200 kW TE_{0p} to $TE_{0,p-1}$ mode transducers ($p = 2$-6) at 140 GHz ($a_0 = 13.9$ mm) are summarised in Thumm (1986) and Kumrić *et al.* (1988). In the case of TE_{02} to TE_{01} transformation, efficiencies of approximately 99% can be achieved with merely two geometrical periods ($N = 2$).

The frequency sensitivity of rippled-wall mode transducers is discussed in Moeller (1982) and Kumrić *et al.* (1988). The conversion efficiency η is approximately given by $\eta = \eta_0/(1+\zeta^2)$, where η_0 is the efficiency at the design frequency f_0 and $\xi \approx 2N(\Delta f/f_0)$. TE_{03} or higher TE_{0p} gyrotron modes (e.g. TE_{04}) can be converted to the basic TE_{01} mode by cascading the corresponding single-step mode transducers resulting in a total conversion efficiency of 99.2% or 98%, respectively; ohmic losses are included (Cu waveguide). Direct TE_{03} to TE_{01} conversion is impossible, because the incoming TE_{03} mode is also very strongly coupled to the undesired TE_{04} mode ($\lambda_B(3,1) \approx \lambda_B(3,4)$, Del Rio *et al.*, 1994), whereas a TE_{04} to TE_{01} conversion efficiency of 98% was also obtained by using a direct TE_{04} to TE_{01} mode transformer ($N = 46$, $\delta_0 = 0.0173$, $L = 1.95$ m). However, the bandwidth of such a high N-number converter is inherently narrow ($1/(1+\xi^2) = 0.929$ for $\Delta f/f_0 = 0.3\%$). For a given number N of geometrical periods, the ratio of the coupling coefficient $S_{p,q}$ between TE_{0p} and TE_{0q} modes (for a smooth radius change) and the difference $\Delta h_{p,q} = 2\pi/\lambda_B(p,q)$ of the unperturbed wavenumbers remains constant. As a result the efficiency of 70 GHz and 140 GHz TE_{0p} to $TE_{o,p-1}$ transducers with $a_0 = 13.9$ mm are almost the same. As an example, results of calculations for improved eight main period TE_{03} to TE_{02} and TE_{02} to TE_{01} transducers for the conversion of the output mixture of VARIAN 70 GHz gyrotrons (92% to 95% TE_{02}), the remaining power being equally distributed between TE_{03} and TE_{01} (Thumm, 1986) into the fundamental TE_{01} mode are summarised in Table 6.1 along with the geometrical converter data. The waveguide radius of these advanced mode transducers is given by

$$a(z) = a_0\left[1-\delta_{01}\cos(\Delta h_w(p,p-1)\cdot z)-\delta_{02}\cos(2\cdot\Delta h_w(p,p-1)\cdot z)\right]/(1-\delta_{01}-\delta_{02}) \qquad (6.5)$$

In this case of symmetrical cosine perturbations, the conversion efficiencies from TE_{01} to TE_{02} and from TE_{02} to TE_{01} are identical.

Table 6.1 *Geometrical data, computed and measured conversion efficiencies η_o (ohmic attenuation is included), power transmission efficiencies (P_{sum}) and normalised spurious mode levels at the output of improved eight main period TE_{01} to TE_{02} and TE_{02} to TE_{03} converters (a_0 = 13.9 mm) at 70 GHz (Thumm, 1986)*

	TE_{01} to TE_{02} converter (L = 872 mm)		TE_{02} to TE_{03} converter (L = 449 mm)	
	δ_{01} = 0.0401	δ_{02} = 0.0050	δ_{01} = 0.0237	δ_{02} = 0.0020
mode	calculated	measured	calculated	measured
TE_{01}	0	0	0.002	0.005
TE_{02}	η_o= 0.997	0.995±0.005	0.001	0
TE_{03}	0.001	0.003	η_o= 0.992	0.990±0.007
TE_{04}	0	0	0.002	0.002
P_{sum}	0.998		0.997	

These improved mode transformers were fabricated by direct machining on a numerically controlled lathe and assembled from eight individual sections (aluminium waveguide). The experimental values of conversion efficiencies and parasitic mode levels (low power measurements) are also given in Table 6.1. In both cases measured and predicted converter data are in excellent agreement. Figure 6.1 shows a series of far-field patterns measured by adding 1–8 sections of the TE_{01} to TE_{02} mode converter, each being λ_w long. Complete power transfer is achieved for $L = 8\cdot\lambda_w$. The agreement of these patterns with those that were computed is good and consistent with the calculated efficiencies. In the case of TE_{02} to TE_{01} transformation, theoretical conversion efficiencies $\eta_o \geq 99\%$ were also obtained with considerable shorter improved converters ($N \geq 2$). The number $N = 8$ of main periods was chosen because one geometrical period of such a transducer converts approximately 3.5%, which is the expected power level of TE_{03} and TE_{01} modes emitted by the gyrotron (Thumm, 1986).

A phased-locked gyrotron mode mixture containing ≈ 93% TE_{02}, ≈ 3.5 % TE_{03} and ≈ 3.5% TE_{01} may be cleaned without power losses by using one section of the TE_{03} to TE_{02} transformer at a specified axial position of the longitudinal

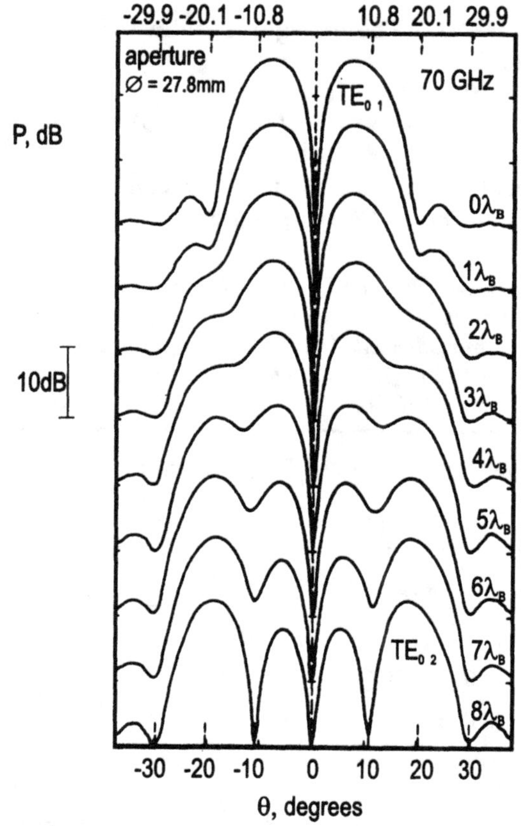

Figure 6.1 Series of measured far-field radiation patterns produced by one to eight geometrical periods (each being λ_w long) of the improved 70 GHz TE_{01} to TE_{02} transducer with $D = 2a_0 = 27.8$ mm (low-power measurements). Separation of transmitting and receiving apertures is $4D^2/\lambda = 720$ mm (Thumm, 1986)

beat structure in the waveguide (phase-matching at the converter input). The resulting TE_{02}/TE_{01} mode composition can finally be converted into an almost pure TE_{01} mode by using only seven sections of the TE_{02} to TE_{01} mode converter, again with phase-matching at the input. At low power levels, the overall conversion efficiency was measured to be 99 ± 1%.

6.3 TE_{01} to TE_{11} mode conversion

Conversion of the circular symmetric TE_{01} mode to the almost linearly polarised TE_{11} mode is achieved by a nonaxisymmetric mode transducer with constant radius (a_0 = 13.9 mm, 200 kW transmission line) and periodically perturbed curvature in one plane (Δm = 1), (Moeller, 1982; Thumm, 1984,

1986). The general expression for the waveguide radius co-ordinate of the actual improved transformers is given by Thumm (1986)

$$a(z,\beta) = a_0\left[1+\left[\delta_{11}\cos(\Delta h_w(01,11)\cdot z) - \delta_{12}\sin(\Delta h_w(01,12)\cdot z) \right.\right.$$
$$\left.\left. -\delta_{21}\sin(\Delta h_w(11,21)\cdot z)\right]\cos\beta\right] \tag{6.6}$$

where β is the azimuthal angle with $\beta = 0,\pi$ in the plane of the wiggles. The E-field plane of the TE_{11} mode is perpendicular to this plane. Six coupled modes were included in the theoretical analysis: TE_{01}, TE_{11}, TE_{12}, TE_{21}, TM_{11} and TM_{21}. Ohmic attenuation and possible ellipticity coupling, introduced by the manufacturing procedure, are included in the coupling matrices. The influence of TM_{11} and TM_{21} turned out to be negligible because there is a continuous and coherent conversion and reconversion (with no change in phasing) between TE_{01} and its degenerate TM_{11} mode and because the TM_{21} mode is only weakly coupled.

The TE_{01} mode is also strongly coupled to the unwanted TE_{12} mode because $\Delta h(01,11)$ is quite close to $\Delta h(01,12)$ and $S(01,12) > S(01,11)$. In addition, the TE_{21} wave is tightly coupled to the generated TE_{11} mode (large value of $S(11,21)/\Delta h(11,21)$). In order to get a conversion efficiency of approximately 93%, it turned out to be necessary to increase the number N of geometrical periods with simple cosine-shaped perturbation to ten (Thumm, 1984). The overall length of this converter is L = 3.12 mm (a_0 = 13.9 mm, δ_{11} = 0.049, $\delta_{12} = \delta_{21} = 0$; $\Delta h(01,11)$ = 20.15 m^{-1}).

The results of numerical computations and the geometrical data of various 70 GHz TE_{12} to TE_{11} mode transducers with eight geometrical periods (N = 8) are given in Thumm (1986). The conversion efficiency of a simple cosine-shaped transducer is quite low (η_0 = 86.2%) and the levels of parasitic modes are rather high (Figure 6.2*a*). The multimode calculations show that the conversion efficiency can be improved to η_0 = 94.0% by a 1.6% increase in the perturbation period (Δh_w = 19.87 m^{-1}). This reduces the unwanted TE_{12} mode level and causes a continuous rematching of the required phase difference between the TE_{01} and TE_{11} modes (Thumm, 1984, 1986), resulting in a reduction of the remaining TE_{01} mode content from 6.9% to 0.9% (Figure 6.2*b*). The power levels of undesired modes are further reduced (Figure 6.2*c*) by superimposing two additional, phase-matched small perturbations, whose periods correspond to the beat wavelengths between TE_{01} and TE_{12} (Δh_w (01,12) = 24.84 m^{-1}) and between TE_{11} and TE_{21} (Δh_w (11,21) = 9.94 m^{-1}). In this case, eight main TE_{01}/TE_{11} periods correspond to ten TE_{01}/TE_{12} periods and to four TE_{11}/TE_{21} periods ($8\cdot\Delta h_w(01,11) = 10\cdot\Delta h_w(01,12) = 4\cdot\Delta h_w(11,21)$). The theoretical conversion efficiency of this improved TE_{01} to TE_{11} mode converter is η_0 = 97.4% (including ohmic attenuation of 1.4% in a Cu waveguide). The bandwidth factor for a frequency deviation of 0.3% is 0.997 (0.3% reduction of the conversion efficiency). Measurements with a wavenumber spectrometer (Kasparek and Müller, 1988) revealed an experimental efficiency of η_0(exp) = (97.2 ± 0.5)% which is in excellent agreement with the predicted value.

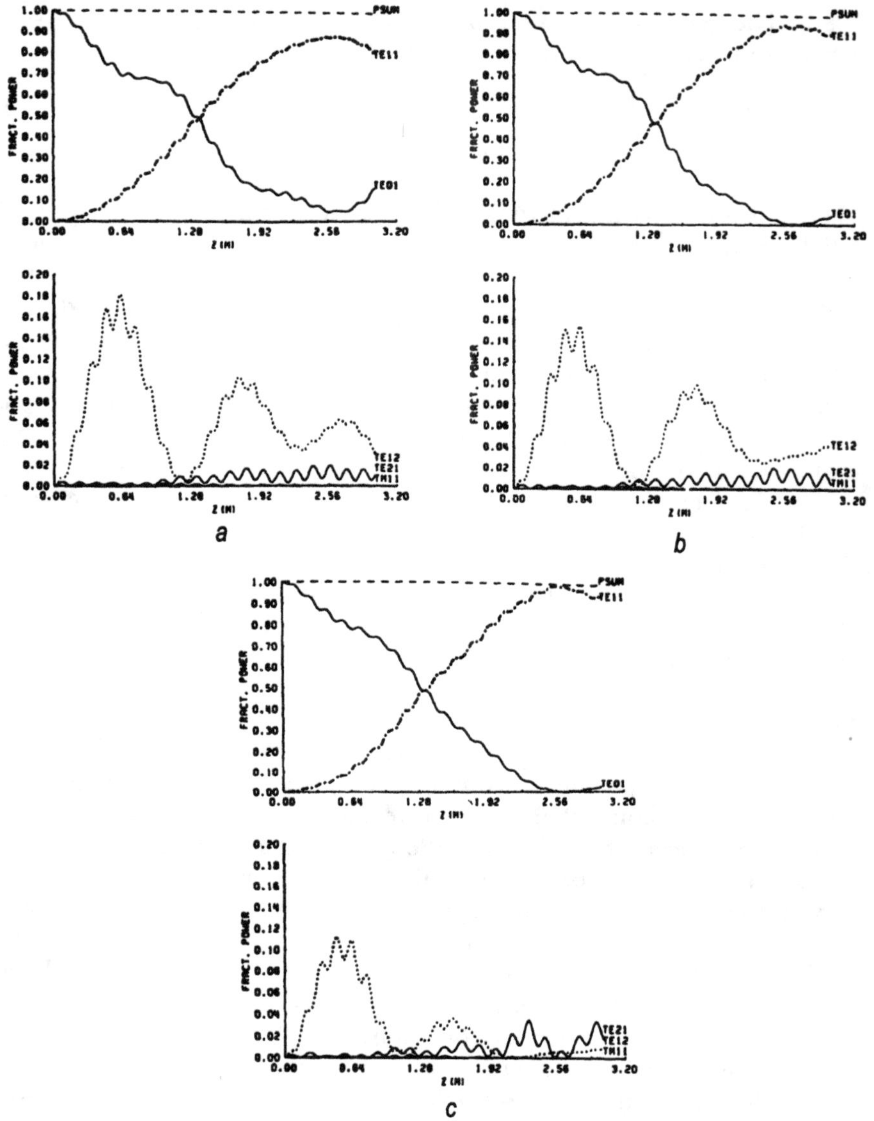

Figure 6.2 *Calculated fractional power in each mode along 70 GHz TE_{01} to TE_{11} converters (a_0 = 13.9 mm, eight geometrical periods) with: (a) simple cosine-shaped perturbation, (b) as a but slightly changed geometrical period, (c) slightly changed geometrical period and superposition of three different, phase matched periodic perturbations (Thumm, 1986)*

6.4 Step-type coupling

Vinogradov and Denisov (1991) have developed the principle of very short waveguide mode converters with step-type coupling described by the coupling coefficient

$$S(z) = \begin{cases} S_0 & 0 \le z < z^* \\ -S_0 & z^* \le z \le L \end{cases} \tag{6.7}$$

and $|2S_0/\Delta h| = 1$. Based on this concept, a number of millimetre wave mode converters were designed and fabricated (TE_{11}-TM_{01}, with TE_{11} polarisation in the plane of curvature; TE_{51}-TE_{22}; $TE_{26,1}$-$TE_{16,4}$). These components have high efficiency $\eta = 0.97$-0.99 and broad frequency band $\Delta f \approx 10$ % and have been successfully used in high power microwave devices (e.g. the TE_{11}-TM_{01} converter in Cherenkov amplifiers) and for cold-test measurements.

6.5 Rippled-wall mode converters for rotating modes

Wall deformations with suitable azimuthal symmetry (eqn. 6.3) that rotate around the waveguide axis (cork-screw type) may be used to directly convert a circular electric TE_{op} mode to a desired rotating TE_{mq} mode (Garin *et al.*, 1987). At the waveguide wall, the phases of two modes with wavenumbers h_1 and h_2 and azimuthal mode indices m_1 and m_2 are at any time equal on a helix of the Δmth order such as (see eqn. 6.1)

$$\Delta h \cdot z + \Delta m \cdot \beta = \text{const} \tag{6.8}$$

Numerical optimisation calculations using coupled-wave equations on mode converters to generate pure, rotating $TE_{22,2}$ (at 110 GHz), $TE_{12,2}$ (at 120 GHz), $TE_{15,2}$ (at 140 GHz) and $TE_{10,4}$ (at 140 GHz) for conversion efficiency optimisation studies on high power quasioptical gyrotron output transducers are presented in Thumm and Jacobs (1988). Two-stage TE_{01} to TE_{op} to TE_{m2} conversion sequences using periodic radius and m-fold helical wall deformations, respectively, provide output purities of almost 99 %. The choice of the intermediate TE_{op} mode ($\mu_{op} \approx \mu_{m2}$) is totally determined by the requirements of the second conversion stage since highly efficient rippled-wall TE_{01} to TE_{op} mode transducers are feasible (Section 6.1).

6.6 Adiabatic surface impedance mode converters

Adiabatic conversion of the almost linearly polarised TE_{11} mode to the perfectly polarised HE_{11} hybrid mode is achieved in a straight, circumferentially corrugated waveguide section with a slot depth which is nonlinearly tapered from a half wavelength to a quarter wavelength (Kerzhentseva, 1971; Thumm *et al.*, 1991), as can be understood from Figure 5.16. A corresponding TM_{11} to HE_{11} transducer consists of a straight corrugated waveguide having a groove depth which changes gradually from zero to a quarter wavelength (Kerzhentseva, 1971; Doane, 1982; Thumm, 1986). These converters can be optimised either by coupled-wave equations (Doane, 1982; Li and Thumm, 1991b) or by the scattering matrix (mode matching) method (Thumm *et al.*, 1991). Similar adiabatic surface impedance mode converters have been employed very efficiently as mode filters in high-power TE_{01} and TE_{02} transmission lines (Thumm *et al.*, 1986).

In conclusion of this chapter on mode converters, Figure 6.3*a* shows the sequence of mode conversion TE_{02}-TE_{01}-TE_{11}-HE_{11} measured at high power (200 kW, 1 ms) for 28 and 70 GHz by thermographic pattern analysers. The principle of periodically perturbed waveguide wall and circumferentially corrugated wall mode converters and the measured efficiencies are summarised in Figure 6.3*b*.

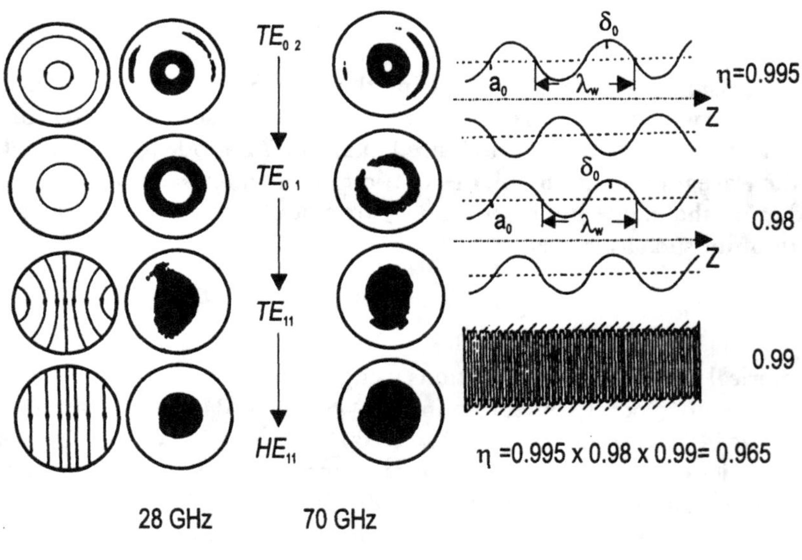

a b

Figure 6.3 (a) *Sequence of mode conversion TE_{02}-TE_{01}-TE_{11}-HE_{11} measured at high power (200 kW, 1 ms) for 28 and 70 GHz by thermographic pattern analysers.*
(b) *Principle of periodically perturbed waveguide wall and circumferentially corrugated wall mode converters. The measured efficiencies are given (Thumm, 1993)*

References

ALPERT, YA. L. (1940): 'To the problem of electromagnetic waves propagation in tubes' (in Russian), *Zh.Technicheskoi Physiki* (*J. Tech. Phys.*), **10**, p. 1358

ANDREASEN, M. G. (1958a): 'Ausbreitung von Grundwellen in kreisrunden und quadratischen gebogenen Hohlleitern konstanter Kruemmung', *Arch. Elektr. Uebertrag.*, **12**, pp. 414–418

ANDREASEN, M. G. (1958b): 'Synthese eines gebogenen Hohlleiters mit stetigen Verlauf der Krümmung', *Archiv. Elektr. Üebertrag.*, **12**, pp. 463–471

ARNDT, F., BEYER, R., REITER, J.M., SIEVERDING, T., and WOLF, T. (1997): 'Automated design of waveguide components using hybrid mode-matching/numerical EM building blocks in optimization-oriented CAD frameworks—state-of-the-art and recent advances', *IEEE Trans. on Microwave Theory and Techniques*, **MTT-45**, (5), pp. 747–760

BARLOW, F. J., RAMSKOV-HANSEN, J. J., and PAYNE, D.N. (1981): 'Birefringence and polarization mode-dispersion in spun single-mode fibers', *Appl. Opt.*, **20**, pp. 2962–2968

BARLOW, H. E. M. (1957a): 'Propagation of the circular H_{01} low-loss wave mode around bends in tubular metal waveguide', *Proc. IEE*, **104B**, pp. 403–409

BARLOW, H. E. M. (1957b): 'Propagation around bends in waveguides', *Proc. IEE*, **106C**, pp. 11–15

BARSUKOV, K. A. (1959): 'On inhomogeneous gyrotropic waveguide filling', *Radio Eng. Electron.*, **4**, (9), pp. 255–256

BEHM, K., and JENSEN, E. (1986): '70 GHz gyrotron development at Valvo'. Conference digest, 11th international conference on *Infrared and millimeter waves*, Pisa, pp. 218–220

BICHUTSKAYA, T. I. and NOVIKOV, V. V. (1979): 'Interaction of normal modes in a plane irregular waveguide with a point of degeneracy', *Radiophys. Quantum Electron.*, **22**, (7), pp.596–603

BORIE, E., GANTENBEIN, G., JÖDICKE, B., DAMMERTZ, G., DUMBRAJS, O., GEIST, T., HOCHSCHILD, G., KUNZE, M., NICKEL, H.-U., PIOSCZYK, B., and THUMM, M. (1992): 'Mode competition using TE_{03} gyrotron cavities', *Int. J. Electron.*, **72**, pp.687–720

BREKHOVSKIKH, L. M. and IVANOV, I.D. (1952): 'On extension of limits for application of ray theory to studying of wave propagation in layered media' (in Russian), *Dokl. AN USSR*, **83**, p. 545

BREKHOVSKIKH, L. M. (1980): 'Waves in layered media' (New York Academic Press, 2nd edn.)

BRODSKIY, V. B. (1958): 'Propagation in waveguides filled by a medium having variable dielectric constant and magnetic permeability', *Radio Eng. Electron.*, **3**, (5), pp. 58–65

BUCKLEY, M. J., STEIN, D. A., and VERNON, R. J. (1991): 'A single-period TE_{02}-TE_{01} mode converter in a highly overmoded circular waveguide', *IEEE Trans. Microw. Theory Tech.*, **MTT-39**, pp. 1301–1306

CHONG, C. K., MCDERMOTT, D. B., RAZEGHI, M. M., LUHMANN, JR., N. C., PRETTEREBNER, J., WAGNER, D., THUMM, M., CAPLAN, M., and KULKE,B. (1992): 'Bragg reflectors', *IEEE Trans. Plasma Sci.*, **PS-20**, pp. 393–402

CLARRICOATS, P. J. B., and OLVER, A. D. (1984): 'Corrugated horns for microwave antennas' (Peter Peregrinus Ltd, London, IEE Electromagnetic Waves Series)

COLLIN, R. E. (1991): 'Field theory of guided waves' (IEEE Press, New York, 2nd edn.)

COMTE, G., CARFORT, F., PONTHUS, A., and PARIS, J. (1957): 'Utilisation de guide d'ondes circulaires pour la transmission á grande distance d'onde centimetriques et millimetriques', *Câbles Transm.*, **11**, p. 342

COMTE, G. (1957): 'Perfectionement aux dispositits de franchissement des coudes par les ondes electromagnetiques du type TE_{01} circulaire'. French patent, class 12, 4, 1142076, 13.09.57

DEL RÍO, C., NET, J. M., PADULLÉS, J., SANCHO, T., SOROLLA, M., THUMM, M., and WIEN, A. (1994): 'Undesired resonances in oversized rippled-wall mode converters', *Int. J. Electron.*, **77**, pp. 1101–1111

DIKAREV, V. A. (1974): 'Asymptotic representations of solutions of a generalized system of telegraphic equations'. Abstract in *Radio Eng. Electron. Phys.*, **19**, pp. 85–91; full text in Russian in: *Radiotekh. Elektron.*, **19**, pp. 2349–2356

DIKAREV, V. A. (1980): 'Waves in multimode systems with turning points', *Radio Eng. Electron. Phys.*, **25**, (4), pp. 5–9

DMITRIEV, YU. (1959): 'Reversal points in waveguides of variable cross section', *Radio Eng. Electron.*, **4**, (6), pp. 78–85

DOANE, J. L. (1982): 'Mode converters for generating the HE_{11} (gaussian-like) mode from TE_{01} in a circular waveguide', *Int. J. Electron.*, **53**, pp. 573–585

DOANE, J. L. (1984a): 'Parabolic tapers for overmoded waveguides', *Int. J. Infrared Millim. Waves*, **5**, pp. 737–751

DOANE, J. L. (1984b): 'Hyperbolic secant coupling in overmoded waveguide', *IEEE Trans. Microw. Theory Tech.*, **MTT-32**, pp. 1362–1371

DOANE, J. L. (1985): 'Propagation and mode coupling in corrugated and smooth-wall circular waveguides', *in* BUTTON, K. J. (Ed.); 'Infrared and millimeter waves', (New York: Academic Press) **13**, pp. 123–170

DOANE, J. L. (1986): 'Polarization converters for circular waveguide modes', *Int. J. Electron.*, **61**, pp. 1109–1133

DOANE, J. L. (1988): 'Low-loss twists in oversized rectangular waveguide', *IEEE Trans. Microw. Theory Tech.*, **MTT-36**, pp. 1033–1042

DOANE, J. L., and ANDERSON, T. N. (1989): 'Oversized rectangular waveguides with mode-free bends and twists for broadband applications', *Microw. J.*, **32**, (3) pp. 153–160

DUMBRAJS, O., THUMM, M., PRETTEREBNER, J., and WAGNER, D. (1992): 'A cavity with reduced mode conversion for gyrotrons', *Int. J. Infrared Millim. Waves*, **13**, pp. 825–840

EDGCOMBE, C. J. (Ed.) (1993): 'Gyrotron oscillators: their principles and practice' (Taylor and Francis, London)

EMELIN, B. F. (Yemelin) (1958): 'Waveguide equations for an irregular waveguide', *Radio Eng. Electron.*, **3**, (5), pp. 40–57

GAPONOV-GREKHOV, A. V.,and GRANATSTEIN, V. L. (Eds.) (1994): 'Applications of high-power microwaves' (Artech House, Boston)

GARIN, P., JEDAR, E., JENDRZEJCZAK, G., MOURIER, G., PAYEN, F., and TEYSSIER, L. (1987): 'Symmetric and non-symmetric modes in a 200 kW, 100 GHz gyrotron'. Conference digest, 12th international conference on *Infrared and millimeter waves*, Lake Buena Vista, Orlando, US, pp. 194–195

GOLOVANOV, O. A. (1991): 'A numerical algorithm for solving diffraction problems for waveguide based microwave devices containing nonlinear media', *Sov. J. Commun.Technol. Electron.*, **36**, (1), pp. 50–59

GOLUBEV, S. V., PLATANOV, YU. YA., RAZIN, S. V., and ZORIN, V. G. (1996): 'Soft x-ray emission from millimeter-wave electron cyclotron resonance discharge', *J. X-Ray Sci. Technol.*, **6**, pp. 244–248

GOUBAU, G., and CHRISTIAN, J. (1961): 'Experimental studies on a beam waveguide for millimeter waves', *IRE Trans. Antennas Propag.*, **AP-9**, p. 256

GRANATSTEIN, V. L., and LAWSON, W. (1996): 'Gyro-amplifiers as candidate RF drivers for TeV linear colliders', *IEEE Trans. Plasma Sci.*, **PS-24**, pp.648–665

GUTMAN, A. L. (1957): 'Calculation of waveguides with smoothly varying cross section' (in Russian), *Radiotekhnika*, **12**, (9), p. 20

GUTMAN, A. L. (1958): 'Calculation of transitions from rectangular waveguides to single and double ridge waveguides' (in Russian), *Radiotekhnika*, **13**, (12), p. 11

GUTMAN, A. L. (1959a): 'Resonant regions in waveguides with continuously variable cross section', *Radio Eng. Electron.*, **4**, (12), pp. 97–113

GUTMAN, A. L. (1959b): 'Application of asymptotic integration of the wave equation to the solution of certain waveguide and resonator problems', *Sov. Phys.-Dokl.*, **4**, pp. 456–459

HEYN, E. (1955): 'Beweis der Vollständigkeitsrelation für die charakteristischen Vektorfelder der Hohlleitertheorie', *Mathemat. Nachricht.*, **13**, p. 25

HUANG, H.-C. (1960): 'Generalized theory of coupled local normal modes in multi-wave guides', *Sci. Sin.*, **9**, (1), pp. 142–154

IIGUCHI, S. (1961): 'Mode conversion in the excitation of TE_{01} waves in a TE_{01}-mode transducer (rectangular- sector portion-circular)', *Rev. Electr. Commun. Lab.*, **9**, pp. 725–732

ISAENKO, YU. M. (1959): 'Continuous exciter of H_{01}-wave in a circular waveguide', *Radio Eng. Electron.*, **4**, (8), pp. 253–260

ISAKOVICH, M. A. (1957): 'Scattering of acoustic waves at small inhomogeneities in a waveduct', *Sov. Phys.-Acoust.*, **3**, (1) pp. 35–45

JOHNSON, R. (1959): 'Design of linear double tapers in rectangular waveguides', *IRE Trans. Microw. Theory Tech.*, **MTT-7**, p. 374

JOUGUET, M. (1947a): 'Les effets de la courbure et des discontinuites de courbure sur la propagation des ondes dans les guides a section rectangulaire', *Câbles Transm.*, **1**, pp. 39–60

JOUGUET, M. (1947b): 'Effets de la courbure dans un guide á section circulaire', *Câbles Transm.*, **1**, pp. 133–153, and (1948), ibid., **2**, pp.257–284

KASPAREK, W., and MÜLLER, G. A. (1988): 'The wavenumber spectrometer—an alternative to the directional coupler for multimode analysis in oversized waveguides', *Int. J. Electron.*, **64**, pp. 5–20

KATSENELENBAUM, B. Z. (1953): 'Waveguides with non-perfect walls' (in Russian), *Dokl. AN SSSR*, **88**, (1), p. 37

KATSENELENBAUM, B. Z. (1954): 'Junction of two waveguides of nearly the same cross-sections' (in Russian), *Zhurnal Technicheskoi Physiki* (*J. Tech. Phys.*), **24**, p. 1892

KATSENELENBAUM, B. Z. (1955a): 'Irregular waveguides with slowly changed parameters' (in Russian), *Dokl. Akad. Nauk SSSR*, **102**, pp. 711–714

KATSENELENBAUM, B. Z. (1955b): 'Degenerated oscillations in waveguides' (in Russian), *Izv. AN USSR*, Division of Technical Sciences, (7), p. 19

KATSENELENBAUM, B. Z. (1956a): 'Bent waveguides of constant cross section' (in Russian), *Radiotekh. Elektron.*, **1**, (2), p. 171

KATSENELENBAUM, B. Z. (1956b): 'Symmetrical dielectric taper in a circular cross section waveguide for the H_{01} mode' (in Russian), *Radiotekh. Elektron.*, **1**, (3), p. 339

KATSENELENBAUM, B. Z. (1957a): 'On general theory of irregular waveguides' (in Russian), *Dokl. AN SSSR*, **116**, pp. 203–206

KATSENELENBAUM, B. Z. (1957b): 'Long symmetrical waveguide junction for H_{01} wave', *Radio Eng. Electron.*, **2**, (5), pp. 21–42

KATZENELENBAUM, B. Z. (1957c): 'On the theory of non-uniform waveguides with slowly changing parameters', Proceedings of international conference on *Ultra high-frequency circuits and antennas*, Paris, October

KATSENELENBAUM, B. Z. (1958a): 'Bent waveguides with inhomogeneous filling', *Radio Eng. Electron.*, **3**, (5), pp. 66–74

KATSENELENBAUM, B. Z. (1958b): 'Irregular waveguides with variable dielectric filling', *Radio Eng. Electron.*, **3**, (7), pp. 48–55

KATSENELENBAUM, B. Z., and MALIN, V. V. (1958c): 'The formation of side-flow in a long waveguide line — part 1', *Radio Eng. Electron.*, **3**, (6), pp. 32–40 and 'The formation of secondary flux in a long wave guide line — part 2', *Radio Eng. Electron. Phys.*, **3**, (11), pp. 93–106

KATSENELENBAUM, B. Z. (1958d): 'Critical cross sections in irregular waveguides', *Sov. Phys.-Dokl.*, **3**, pp. 1124–1127

KATSENELENBAUM, B. Z. (1959a): 'Twisted waveguides', *Radio Eng. Electron.*, **4**, (9), pp. 46–52

KATSENELENBAUM, B. Z. (1959b): 'Waveguide transitions', *Radio Eng. Electron.*, **4**, (9), pp. 247–254

KATSENELENBAUM, B. Z. (1960a): 'Wave impedance of rectangular waveguide' (in Russian), *Radiotekhnika*, **15**, (5)

KATSENELENBAUM, B. Z. (1960b): 'Problem of normal incidence of a plane electromagnetic wave on a periodic boundary separating two dielectrics', *Radio Eng. Electron.*, **5**, (12), pp. 77–82

KATSENELENBAUM, B. Z. (1961a): 'On the theory of irregular acoustic waveguides', *Sov. Phys.-Acoust.*, **7**, (2), p. 159–164

KATSENELENBAUM, B. Z., and MALINA, Z. A. (1961b): 'Transition calculations for a symmetrical magnetic wave in a circular waveguide', *Radio Eng. Electron. Phys.*, **6**, (2), pp. 198–203

KATSENELENBAUM, B. Z. (1961c): 'Theory of non-uniform waveguides with slowly varying parameters' (in Russian). Published by the former Soviet Union Academy of Sciences, Izd. Akad. Nauk SSSR, Moscow

KATSENELENBAUM, B. Z., KOLESNICHENKO, YU. V., FRANTSESSON, A. V., and SHEVCHENKO, V. V. (1996): 'Twisted dielectric waveguides: macro- and micro-chirality and the rejection band', *J. Commun. Technol. Electron.*, **41**, (6), pp. 489–496

KAZANTSEV, YU. N. (1957): 'Calculation of symmetrical transition sections for H_{0n} waves in circular guide', *Radio Eng. Electron.*, **2**, (2), pp. 29– 38

KAZNACHEEV, YU. I. (1959): 'Broadband telecommunications using waveguides' (in Russian). Publication of the Academy of Sciences of USSR, Moscow

KERZHENTSEVA, N. P. (1958): 'Propagation of electromagnetic waves in bent circular waveguides', *Radio Eng. Electron.*, **3**, (5), pp. 86–102

KERZHENTSEVA, N. P. (1959): 'The passage of a H_{01} wave through a curved helical waveguide', *Radio Eng. Electron.*, **4**, pp. 249–254

KERZHENTSEVA, N. P. (1960): 'Waveguide bend with varying curvature', *Radio Eng. Electron.*, **5**, pp. 40–49

KERZHENTSEVA, N. P. (1971): 'Conversion of wave modes in a waveguide with smoothly varying impedance of the walls', *Radio Eng. Electron. Phys.*, **16**, pp. 28–31

KIENLIN, U., and KÜRZL, A. (1958): 'Reflexionen an Hohlleiter-Flanschverbindungen', *Nachr.techn. Z.*, (11), p. 561

KING, A. P., and MARCATILI, E. A. (1956): 'Transmission loss due to resonance of loosely-coupled modes in a multi-mode system', *Bell Syst. Tech. J.*, **35**, pp.899–906

KISUN'KO, G. V. (1949): 'Electrodynamics of hollow systems' (in Russian) (Publishing House VKAS, Leningrad)

KLOPFENSTEIN, R. W. (1956): 'A transmission line taper of improved design', *Proc. IRE*, **44**, pp. 31–35

KOVALEV, N. F., ORLOVA, I. M., and PETELIN, M. I. (1968): 'Wave transformation in a multimode waveguide with corrugated walls', *Radiophys. Quantum Electron.*, **11**, pp. 449–450

KOVALEV, N. F. (1986): 'Theoretical basis for the transverse-section method in the theory of irregular waveguides', *Sov. J. Commun. Technol. Electron.*, **31**, (1), pp. 60–64

KRASNUSHKIN, P. E. (1945): 'The waves in curved tubes' (in Russian). Uchennye Zapiski MGU (Scientific Reports of Moscow State University). **75-Phys.**, book 2, part II, p. 9

KUMRIĆ, H., AND THUMM, M. (1986): 'Optimized overmoded TE_{01} to TM_{11} mode converters for high-power millimeter wave applications at 70 and 140 GHz', *Int. J. Infrared Millim. Waves*, **7**, pp. 1439–1463

KUMRIĆ, H., THUMM, M., and WILHELM, R. (1988): 'Optimization of mode converters for generating the fundamental TE_{01} mode from TE_{06} gyrotron output at 140 GHz', *Int. J. Electron.*, **64**, pp. 77–94

LEONARD, D., and YEN, L. (1957): 'Junction of smooth flared wave guides', *J. Appl. Phys.*, **28**, p. 1441

LEWIN, L. (1955a): 'Theory of waveguides' (John Wiley, New York) and (1975): 'Theory of waveguides: techniques for the solution of waveguide problems' (Newnes-Butterworths, London)

LEWIN, L. (1955b): 'Propagation in curved and twisted waveguides of rectangular cross-section', *Proc. IEE*, **102B**, p. 75

LEWIN, L. (1959): 'A note on quasi-optical method at millimeter waves'. Proceedings of the symposium on *Millimeter waves*, New York, p. 469

LI, H., and THUMM, M. (1991a): 'Mode conversion due to curvature in corrugated waveguide', *Int. J. Electron.*, **71**, pp. 333–347

LI, H., and THUMM, M. (1991b): 'Mode coupling in corrugated waveguides with varying wall impedance and diameter change', *Int. J. Electron.*, **71**, pp.827–844

LYKHNE, A. M. (1959): 'The theory of horns' (English translation, in Russian version the author was Dykhne), *Sov. Phys.–Dokl.*, **4**, pp. 651–652

LYUDBARSKI, G. YA., and POVZNER, A. YA. (1957): 'On the theory of wave propagation in variable-cross-section waveguides', *Sov. Phys.-Tech. Phys.*, **4**, (2), pp. 146–154

MALYGIN, S. A., PAVELYEV, V. G., and TSIMRING, SH. E. (1983): 'Resonant transformation of modes in oversized electrodynamic systems', *Radiophys. Quantum Electron.*, **26**, (9), pp. 827–832

MARCATILI, E. (1961): 'A circular-electric hybrid junction and some channel-dropping filters', *Bell Syst. Tech. J.*, **40**, p. 185

MARCUSE, D. (1974): 'Theory of dielectric optical waveguides' (Academic Press, New York)

MARCUSE, D. (1989): 'Light transmission optics' (F.L. Krieger, Melbourne, 2nd edn.) and (Van Nostrand, New York, 1972, 1st edn.)

MARIE, P. (1959): 'A bend for TE_{01}-mode propagation in circular waveguide', *Proc. IEE*, **106B**, suppl. (13), p. 108

MAR'IN, N. P. (1959a): 'Butt joint of different waveguide types', *Radio Eng. Electron.*, **4**, (1), pp. 1–14

MAR'IN, N. P. (1959b): 'Applying the conform transformations method to the solution of the problem of a radial taper between two rectangular cross section waveguides' (in Russian), *Radiotekhnika*, **14**, (6), p. 17

MATSUMARU, K. (1959): 'Reflection of a pyramidally tapered rectangular waveguide', *IRE Trans. Microw. Theory Tech.*, **MTT-7**, p. 192

MILLER, S. E. (1952): 'Notes on methods of transmitting the circular electric wave around bends', *Proc. IRE*, **40**, pp. 1104–1113

MILLER, S. E. (1954): 'Waveguide as a communication medium', *Bell Syst. Tech. J.*, **33**, pp. 1209–1265

MÖBIUS, A., and THUMM, M. (1993): 'Gyrotron output launchers and output tapers', *in* EDGCOMBE, C. J. (Ed.): 'Gyrotron oscillators' (Taylor and Francis, London) chapter 7, pp. 179–222

MOELLER, C. (1982): 'Mode converters used in the Doublet III ECH microwave system', *Int. J. Electron.*, **53**, pp. 587–593

MORGAN, S. P. (1957): 'Theory of curved circular waveguide containing an in-homogeneous dielectric', *Bell Syst. Tech. J.*, **36**, pp. 1209–1251

NISHIHARA, H., HARUNA, M. and SUHARA, T. (1985): 'Optical integrated circuits' (McGraw-Hill, New York)

OGUCHI, B., and KATO, M. (1958): 'The effects of circular TE_{1m} waves on the propagation of circular TE_{10}-wave in a curved waveguide', *Onde Electr.*, suppl. special, **376**, pp. 185–192

OUDIN, J., and COMTE, G. (1959): 'Perfectionement aux dispositits de guide d'ondes comportant des miroirs'. French patent, class 12, 4, 1190284, 12.10.59

PAVELYEV, V. G., TSIMRING, SH. YE., and ZAPEVALOV, V.E. (1987): 'Coupled cavities with mode conversion in gyrotrons', *Int. J. Electron.*, **63**, pp. 379–391

PAVELYEV, V. G., and TSIMRING, SH. YE. (1988): 'On the method of planar cross sections in the theory of highly irregular oversize electromagnetic waveguides', *Radio Eng. Electron. Phys.*, **32**, pp. 173–176

PAVELYEV, V. G., and STAROV, V. G. (1995): 'Planar cross-section method in the theory of axisymmetrical structures, containing segments of irregular coaxial and circular waveguides', *J. Commun. Technol. Electron.*, **40**, (4), pp. 1–5

PERSIKOV, M. YU., KAZANTSEV, YU. N., and KOZELEV, A. I. (1960): 'Field indicator in a circular cross section waveguide' (in Russian), *Elektrosvyaz*, (5), p. 38

PIEFKE, G. (1957): 'Reflexion beim Übergang vom Rechteck-Hohlleiter zum Zektor-Horn', *Arch. Elektr. Üebertrag.*, **11**, (3), pp. 123–135

POCHELON, A., GOODMAN, T., WHALEY, D., TRAN, M. Q., PERRENOUD, A., REINHARD, D., JÖDICKE, B., MATHEWS, H. G., KASPAREK, W., and THUMM, M. (1991): '39 GHz ECRH system for breakdown studies on the TCA tokamak' (Elsevier Science Publishers B. V., Fusion Technology, 1990, KEEN, B. E., HUGUET, M., and HEMSWORTH, R. (Eds.)) pp. 1075–1079

POKROVSKII, V. L., ULINICH, F. R., and SAVVINYKH, S. K. (1958): 'Local reflection in waveguides of variable cross section', *Sov. Phys.-Dokl.*, **3**, pp. 580–583

POKROVSKII, V. L., ULINICH. F. R., and SAVVINYKH, S. K. (1959a): 'Nonlocal reflection in waveguides of varying cross section', *Sov. Phys.-Dokl.*, **4**, pp. 108–110

POKROVSKII, V. L., ULINICH, F. R., and SAVVINYKH, S. K. (1959b): 'The theory of tapered waveguides', *Radio Eng. Electron.*, **4**, (2), pp. 10–23

RAMO, S., WHINNERY, J. R., and VAN DUZER, T. (1994): 'Fields and waves in communication electronics' (John Wiley, New York, 3rd edn.)

REITER, G. (1959): 'Generalized telegraphist's equation for waveguides of varying cross section', *Proc. IEE*, **106B**, suppl. (13), pp. 54–57

RICE, O. (1949): 'Reflection from corners in rectangular waveguides-conformal transformation', *Bell Syst. Tech. J.*, **28**, pp. 104–134

ROZHDESTVENSKIY, B. L., and CHETAYEV, D. N. (1951): 'On the problem of reflection diminishing in a varying cross section waveguide' (in Russian), *Dokl. AN USSR*, **79**, (3)

ROZHDESTVENSKIY, B. L. (1953): 'On the theory of a planar horn' (in Russian), *Zhurnal Technicheskoi Physiki* (*J. Tech. Phys.*), **23**, p. 1609

RYVKIN, M. S. (1959): 'Rectangular waveguide of variable cross-section', *Radio Eng. Electron.*, **4**, (9), pp. 78–92

SANDSMARK, P. I. (1955): 'Effect of ellipticity of mode-dominant axial ratio in nominally circular waveguides', *IRE Trans. Microw. Theory Tech.*, **MTT-3**, (5), pp. 15–20

SAVVINYKH, S. K. (1959): 'On the theory of waveguides of variable circular cross-section', *Radio Eng. Electron.*, **4**, (6), pp. 86–97. The same author (1960): 'Regarding the paper 'On the theory of waveguides of variable circular cross section'' (in Russian), **5**, p. 260–261

SCHAFFELD, W., and BAYER, H. (1956): 'Über das Verhalten elektromagnetischer Wellen in kreiszylindrischen Hohlleitern im Bereich der Grenzfrequenz unter Beruecksichtigung der endlichen Wandleitfaehigkeit', *Arch. Elektr. Üebertrag.*, **10**, (3), pp. 89–97

SCHELKUNOFF, S. A. (1952): 'Generalized telegraphist's equations for waveguides', *Bell Syst. Tech. J.*, **31**, pp. 784–801

SCHELKUNOFF, S. A. (1955): 'Conversion of Maxwell's equations into generalized telegraphist's equations', *Bell Syst. Tech. J.*, **34**, pp. 995–1043

SCHNETZLER, K. (1960a): 'Die Mitführung der Polarisation bei der Verwindung eines quadratischen Hohlleiters', *Frequenz*, **14**, p. 123

SCHNETZLER, K. (1960b): 'Die Reflexion der Grundwelle an den Knickstellen eines Hohlleiters, insbesondere bei einem stetigen Übergang von einem recheckigen auf einem runden Hohlleiter', *Arch. Elektr. Üebertrag.*, **14**, pp. 177–182

SHEVCHENKO, V. V. (1961): 'Irregular acoustic waveguides', *Sov. Phys.-Acoust.*, **7**, pp. 392–397

SHEVCHENKO, V. V. (1962): 'A waveguide with wall impedance as a surface compensator', *Radio Eng. Electron. Phys.*, **7**, pp. 1034–1039

SHEVCHENKO, V. V. (1967): 'Coupling coefficients of waves in a variable cross section waveguide with an impedance', *Radio Eng. Electron. Phys.*, **12**, pp. 147–149

SHEVCHENKO, V. V. (1969): 'Graphic classification of modes guided by regular open waveguides', *Radio Eng. Electron. Phys.*, **14**, pp. 1530–1533

SHEVCHENKO, V. V. (1971): 'Continuous transitions in open waveguides: introduction to the theory' (The Golem Press, Boulder, Colorado)

SHEVCHENKO, V. V. (1996): Private communication

SHIMIZU, Y. (1957): 'Theory of transmitting circular electric wave around bends'. Proceedings of international conference on *Ultra high-frequency circuits and antennas*, Paris, October

SHUSHPANOV, O. E. (1961): 'Coupling coefficients for a bent coaxial waveguide of round cross section', *Radio Eng. Electron. Phys.*, **6**, (10), pp. 1573–1574

SOHDA, M. S., and GHATAK, A. K. (1977): 'Inhomogeneous optical waveguides' (Plenum Press, New York)

SOLYMAR, L. (1958): 'Design of a conical taper in circular waveguide system supporting H_{01} mode', *Proc. IRE*, **46**, (3), pp. 618–619

SOLYMAR, L. (1959a): 'Spurious mode generation in nonuniform waveguide', *IRE Trans. Microw. Theory Tech.*, **MTT-7**, pp. 379–383

SOLYMAR, L. (1959b): 'Mode conversion in pyramidal tapered waveguides', *Electron. Radio Eng.*, **36**, pp. 461–463

SOLYMAR, L. (1959c): 'Design of a two-section conical taper in circular waveguide system supporting the H_{01} mode', *Proc. IEE*, **106 B**, suppl. (13), pp. 119–120

SOLYMAR, L. (1959d): 'Monotonic multi-section tapers for overmoded circular waveguides', *Proc. IEE*, **106B**, suppl. (13), pp. 121–128

SPORLEDER, F., and UNGER, H. G. (1979): 'Waveguide tapers, transitions and couplers' (Peter Peregrinus Ltd., London)

STEVENSON, A. (1951a): 'General theory of electromagnetic horns', *J. Appl. Phys.*, **22**, pp. 1447–1454

STEVENSON, A. (1951b): 'Exact and approximate equations for wave propagation in acoustic horns', *J. Appl. Phys.*, **22**, p. 1461

SVESHNIKOV, A. G. (1956a): 'Approximate method for computation of a slightly irregular waveguide' (in Russian), *Dokl. AN SSSR*, **110**, (2), pp. 197–199

SVESHNIKOV, A. G. (1956b): 'On radio wave propagation in slightly bent waveguides' (in Russian), *Radiotekh. Elektron.*, **1**, p. 1222

SVESHNIKOV, A. G. (1958): 'Waves in bent tubes', *Radio Eng. Electron.*, **3**, (5), pp. 75–85

SVESHNIKOV, A. G. (1959a): 'Excitation of irregular waveguides' (in Russian), *Nauchnye Dokl. Vysshei Shkoly* (Scientific Reports of the Highest School), Physical-mathematical sciences, (2), p. 162

SVESHNIKOV, A. G., (1959b): 'Irregular waveguides' (in Russian), *Izv. Vuz. Radiofizika*, **2**, p. 720

TANAKA, K. (1957): 'Mode conversion through the tapered section of circular waveguide system'. Proceedings of international conference on *Ultra high-frequency circuits and antennas*, Paris, October

TATSUKAWA, T., MAEDA, T., SASAI, H., IDEHARA, T., MEKATA, M., SAITO, T., and KANEMAKI, T. (1995): 'ESR spectrometer with a wide frequency range using a gyrotron as a radiation power source', *Int. J. Infrared Millim. Waves*, **16**, pp. 293–305

THUMM, M. (1984): 'High-power millimetre-wave mode converters in overmoded circular waveguides using periodic wall perturbations', *Int. J. Electron.*, **57**, pp. 1225–1246

THUMM, M., ERCKMANN, V., JANZEN, G., KASPAREK, W., MÜLLER, G., SCHÜLLER, P. G., and WILHELM, R. (1985): 'Generation of the gaussian-like HE_{11} mode from gyrotron TE_{0n} mode mixtures at 70 GHz', *Int. J. Infrared Millim. Waves*, **6**, pp. 459–470

THUMM, M., ERCKMANN, V., KASPAREK, W., KUMRIĆ, H., MÜLLER, G. A., SCHÜLLER, P. G., and WILHELM, R. (1986): 'Very high power mm-wave components in oversized waveguides', *Microw. J.*, **29**, (11) pp. 103–121

THUMM, M. (1986): 'High power mode conversion for linearly polarized HE_{11} hybrid mode output', *Int. J. Electron.*, **61**, pp. 1135–1153

THUMM, M., and JACOBS, A. (1988):, 'In-waveguide TE_{01}-to- whispering gallery mode conversion using periodic wall perturbations'. Conference digest, 13th international conference on *Infrared and millimeter waves*, Honolulu, Hawaii, pp. 465–467

THUMM, M., and KUMRIĆ, H. (1989): 'High-power TE_{11} and TM_{11} circular polarizers in oversized waveguides at 70 GHz', *Int. J. Infrared Millim. Waves*, **10**, pp. 1059–1075

THUMM, M., JACOBS, A., and SOROLLA, M. (1991): 'Design of short high power TE_{11}-HE_{11} mode converters in higly overmoded corrugated waveguides', *IEEE Trans. Microw. Theory Tech.*, **MTT-39**, pp. 301–309

THUMM, M. (1993): 'High power microwave transmission systems, external mode converters and antenna technology', *in* EDGCOMBE, C. J. (Ed.): 'Gyrotron oscillators' (Taylor and Francis, London) chapter 13, pp. 365–401

THUMM, M. (1994): 'Progress in the development of high-power millimeter- and submillimeter wave gyrotrons and of free electron masers', *Arch. Elektrotech.*, **77**, pp. 51–55

THUMM, M., and KASPAREK, W. (1995): 'Recent advanced technology in electron cyclotron heating systems', *Fusion Eng. Des.*, **26**, pp. 291–317

TSIMRING, SH. E., and PAVELYEV, V. G. (1982): 'The theory of nonhomogeneous electromagnetic waveguides containing critical cross-sections', *Radio Eng. Electron. Phys.*, **27**, (6), pp. 41–44

UHER, J., BORNEMANN, J., and ROSENBERG, U. (1993): 'Waveguide components for antenna feed systems: theory and CAD' (Artech House, London)

ULINICH, F. P. (1959): 'The theory of horns', *Radio Eng. Electron.*, **4**, (5), pp. 82–91

UNGER, H. G. (1957): 'Normal modes bends for circular electric waves', *Bell Syst. Tech. J.*, **36**, pp. 1292–1307

UNGER, H. G. (1958a): 'Circular waveguide taper of improved design', *Bell Syst. Tech. J.*, **37**, pp. 899–912

UNGER, H. G. (1958b): 'Helix waveguide theory and application', *Bell Syst. Tech. J.*, **37**, pp. 1599–1647

UNGER, H. G. (1977): 'Planar optical waveguides and fibres' (Clarendon Press, London, Oxford and New York)

VAGANOV, R. B. (1960): 'The experimental analysis of the electromagnetic field in waveguide junctions containing critical sections', *Radio Eng. Electron.*, **5**, (5), pp. 31–39

VAN HUAN-CHZHO (VAN KHUAN'-CHZHO) (1960): 'Method of calculating a dielectric insert for matching the H_{01} wave at a junction between two circular waveguides', *Radio Eng. Electron.*, **5**, (7), pp. 57–65

VAN HUAN-CHZHO (WANG HUAN CHO) (1961): 'Calculation of a matching metallic insert in a circular waveguide', *Radio Eng. Electron. Phys.*, **6**, (7), pp. 1067–1069

VESELOV, G. I., SEMENOV, S. G., and BLAGOVESHENSKIY, V. L. (1983): 'Unique features of the propagation of hybrid waves in a circular waveguide with a dielectric rod', *Radio Eng. Electron. Phys.*, **28**, (11), pp. 10–16

VIKTOROVA, S. L., and SVESHNIKOV, A. G. (1958): 'Conversion of H_{01} mode in a space bent waveguide of circular cross section' (in Russian), *Nauchnye Dokl.y Vysshei Shkoly* (Scientific Reports of the Highest School), Physical-mathematical sciences, (2), p. 133

VINOGRADOV, D. V., and DENISOV, G. G. (1991): 'Waveguide mode converters with step type coupling', *Int. J. Infrared Millim. Waves*, **12**, pp. 131–140

VOYTOVICH, N. I., VOYTOVICH, N. N., ZAMOYSKAYA, O. F., and KATSENELENBAUM, B. Z. (1993): 'Calculation of the radiation pattern of flat resonator antenna with variable transmittance and height', *J. Commun. Technol. Electron.*, **38**, (14), pp. 88–96

WAGNER, D. (1996): 'Entwurf und Analyse von korrugierten überdimensionierten Hohlleiterstrukturen für Hochleistungsmillimeterwellensysteme'. Dr. Ing. thesis, Institut für Plasmaforschung, Universität Stuttgart

WEINSTEIN (VAINSTEIN), L. A. (1957): 'The method of approximate separation of variables and its application to boundary value problems of electrodynamics and acoustics', *Sov. Phys.-Tech. Phys.*, **2**, pp. 1955–1973

WEINSTEIN (VAINSTEIN), L. A. (1988): 'Electromagnetic waves' (in Russian) (Publishing House, Radio i Svyaz, Moscow, 2nd edn.)

WILLIS, J., and SINHA, N. (1955): 'Nonuniform transmission lines as impedance matching sections', *Proc. IRE*, **43**, p. 1975

WRAY, D. (1959): 'Frequency compensation for simple stepped waveguide transforming section', *Electr. Eng.*, **31**, (372), pp. 76–79

ZAPEVALOV, V. E., MALYGIN, S. A., PAVELYEV, V. G., and TSIMRING, SH. E. (1984): 'Coupled-resonator gyrotrons with mode conversion', *Radiophys. Quantum Electron.*, **27**, (9), pp. 846–852

ZARUDNEVA, G. I., KALYNOV, YU. K., and MALYGIN, S. A. (1988): 'Radiation mode composition of open resonators in the form of axisymmetric, weakly irregular waveguides', *Radiophys. Quantum Electron.*, **31**, pp. 254–257

Index

www.ingramcontent.com/pod-product-compliance
Lightning Source LLC
LaVergne TN
LVHW011754200926
848484LV00003B/703
* 9 7 8 0 8 5 2 9 6 9 1 8 2 *